Essential Topics in Materials Physics and Chemistry

Volume I

Essential Topics in Materials Physics and Chemistry
Volume I

Edited by **Heather Dale**

NY RESEARCH PRESS
New York

Published by NY Research Press,
23 West, 55th Street, Suite 816,
New York, NY 10019, USA
www.nyresearchpress.com

Essential Topics in Materials Physics and Chemistry: Volume I
Edited by Heather Dale

International Standard Book Number: 978-1-63238-187-3 (Hardback)

Printed in China.

Contents

Preface

Our understanding of the world is rapidly evolving as scientific methods go deeper and deeper into unraveling the universe. Basic physical science fields have also evolved, changed and advanced accordingly and the fields of materials Physics and Chemistry have gained a foothold. The discipline that focuses on the relationship between the atoms, molecules or ions that constitute a material and its overall physical characteristics is called Materials Chemistry. In this field, the material and structure of materials is studied as well as synthesizing new materials. What is surprising is the fact that until perhaps 20 years ago the words 'materials' and 'chemistry' were hardly used together. Advanced technological progress has also been helpful in using computational techniques to predict the material and structure of still undiscovered matter. On the other hand Material physics is instrumental in describing materials in terms of mechanics, heat, light and force. It combines sciences such as chemistry and solid state physics along with solid mechanics. Like materials chemistry has a bit of chemistry in its methodology, so too has physics got chemistry. In recent years such materials sciences have reached a respected and secure place in the scientific community.

This book lists and discusses the research compiled on the advances that materials physics and chemistry have had recently. I am thankful to those scientists and researchers whose toil and hard work help make this book a success as well as those outside the scientific community who have been unwavering in their support

Editor

Effect of Mechanical Treatment Temperature on Electrical Properties and Crystallite Size of PVDF Film

Ambran Hartono[1*], Suparno Satira[2], Mitra Djamal[2], Ramli Ramli[3], Herman Bahar[2], Edi Sanjaya[1]
[1]Department of Physics, Faculty of Science and Technology, Islamic State University Syarif Hidayatullah, Jakarta, Indonesia
[2]Theoretical High Energy Physics and Instrumentation Research Group, Faculty of Mathematics and Natural Sciences, Institut Teknologi Bandung, Bandung, Indonesia
[3]Department of Physics, Faculty of Mathematics and Natural Sciences, Universitas Negeri Padang, Padang, Indonesia

ABSTRACT

Fabrication of PVDF films has been making using Hot Roll Press. Preparation of samples carried out for nine different temperatures. This condition is carried out to see the effect of temperature fabrication on electrical properties and crystallite size of PVDF films. The electrical properties like as surface resistivity are discussion focus in this paper. Surface resistivity properties of PVDF can be improved by mechanical treatment on the varying film thickness and the temperature. To obtain the diffraction pattern of sample characterization is performed using X-Ray Diffraction. Crystallite size of PVDF films calculate from broadening pattern of X-Ray Diffraction. Furthermore, from the diffraction pattern calculated β fraction and crystallite size, for calculation to determine the crystallite size of the sample by using the Scherrer equation. Has been obtained an increase piezoelectric properties of PVDF films that characterized by increasing β fraction. Have been obtained β fraction increased from 25.4% up to 44% for temperatures of 130°C up to 170°C, respectively. Resistivity value has been obtained at temperature 130°C up to 170°C, decreased from 1.23×10^4 Ωm up to 0.21×10^4 Ωm respectively. From the experimental results and the calculation of crystallite sizes obtained for the samples with temperature 130°C up to 170°C respectively are increased from 7.2 nm up to 20.54 nm. These results indicate that mechanical treatment caused increase β fraction and decrease surface resistivity. Increasing temperatures will also increase the size of the crystallite of the sample. This happens because with the increasing temperature causes the higher the degree of crystallization of PVDF film sample is formed, so that the crystallite size also increases.

Keywords: Crystallite Size; PVDF; Roll Hot Press; Scherrer Equation; Surface Resistivity; Temperature Different; XRD

1. Introduction

Recently, the development of manufacturing technology and utilization of polymer films are increasing rapidly, especially polyvinylidene fluoride (PVDF) polymer. PVDF is a material typically used for applications requiring high purity when compared to other fluoropolymer. PVDF easier to melt due to have a low melting point [1]. PVDF can be synthesized from the gaseous VDF monomers via free radical polymerization process are shown in **Figure 1**.

PVDF polymer has three solid molecule structures that are β-phase, α-phase and γ-phase [2]. Chain agglomerations of α-phase in PVDF are shown in **Figure 2**.

α-phase of PVDF polymer have TGTG type conformation structure [2]. PVDF polymer with β-phase molecule structure is shown in **Figure 3**.

Recently, PVDF with β-phase structure are widely developed due to molecules with this structure provides the greatest piezoelectric effect compared with other phase.

β-phase have planar zig-zag formation which all trans (TTTT) [3] with deflection small because of fluor atoms bonded by nearest monomer.

The electrical properties like as piezoelectric are related to the β fraction PVDF polymer and surface resis-

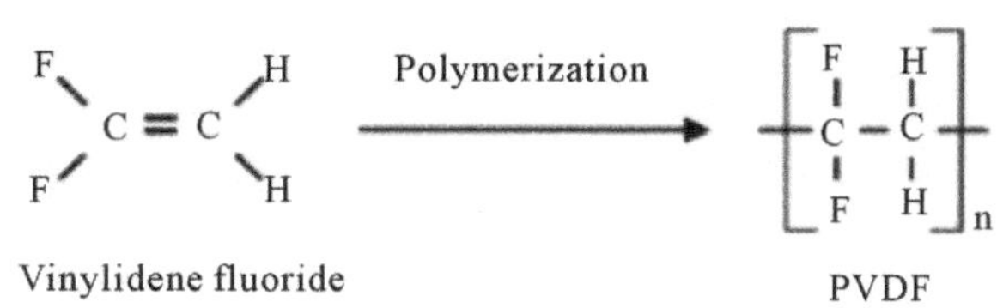

Figure 1. Polymerization process.

*Corresponding author.

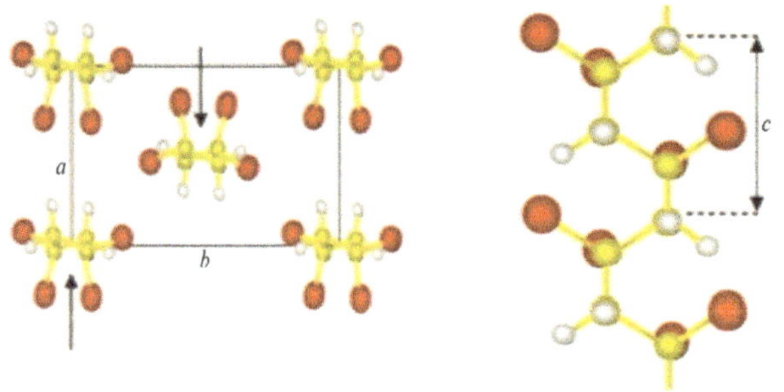

Figure 2. Chain structure of α-phase in PVDF.

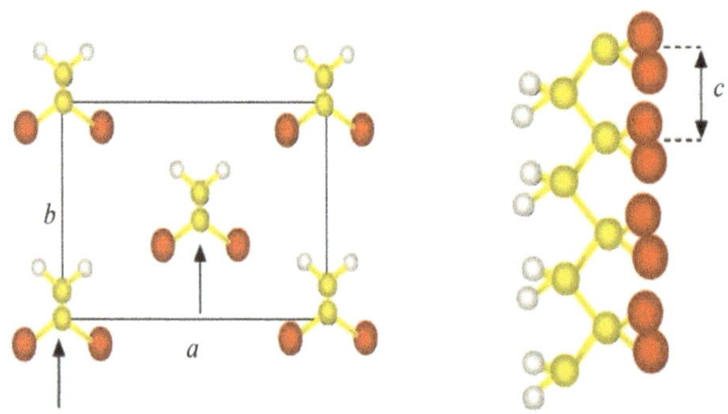

Figure 3. Chain structure of β-phase in PVDF.

tivity. Recently, PVDF with β-phase structure are widely developed due to molecules with this structure provides the greatest piezoelectric effect compared with other phase [4]. The α phase obtained by the crystallization of melt can be transformed to the β phase by mechanical deformation [2]. Drawing at lower temperatures to the natural draw ratio of −4 and at higher temperatures to higher draw ratios (>4) favors the transformation from α to β phase [5-8]. From WAXD and FTIR studies, it is clear that there is α phase transformation during stretching and stretching at lower temperature to draw ratio of 4 or to higher draw ratios at higher temperature results in higher β phase contents [9,10].

The most common polymorph produced during crystallization from the melt is the α-form; it is essentially the sole crystalline obtaoined at all temperatures up to ≈150˚C. Its crystallization kinetics have been studied by Gianotti *et al.* [11] in the range 147˚C - 165˚C, by Nakamura *et al.* [12], between 148˚C - 155˚C and by Mancarella and Martuscelli [13] between 134˚C and 149˚C. The dependence of crystalline and amorphous specific volume on temperature has been systematically studied by Nakagawa and Ishida [14].

Until now research on PVDF piezoelectric properties is still being developed, both in terms of increasing the piezoelectric material and the analysis of physical parameters related to the amount of beta fraction of the sample as well as in its application to sensors. In our research conducted by making a thin layer of material PVDF polymer with Roll Hot Press for several different temperatures. Having obtained samples were then carried out characterization using XRD to analyze the influence of different temperatures on the manufacture of crystallite size of PVDF films [15].

Surface Resistivity value of PVDF film calculated by measurement using I-V meter. Electro active polymers, especially polyvinylidene fluoride, its polyvinylidene fluoride trifluoroethylene copolymer and cellular polymer ferroelectrics'—have interesting and useful piezoelectric and pyroelectric properties [16].

2. Experiment

In the experiments performed and characterization of PVDF film using such equipment; Roll Hot Press machine, micrometer screw, IV meter and X-Ray Diffraction. Initially, the PVDF samples in powder form were placed on the engine cylinder is then heated until it melts and then hot press roll device is operated to obtain a thin film. Fabrication of PVDF films made for variations in temperature from 130˚C to 170˚C. PVDF film thickness was measured with a micrometer screw, the surface resistivity of film obtained by four point probes that was realized using I-V meter, β fraction and the crystalline phase is determined by using the X-Ray Diffraction. Roll hot press machine are shown in **Figure 4**.

The scheme of four point probes is shown in **Figure 5**.

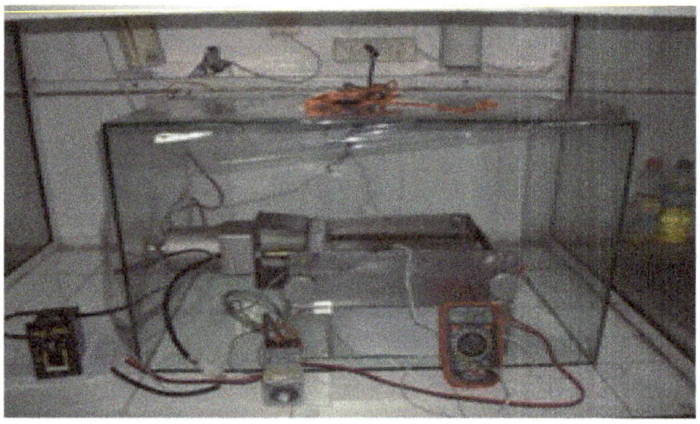

Figure 4. Roll hot press machine.

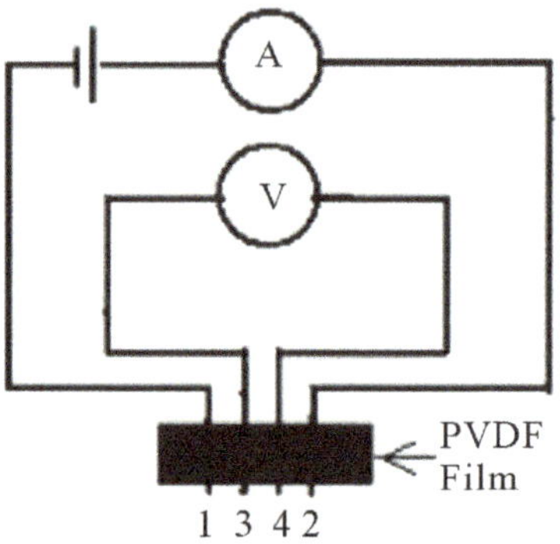

Figure 5. The point four probe scheme.

3. Results and Discussion

PVDF films have been obtained with thicknesses of 13 µm as shown in **Figure 6**. Characterization results performed on samples by using XRD.

X-Ray Diffraction characterization results of sample for temperature variation from 130˚C to 170˚C as shown in **Figure 7**.

XRD characterization results of sample with temperature at 130˚C as shown in **Figure 8**.

While the results for the characterization of the sample for temperature at 135˚C as shown in **Figure 9**.

Calculation to determine the amount of beta fraction of each sample using the Equation (1):

$$F(\beta)=\frac{A_\beta}{1.26A_\alpha+A_\beta}\times100\% \quad (1)$$

From this characterization obtained β fraction of sample for temperature variation 130˚C to 170˚C are shown in **Table 1**. Graph of increasing β fraction with increasing temperature is shown in **Figure 10**.

Figure 6. PVDF film produced.

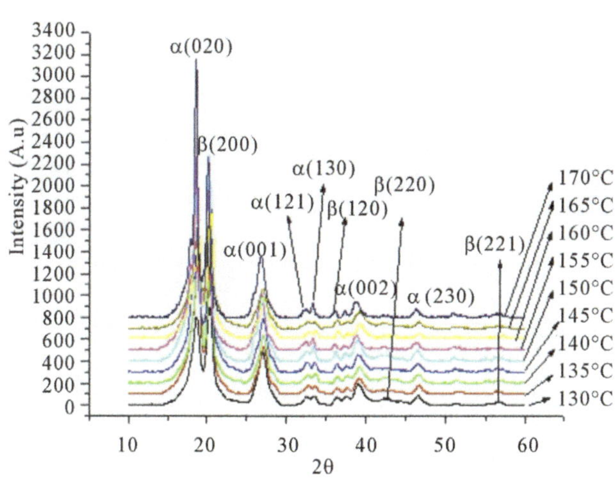

Figure 7. XRD results for sample at 130˚C to 170˚C.

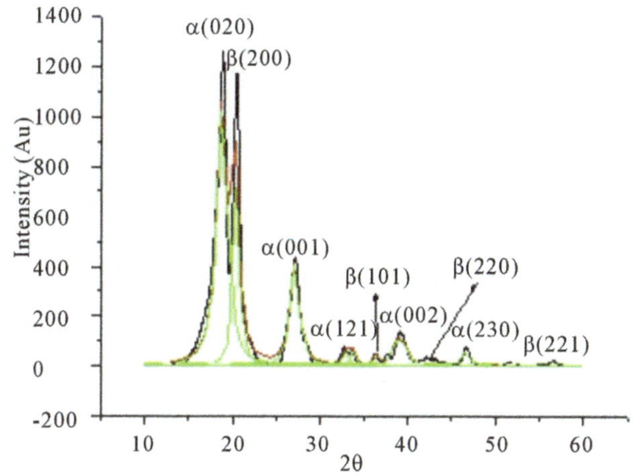

Figure 8. XRD results for sample at 130˚C.

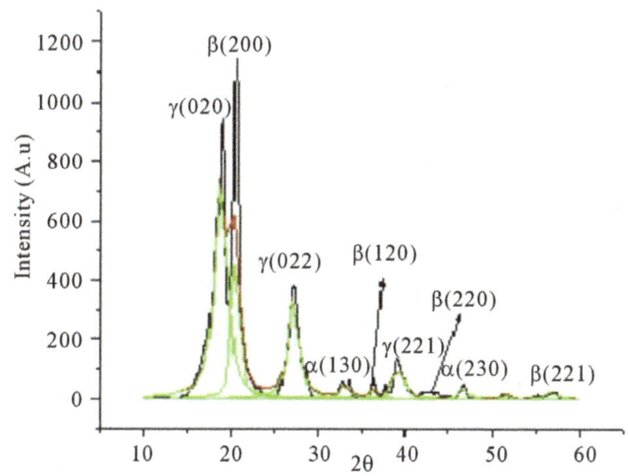

Figure 9. XRD results for sample at 135˚C.

The increase in temperature causes an increase in polarity due to the dipole orientation resulting in the transformation of the structure occurs. It is characterized by the addition of β fraction [17].

The results of current-voltage measurements for the same thickness of 13 nm with a temperature variation of 130˚C to 170˚C are shown in **Figure 11**.

The results of resistivity measurements show that the applied voltage is proportional to the increase in current, in accordance with Ohm's law.

$$R=\frac{V}{I} \quad (2)$$

where R is resistance (Ω), V is applied voltage (V) and I is current (A).

The value of R can obtained from I-V curve gradient. Resistivity of sample calculated by Equation (3):

$$\rho=2\pi RS \quad (3)$$

Table 1. β **fraction of sample for temperature variation from 130˚C up to 170˚C.**

No	Temperature (˚C)	β Fraction (%)
1	130	25.4
2	135	26.34
3	140	27.51
4	145	29.23
5	150	29.56
6	155	35.5
7	160	39.13
8	165	41.07
9	170	44

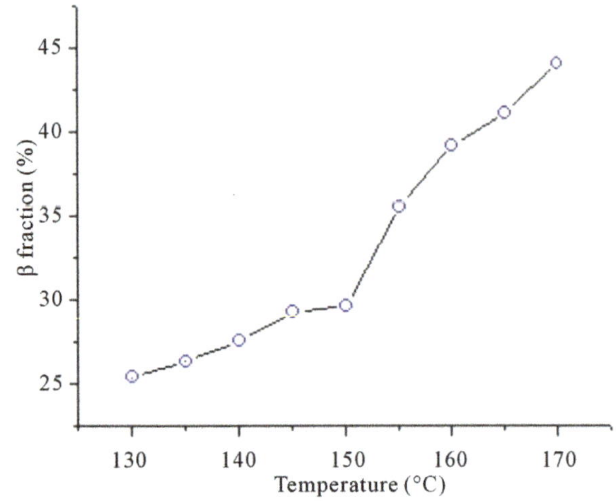

Figure 10. Graph of β **fraction increase to the temperature from 130˚C to 170˚C.**

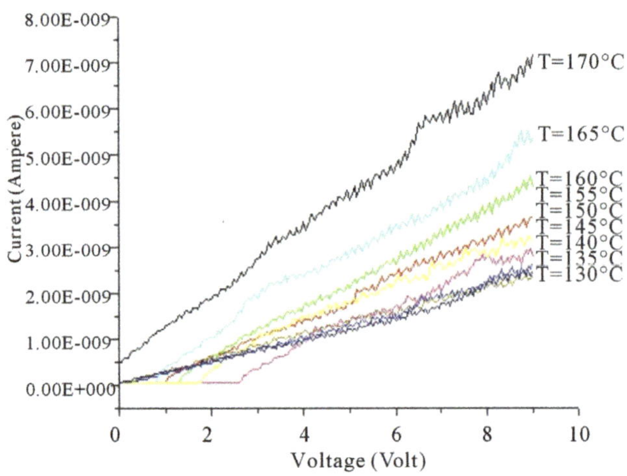

Figure 11. I-V curve of sample for variation temperature from 130˚C to 170˚C.

where ρ is resistivity (Ωm) and S is distance between the electrodes (5 mm). Surface Resistivity values of PVDF film for temperature 130˚C to 170˚C are decreased from 1.23×10^4 Ωm to 0.21×10^4 Ωm respectively. Graph of surface resistivity versus temperature are shown in **Figure 12**.

While the temperature increase causes a decrease in the resistivity value, which indicates that there has been an increase in piezoelectric properties.

Crystallite size of PVDF film is determined by using Scherrer equation. Scherrer has derived an expression for broadening of x-ray diffraction peaks due only to small crystallite sizes [18]:

$$B_{crist} = \frac{k\lambda}{L\cos\theta} \quad (4)$$

where λ is the wavelength of the x-ray used, θ is the Bragg angle, L is the "average" crystallite size measured in a direction perpendicular to the surface of the specimen, and k is a constant. Equation (4) is commonly known as the Scherrer equation and was derived based on uniform size. However, this equation is now frequently used to estimate the crystallite sizes of both cubic and noncubic materials. The constant k has been determined to vary between 0.89 and 1.39, but is usually taken as close to unity, the assumption that $k = 1.0$ is generally justifiable.

In experiment for fabrication temperature is 130˚C, this pattern was indexed, and we found that these three reflections have indices (020), (200) and (001) at the 2θ value of 18.67, 20.405, and 27.09, respectively. From this characterization and calculation has produces crystallite size is 7.2 nm.

From second experiment for fabrication temperature is 135˚C, we found that these three reflections have indi-

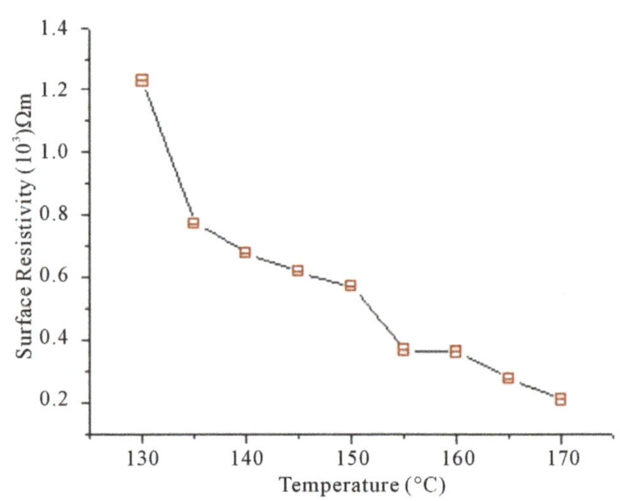

Figure 12. Graph resistivity as a function of temperature from 130˚C to 170˚C.

ces (020), (200) and (002) at the 2θ value of 18.72, 20.51, and 27.19, respectively. From this characterization and calculation has produces crystallite size is 7.9 nm. From the difraction pattern for temperature 140˚C, we found that these three reflections have indices (020), (200) and (022) at the 2θ value of 18.63, 20.40, and 27.08 respectively. From this characterization and calculation has produces crystallite size is 19 nm [15].

Completely experimental results from this experiment are shown in **Table 2**.

These results show that increasing temperatures will also increase the size of the crystallite of the sample. This happens because with the increasing temperature causes the higher the degree of crystallization of PVDF film sample is formed, so that the crystallite size also increases.

Table 2. Crystallite size for each sample.

Temperature	2θ	hkl	B_c (rad) Cosθ ($\times 10^{-2}$)	Crystallite size (nm)
	18.67	020	2.27	
130˚C	20.405	200	1.76	7.2
	27.09	001	2.42	
	18.72	020	2.25	
135˚C	20.51	200	2.09	7.9
	27.19	002	1.5	
	18.63	020	2.41	
140˚C	20.40	200	1.61	8.7
	27.08	022	1.57	
	18.72	020	2.032	
145˚C	20.51	200	1.093	10.93
	27.18	022	1.102	
	18.49	020	0.84	
150˚C	20.225	200	1.241	14
	26.92	022	1.181	
	18.57	020	1.436	
155˚C	20.37	200	0.809	14.67
	27.00	022	0.901	
	18.50	020	0.510	
160˚C	20.31	200	0.335	19
	26.99	022	1.677	
	18.80	020	0.414	
165˚C	20.44	200	0.99	19.7
	27.08	022	1.065	
	17.75	110	0.276	
170˚C	18.56	020	1.253	20.54
	20.12	200	0.731	

4. Conclusion

Has been successfully produced PVDF films with using roll hot press. XRD characterization showed increase in the β fraction when the temperature is raised. PVDF films showed a decrease in resistivity for every increase of temperature. This indicates that mechanical treatment to produce film thickness and temperature has a strong influence on the β fraction of PVDF films. Increasing of temperatures will also increase the size of the crystallite of the sample. This happens because with the increasing temperature causes the higher the degree of crystallization of PVDF film sample is formed, so that the crystallite size also increases.

5. Acknowledgements

The authors thank to Mr. Priambodo as laborant at Integrated Centre Laboratory of State Islamic University (UIN JAKARTA) for support in X-Ray Diffraction Characterization.

REFERENCES

[1] C. L. Radiman, "Kimia Polimer," Penerbit ITB, Bandung, 2004, pp. 5-20.

[2] A. J. Lovinger, "Poly(Vinylidene fluoride)," In: D. C. Bassett, Ed., *Development in Crystalline Polymers*, Applied Science Publisher, London, 1982, pp. 195-261.

[3] A. J. Lovinger, "Poyivinylidene Fluoride," Applied Science Publisher, London, 1981, p. 202.

[4] A. Hartono, S. Satira, M. Djamal, Ramli and E. Sanjaya, "Effect of Mechanical Treatment Temperature and Layer Thickness on Piezoelctic Properties PVDF Film," *Proceedings of APS*, 2012.

[5] K. Matsushige, K. Nagata, S. Imada and T. Takemura, "The II-I Crystal Transformation of Poly(vinylidene fluoride) under Tensile and Compressional Stresses," *Polymer*, Vol. 21, No. 12, 1980, pp. 1391-1397.

[6] J. C. Mcgrath, "High Effective Draw as a Route to Increased Stiffness and Electrical Response in Poly(vinylidene fluoride)," *Polymer*, Vol. 21, No. 8, 1980, pp. 855-857.

[7] J. Humphreys, *et al.*, "A Study of the Drawing Behavior of Polyvinylidene Fluoride," *Journal of Applied Polymer Science*, Vol. 30, No. 10, 1985, pp. 4069-4079.

[8] T. Hsu and P. J. Geil, "Deformation and Transformation Mechanisms of Poly(vinylidene fluoride) (PVF_2)," *Journal of Materials Science*, Vol. 24, No. 4, 1989, pp. 1219-

1232.

[9] V. Sencadas, V. M. Moreira, S. Lanceros-Mendez, A. S. Pouzada and R. Gregorio Jr., "α- to-β Transformation on PVDF Films Obtained by Uniaxial Stretch," *Materials Science Forum*, Vol. 514-516, 2006, pp. 872-876.

[10] R. P. Vijayakumar, D. V. Khakhar and A. Misra, "Studies on A to B Phase Transformations in Mechanically Deformed PVDF Films," *Journal of Applied Polymer Science*, Vol. 117, 2010, pp. 3491-3497.

[11] G. Gianotti, A. Capizzi and V. Zamboni, *Chim. e Industr.*, Vol. 55, 1973, 1973, p. 501.

[12] S. Nakamura, T. Sasaki, J. Funamoto and K. Matsuzaki, "Dynamic Cooling and Isothermal Crystallizations of Poly (vinylidene fluoride) from the Melt," *Die Makromolekulare Chemie*, Vol. 176, No. 11, 1975, pp. 3471-3481.

[13] C. Mancarella and E. Martuscelli, "Crystallization Kinetics of Poly(vinylidene fluoride)," *Polymer*, Vol. 18, No. 12, 1977, pp. 1240-1242.

[14] K. Nakagawa and Y. Ishida, "Estimation of Amorphous Specific Volume of Poly(vinylidene fluoride) as a Function of Temperature," *Kolloid-Zeitschrift und Zeitschrift für Polymere*, Vol. 251, No. 2, 1973, pp. 103-107.

[15] A. Hartono, S. Satira, M. Djamal, R. Ramli, H. Bahar and E. Sanjaya, "Effect of Fabrication Temperature on Crystallite Size of PVDF Film," *Proceedings of ICMNS*, 2012.

[16] S. B. Lang and S. Muensit, "Review of Some Lesser-Known Applications of Piezoelectric and Pyroelectric Polymers," *Applied Physics A*, Vol. 85, No. 2, 2006, pp. 125-134.

[17] A. Hartono, R. Ramli and M. D. dan S. Satira, "Pengaruh Temperatur Perlakuan Mekanik Terhadap Jumlah Fraksi β dan Resistivitas Permukaan Lapisan PVDF," *BKS PTN-BMIPA*-2012, Universitas Negeri Medan, Medan, 2012, pp. 17-22.

[18] C. Suryanarayana and M. G. Norton, "X-Ray Diffraction a Practical Approach," Plenum Press, New York, 1998.

Structure, Photochromism and Liquid Crystal Properties of 1-Alkyl-2-(Arylazo)Imidazoles (Raai-C_nH_{2n+1}, n (Even) = 10 - 22)

Avijit Nandi[1], Chandana Sen[1], Debashis Mallick[2], Randhir K. Sinha[1], Chittaranjan Sinha[1*]
[1]Department of Chemistry, Inorganic Chemistry Section, Jadavpur University, Kolkata, India
[2]Department of Chemistry, Mrinalini Datta Mahavidyapith, Birati, India

ABSTRACT

Photochromism is the reversible structural change of the molecules upon light irradiation having different spectral pattern before and after irradiation. Liquid crystal (LC) is a long-range orientational order of the molecules in between solid and isotropic liquid phases which share the properties of the crystal and liquid phases. 1-Alkyl-2-(arylazo)imidazoles, (Raai-C_nH_{2n+1}, R = H, Me and n = 10, 12, 14, 16, 18, 20, 22) exhibits photochromism and two of them (Haai-$C_{18}H_{37}$ and Haai-$C_{22}H_{45}$) show Liquid crystal (LC) properties. One of these compounds, Meaai-$C_{16}H_{33}$ is structurally characterized by single crystal X-ray diffraction study. The compounds upon irradiation with UV light show E-to-Z isomerisation. Quantum yields ($\phi_{E\rightarrow Z}$) of E-to-Z isomerisation vary with the molar mass of the compounds. The reverse transformation, Z-to-E, is recorded in thermal condition. The activation energy (E_a) of isomerisation is determined by the controlled temperature reaction. Liquid crystal property of the molecules is examined by polarizing optical microscopy (POM) and DSC. The carbon chain length n = 18 shows nematic (N)-isotropic (I) phase transition and an unknown Smectic (Sm) phase is observed with n = 22 on heating cycle.

Keywords: Arylazoimidazoles; Structure; Photochromism; Liquid Crystal; Nematic-Isotropic Phase Transition; Smectic Phase

1. Introduction

Smart molecules can respond to a specific external stimulus in a specific manner and are of increasing interest in areas as diverse as microbiology [1], microfluidics [2], organic electronics [3,4], high density optical data storage devices [5,6] etc. Using of light as external stimulating agent has manifold advantages. Fritzsche (1867) observed the bleaching of an orange colored tetracene solution in daylight and then re-colored at night [7]. However, during the 1950s, the research groups of Hirshberg and Fischer carried out comprehensive studies on light irradiated color changes and the subsequent physical and chemical changes of different organic molecules [8]. The research field has received a renewable thrust in the 1980s when the photo-induced reversible transformation was observed for organic molecules like spirooxazines, diarylethenes etc. without photodegradation. This process is known as photochromism. It is defined as "reversible transformation of a chemical species induced in one or both directions by the absorption of electromagnetic radiation between two forms having different absorption spectra" [9]. The photochromic compounds have become appealing for their potential advantages in the construction of photo-switching devices and supramolecular architectures [9-15]. Various organic compounds show cis(Z)/trans(E) isomerization, pericyclic reactions, H atom transfer reactions, and dissociation processes. The molecules have been used as photoresponsive functional devices utilized as smart polymers [16], liquid crystals [17], molecular switches [18] and machines [19].

Azobenzene is used for a wide range of applications, from dye indicators and nonlinear optical chromophores to quantum interferences [20]. The molecule shows photoinduced E/Z isomerization by UV or visible light. More recently the synthesis and the characterization of heterocyclic azo dyes such as arylazopyridines have been extensively reported [21-24]. The prospect of azobenzene has encouraged us for synthesizing arylazoimidazoles

*Corresponding author.

[25] which acts as a potential switching group in biology and in coordination chemistry, since imidazole is a ubiquitous and essential group in biology, especially as a metal coordinating. The photochromism of 1-alkyl-2-(arylazo)imidazole (RaaiR) [26,27] and some of their complexes [28-31] are also reported.

The liquid crystal (LC) is a long-range orientational order and lies in between solid and isotropic liquid phases. The LC phase (mesophase) shares properties of the crystal and liquid phases. Some examples of anisotropic properties are birefringence, alignment in electric and magnetic fields (electrical permittivity and magnetic susceptibility), elasticity, viscosity, and conductivity [32,33]. Liquid crystals are widely used in information display technology and the anisotropic fluid states of rigid polymers are used for processing of high strength fibers. Liquid crystals (LCs) should also be useful in bio-related fields because the self-organized structures of lyotropic liquid crystals are compatible with those in living systems [34].

Comparing with azobenzene photochrome, arylazoimidazoles have several advantages of possessing both the photoactivity related to the azo unit, the capability of self-assembly through the imidazolyl group and coordinating ability to metal ions. Azoimidazoles display surprisingly a strong tendency toward self-assembly, which offers a new and efficient way to make photoactive liquid crystal (LC) materials. Attachment of long chain alkyl groups to aromatic molecules may develop liquid crystal (LC) property. The E form of azo derivatives is a rod and is nematic LCs (NLCs) while the Z is bent and tends to destabilize the phase structure. Therefore, E/Z photoisomerization of the azobenzene in the N phase can disorganize the phase structure of NLCs, resulting in a nematic-isotropic (N-I) isothermal phase transition (photochemical phase transition). In this article, we describe a preliminary account on photochromic LC (PLC) activity of some long chain alkyl group substituted 1-alkyl-2-(arylazo)imidazoles, Raai-C_nH_{2n+1} (n = 10, 12, 14, 16, 18, 20, 22).

2. Experimental

2.1. Materials

1-Alkyl-2-(arylazo)imidazoles were synthesized by reported procedure [27]. 1-Bromo-n-alkanes, n-$C_{10}H_{21}$-1-Br, n-$C_{12}H_{25}$-1-Br, $C_{14}H_{29}$-1-Br and 1-chloro-n-alkanes, $C_{16}H_{33}$-1-Cl, $C_{18}H_{37}$-1-Cl, $C_{20}H_{41}$-1-Cl and $C_{22}H_{45}$-1-Cl were purchased from Sigma-Aldrich and used as such. All other chemicals and solvents were analytical reagent grade as received.

2.2. Physical Measurements

Microanalytical data (C, H, N) were collected on Perkin-Elmer 2400 CHNS/O elemental analyzer. Spectroscopic data were obtained using the following instruments: UV-Vis spectra from a Perkin Elmer Lambda 25 spectrophotometer; IR spectra (KBr disk, 4000 - 400 cm^{-1}) from a Perkin Elmer RX-1 FTIR spectrophotometer; photo excitation has been carried out using a Perkin Elmer LS-55 spectrofluorimeter and ^1H NMR spectra were recorded from a Bruker (AC) 300 MHz FTNMR spectrometer. The liquid crystalline properties were established by thermal microscopy (Nikon polarizing microscope LV100POL attached with Linkam Temperature controlled stage model Examina-THMS600, and the phase transitions were confirmed by differential scanning calorimetry (Perkin-Elmer Diamond DSC Pyris1 system).

2.3. Synthesis of Compounds

2-(Arylazo)imidazoles were synthesized by the diazotization of aniline or *p*-toluidine in $NaNO_2$/HCl at low temperature (0˚C - 5˚C) and followed by coupling with imidazole in aqueous sodium carbonate solution (pH-7). The alkylation was carried out by adding alkyl halide in dry THF solution to the corresponding 1-alkyl-2-(arylazo) imidazole in presence of NaH (**Scheme 1**). Long chain alkyl groups are appended at imidazole motif to synthesise present series of molecules. The compounds are 1-alkyl-2-(arylazo)imidazole [Raai-C_nH_{2n+1}, where R = H (**a**), Me (**b**); C_nH_{2n+1}—where n = 10 ($C_{10}H_{21}$, **1**), 12 ($C_{12}H_{25}$, **2**), 14 ($C_{14}H_{29}$, **3**), 16 ($C_{16}H_{33}$, **4**), 18 ($C_{18}H_{37}$, **5**), 20 ($C_{20}H_{41}$, **6**), 22 ($C_{22}H_{45}$, **7**)]. Detail synthesis for Me-aai-$C_{10}H_{21}$ (**1b**) is given below.

To dry THF solution (75 ml) of 2-(*p*-tolylazo)imidazole (2 g, 10.8 mmol), NaH (50% paraffin) (0.56 g) was added in small portion and stirred at cold condition on ice bath for 0.5 h. 1-Bromodecane (n-$C_{10}H_{21}$-1-Br; 2.50 g, 11.31 mmol) in dry THF (25 ml) was added slowly through pressure equalizing system under stirring condition for a period of 1 h and then warm for another 1 h on steam bath. The solution was evaporated to dryness, extracted with CH_2Cl_2, washed with NaOH solution (10%, 10 ml × 3) and finally by distilled water (20 ml × 3). The CH_2Cl_2 extract was chromatographed over silica gel-column (45 × 1 ml) prepared in petroleum ether (40˚C - 60˚C). The elution was performed by ethylacetate-pe-

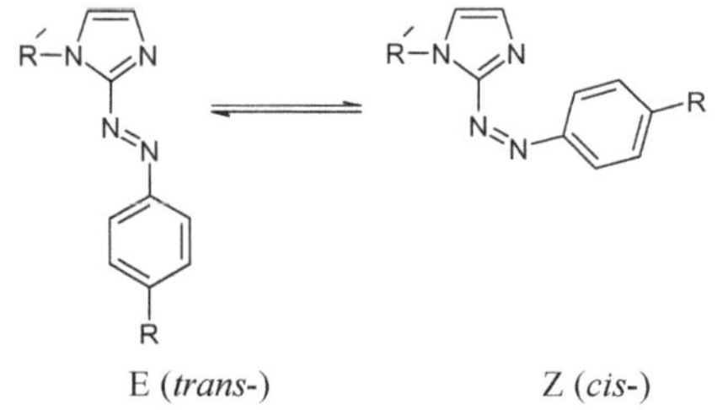

Scheme 1. Isomerization of 1-alkyl-2-(arylazo)imidazole.

troleum ether (1:3 v/v) mixture and evaporated under reduced pressure. An orange crystalline product was obtained and then dried over P_4O_{10} under vacuum. Yield of the product was 2.28 g (65%).

All other compounds were prepared and purified by identical procedure and the yield was varied 55% - 70%. The microanalytical data of the compounds are as follows: Haai-$C_{10}H_{21}$ (**1a**), 54˚C ± 1˚C, *Anal.* Found: C, 73.00; H, 9.00; N, 18.00 Calc. for $C_{19}H_{28}N_4$: C, 73.08; H, 8.97; N, 17.95. Meaai-$C_{10}H_{21}$ (**1b**), 53˚C ± 1˚C, *Anal.* Found: C, 73.70; H, 9.13; N, 17.23 Calc. for $C_{20}H_{30}N_4$: C, 73.62; H, 9.20; N, 17.18. Haai-$C_{12}H_{25}$ (**2a**), 54˚C ± 1˚C, *Anal.* Found: C, 74.20; H, 9.36; N, 16.40. Calc. for $C_{21}H_{32}N_4$: C, 74.12; H, 9.41; N, 16.47. Meaai-$C_{12}H_{25}$ (**2b**), 52˚C ± 1˚C, *Anal.* Found: C, 74.64; H, 9.55; N, 15.90 Calc. for $C_{22}H_{34}N_4$: C, 74.58; H, 9.60; N, 15.82. Haai-$C_{14}H_{29}$ (**3a**), 54˚C ± 1˚C, *Anal.* Found: C, 74.94; H, 9.84; N, 15.30 Calc. for $C_{23}H_{36}N_4$: C, 75.00; H, 9.78; N, 15.22. Meaai-$C_{14}H_{29}$ (**3b**), 53˚C ± 1˚C *Anal.* Found: C, 75.47; H, 9.92; N, 14.72 Calc. for $C_{24}H_{38}N_4$: C, 75.39; H, 9.95; N, 14.66. Haai-$C_{16}H_{33}$ (**4a**), 53 ± 1˚C, *Anal.* Found: C, 75.70; H, 10.05; N, 14.20 Calc. for $C_{25}H_{40}N_4$: C, 75.76; H, 10.10; N, 14.14. Meaai-$C_{16}H_{33}$ (**4b**), 54˚C ± 1˚C, *Anal.* Found: C, 76.12; H, 10.20; N, 13.60 Calc. for $C_{26}H_{42}N_4$: C, 76.09; H, 10.24; N, 13.66. Haai-$C_{18}H_{37}$ (**5a**), 59˚C ± 1˚C, *Anal.* Found: C, 76.50; H, 10.35; N, 13.30. Calc. for $C_{27}H_{44}N_4$: C, 76.42; H, 10.38; N, 13.21. Meaai-$C_{18}H_{37}$ (**5b**), 56˚C ± 1˚C, *Anal.* Found: C, 76.80; H, 10.55; N, 12.72 Calc. for $C_{28}H_{46}N_4$: C, 76.71; H, 10.50; N, 12.79. Haai-$C_{20}H_{41}$ (**6a**), 65˚C ± 1˚C, *Anal.* Found: C, 76.94; H, 10.66; N, 12.30 Calc. for $C_{29}H_{48}N_4$: C, 76.99; H, 10.62; N, 12.39. Meaai-$C_{20}H_{41}$ (**6b**), 63˚C ± 1˚C, *Anal.* Found: C, 77.31; H, 10.76; N, 12.06 Calc. for $C_{30}H_{50}N_4$: C, 77.25; H, 10.73; N, 12.02. Haai-$C_{22}H_{45}$ (**7a**), 65˚C ± 1˚C, *Anal.* Found: C, 77.60; H, 10.86; N, 11.70 Calc. for $C_{31}H_{52}N_4$: C, 77.51; H, 10.83; N, 11.67. Meaai-$C_{22}H_{45}$ (**7b**), 63˚C ± 1˚C, *Anal.* Found: C, 77.80 H, 10.88; N, 11.30 Calc. for $C_{32}H_{54}N_4$: C, 77.73; H, 10.93; N, 11.34%.

2.4. X-Ray Diffraction Study

The crystallographic data are shown in **Table 1**. Suitable single crystal of Meaai-$C_{16}H_{33}$ (**4b**) was mounted on a Siemens CCD diffractometer equipped with graphite monochromated Mo-K_α (λ = 0.71073 Å) radiation. The unit cell parameters and crystal-orientation matrices were determined for two complexes by least squares refinements of all reflections. The intensity data were corrected for Lorentz and polarisation effects and an empirical absorption correction were also employed using the SAINT program. Data were collected applying the condition $I > 2\sigma(I)$. All these structures were solved by direct methods and followed by successive Fourier and difference Fourier syntheses. Full matrix least squares refinements on F^2 were carried out using SHELXL-97 with anisotropic displacement parameters for all non-hydrogen atoms. Hydrogen atoms were constrained to ride on the respective carbon or nitrogen atoms with an isotropic displacement parameters equal to 1.2 times the equivalent isotropic displacement of their parent atom in all cases. Complex neutral atom scattering factors were used throughout for all cases. All calculations were carried out using SHELXS 97 [35], SHELXL97 [36] and PLATON 99 [37].

2.5. Computational Study

The direction of the dipole of the compound **5a** was calculated by Density Functional Theory (DFT), which may determine the presence of dipolar interaction within the molecules. All calculations have been performed by the Chem3D (version 10) [38] with GAUSSIAN 03 Interface [39]. The first optimization was done by the MM2 method, and finally DFT was performed as stated earlier.

3. Results and Discussion

3.1. Synthesis and Formulation

1-Alkyl-2-(arylazo)imidazole (Raai-C_nH_{2n+1}) are synthesized by controlled alkylation of 2-(arylazo)imidazole (RaaiH) by adding alkyl halide ($C_nH_{2n+1}X$) in dry THF using NaH. RaaiH are synthesized by diazo reaction of aryldiazonium salt with imidazole at pH 7 (**Scheme 2**). Alkyl halides of varying carbon chain (C_{10} to C_{22}) are used to synthesize Raai-C_nH_{2n+1} (n = 10 ($C_{10}H_{21}$, **1**), 12 ($C_{12}H_{25}$, **2**), 14 ($C_{14}H_{29}$, **3**), 16 ($C_{16}H_{33}$, **4**), 18 ($C_{18}H_{37}$, **5**), 20 ($C_{20}H_{41}$, **6**), 22 ($C_{22}H_{45}$, **7**)). The compounds are purified by chromatographic separation in silica gel column using acetonitrile-benzene (8:2, v/v). The elemental composition is supported by microanalytical data and they are low melting solid (*vide Experimental section*).

3.2. Spectral Studies

A moderately intense stretching for ν(C=N) is observed at 1585 - 1615 and ν(N=N) at 1370 - 1374 cm^{-1} (**Table 2**).

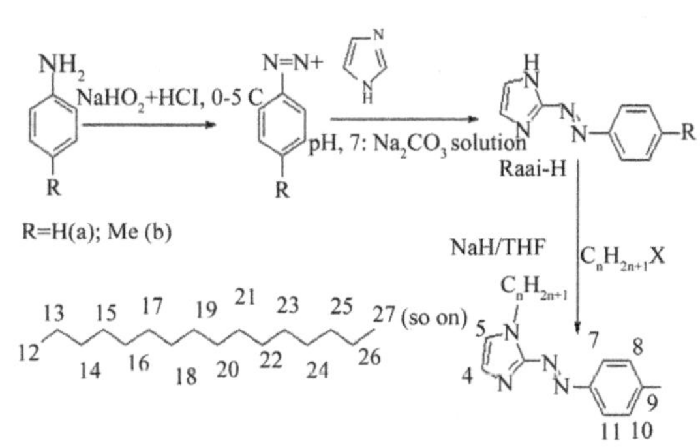

C_nH_{2n+1} n = 10 ($C_{10}H_{21}$, **1**), 12 ($C_{12}H_{25}$, **2**), 14 ($C_{14}H_{29}$, **3**), 16 ($C_{16}H_{33}$, **4**), 18 ($C_{18}H_{37}$, **5**), 20 ($C_{20}H_{41}$, **6**), 22 ($C_{22}H_{45}$, **7**)

Scheme 2. Synthetic process of Raai-C_nH_{2n+1} (1 - 7).

Table 1. Summarized crystallographic data for Meaai-$C_{16}H_{33}$ (4b).

	4b
Empirical formula	$C_{26}H_{42}N_4$
Formula weight	410.64
Temperature (K)	296
Crystal system	Monoclinic
Space group	P21/c
Crystal size $(mm)^3$	0.10 × 0.35 × 0.40
a(Å)	24.405(6)
b(Å)	11.049(3)
c(Å)	9.927(3)
β(°)	93.179(7)
$V(Å)^3$	2672.7(13)
Z	4
μ (MoK_{α}) (mm^{-1})	0.060
θ range	1.7 - 25.1
hkl range	−29 < h < 28; −13 < k < 12; −8 < l < 11
D_{calc} ($mg \cdot m^{-3}$)	1.020
Refine parameters	273
Total reflections	15,392
Unique reflections	4696
R_1^a [$I > 2\sigma(I)$]	0.0647
wR_2^b	0.1409
Goodness of fit	1.039

$^aR = \Sigma||F_o|-|F_c||/\Sigma|F_o|$; $^bwR_2 = [\Sigma w(F_o^2-F_c^2)^2/\Sigma w(F_o^2)^2]^{1/2}$, $w = 1/[\sigma^2(F_0)^2 + (0.0986P)^2]$ for **4b**; where $P = (F_0^2 2^2 + 2Fc^2 2^2)/3$.

The stretching frequency of –N=N- (exocylic) are insignificantly affected on comparing with –C=N- (imidazolyl) group. This may support effect of alkylation at N(1)-imidazolyl position of the compounds [27].

The absorption spectra of the compounds were recorded in MeOH solution in the wavelength range 200 - 900 nm (**Figure 1**, **Table 2**). The absorption spectra of Raai-C_nH_{2n+1} show main absorption band at 355 - 375 nm with a molar absorption coefficient in the order of 10^4

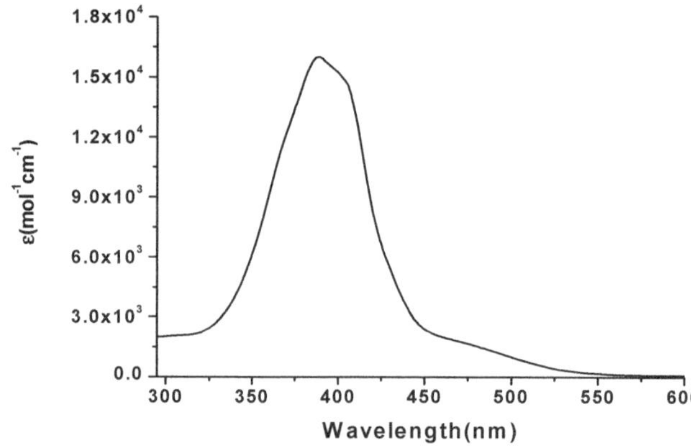

Figure 1. UV-Vis spectrum of Haai-$C_{20}H_{41}$ (6a) in methanol.

$M^{-1} \cdot cm^2$ and a weak band at 260 - 280 nm and 450 - 455 nm. The intense absorption band corresponds to π-π^* transitions, while the tail portion is due to n-π^* transition. With increasing chain length the absorption maxima have been shifted to longer wavelength which supports the enhancement of inductive effect of substituent with increase in number of bonded carbon centres at N(1)-imidazolyl group. The –N=N- is a part of a conjugated system of double bonds, both the **n**$\rightarrow\pi^*$ and the $\pi\rightarrow\pi^*$ bands are shifted to longer wavelength [40].

The ^1H NMR spectra of Raai-C_nH_{2n+1} (**1-7**) are recorded in $CDCl_3$. The atom numbering is shown in **Scheme 2**. The spectral data are set out in **Table 3**. The proton signals are assigned on the basis of spin-spin interaction and effect of substitution. The alkylation of imidazole is supported by the disappearance of δ(N–H) at ~10.30 ppm and the appearance of 1-alkyl signal at 0.85 - 4.40 ppm for -N–CH_2–(CH_2)n–CH_3. A triplet signal for -N–CH_2– appears at 4.40 ppm; and multiplet is observed for –(CH_2)n– at 0.80 - 1.50 ppm. Imidazolyl 4- and 5-H appear as broad singlet at 7.24 - 7.28 and 7.14 - 7.17 ppm, respectively. Broadening may be due to rapid proton exchange between these imidazolic protons. The aryl protons (7-H to 11-H) are upfield shifted on going from phenylazo (**a**) to *p*-tolylazo (**b**) which may be due to +I effect of –Me group.

3.3. Molecular Structure

The molecular structure of Meaai-$C_{16}H_{33}$ (**4b**) is shown in **Figure 2**. The bond parameters are listed in **Table 4**. The bond length and angles of the fragment of the structure, (*p*-tolylazo)imidazole, are comparable to the reported structure of Haai-Me [27]. The azo N-N distance is 1.266(4) Å which is slightly longer than the reported value (1.257(6) Å) [27]. Two groups around N=N are not perfectly planar and the dihedral angle between the two least squares planes is *avg.* 32°. The crowding of long chain, 1-$C_{16}H_{33}$ may be the reason for deviation. The

Table 2. FT-IR[a] and UV-VIS[b] spectral data of the compounds.

Compounds	IR in KBr disc (cm^{-1})		UV-Vis spectral data λ_{max}(nm), (10^{-3} ∈ ($dm^3 \cdot mol^{-1} \cdot cm^{-1}$))
	ν(N=N)	ν(C=N)	
Haai-$C_{10}H_{21}$ (1a)	1403	1624	362 (25.9), 374 (24.1), 451 (2.4)
Meaai-$C_{10}H_{21}$ (1b)	1404	1625	363 (25.9), 376 (22.9), 451(2.1)
Haai-$C_{12}H_{25}$ (2a)	1404	1624	363 (27.1), 372 (23.9), 451(2.3)
Meaai-$C_{12}H_{25}$ (2b)	1405	1623	364 (28.9), 376 (22.9) ,452 (2.3)
Haai-$C_{14}H_{29}$ (3a)	1403	1624	365 (25.2), 376 (24.2), 454 (2.1)
Meaai-$C_{14}H_{29}$ (3b)	1403	1625	364 (26.9), 375 (23.6), 455 (2.2)
Haai-$C_{16}H_{33}$ (4[a])	1404	1624	365 (26.2), 375 (24.1), 451 (2.4)
Meaai-$C_{16}H_{33}$ (4b)	1404	1625	367 (25.9), 377 (24.5), 453 (2.1)
Haai-$C_{18}H_{37}$ (5a)	1405	1623	367 (25.8), 378 (24.2), 454 (2.1)
Meaai-$C_{18}H_{37}$ (5b)	1405	1625	368 (26.9), 378 (26.1), 455 (2.2)
Haai-$C_{20}H_{41}$ (6a)	1404	1624	370 (25.2), 380 (22.1), 458 (2.4)
Meaai-$C_{20}H_{41}$ (6b)	1405	1624	372 (25.9), 385 (23.6), 460 (2.1)
Haai-$C_{22}H_{45}$ (7a)	1404	1622	373 (19.3), 386 (17.3), 460 (1.98)
Meaai-$C_{22}H_{45}$ (7b)	1406	1620	373 (2.3), 385 (29.2), 462 (3.6)

[a]In KBr disc; [b]In methanol.

Table 3. ^{1}H NMR spectral data in $CDCl_3$ at room temperature.

Compounds	4-H[s]	5-H[s]	7,11-H[d]	8,10-H	9-R[s]	12-CH_2[t]	$-(CH_2)_{n-1}$-CH_3[*]
Haai-$C_{10}H_{21}$ (**1a**)	7.35	7.16	7.97 (8.2)	7.45[m]	7.43	4.43 (6.7)	1.88 - 1.28
Meaai-$C_{10}H_{21}$ (**1b**)	7.32	7.15	7.88 (8.5)	7.28 (7.0)[d]	2.42	4.40 (6.4)	1.87 - 1.29
Haai-$C_{12}H_{25}$ (**2a**)	7.34	7.17	7.94 (8.0)	7.46[m]	7.43	4.44 (7.1)	1.88 - 1.34
Meaai-$C_{12}H_{25}$ (**2b**)	7.31	7.14	7.90 (7.5)	7.28 (7.0)[d]	2.44	4.40 (6.4)	1.86 - 1.23
Haai-$C_{14}H_{29}$ (**3a**)	7.34	7.17	7.96 (8.1)	7.47[m]	7.46	4.42 (6.6)	1.89 - 1.25
Meaai-$C_{14}H_{29}$ (**3b**)	7.33	7.14	7.89 (8.6)	7.27 (7.0)[d]	2.41	4.41 (7.0)	1.89 - 1.28
Haai-$C_{16}H_{33}$ (**4a**)	7.35	7.16	7.97 (8.0)	7.49[m]	7.49	4.42 (6.6)	1.90 - 1.25
Meaai-$C_{16}H_{33}$ (**4b**)	7.27	7.14	7.96 (7.8)	7.27 (7.4)[d]	2.44	4.43 (7.0)	1.70 - 1.22
Haai-$C_{18}H_{37}$ (**5a**)	7.36	7.17	7.97 (8.2)	7.50[m]	7.49	4.41 (7.0)	1.89 - 1.23
Meaai-$C_{18}H_{37}$ (**5b**)	7.31	7.14	7.88 (8.5)	7.26 (7.4)[d]	2.43	4.40 (7.2)	1.87 - 1.21
Haai-$C_{20}H_{41}$ (**6a**)	7.35	7.16	7.96 (8.1)	7.52[m]	7.51	4.41 (7.0)	1.88 - 1.24
Meaai-$C_{20}H_{21}$ (**6b**)	7.32	7.14	7.89 (8.6)	7.27 (7.0)[d]	2.41	4.43 (7.0)	1.89 - 1.21
Haai-$C_{22}H_{45}$ (**7a**)	7.36	7.16	7.97 (8.2)	7.53[m]	7.52	4.42 (6.6)	1.90 - 1.22
Meaai-$C_{22}H_{45}$ (**7b**)	7.27	7.14	7.88 (8.5)	7.26 (7.0)[d]	2.41	4.40 (7.2)	1.86 - 1.22

[s]Singlet; [d]Doublet; [t]Triplet; [bs]Broad Singlet; [m] Multiplet; [*]protons of $-(CH_2)_{n-1}$-CH_3.

Table 4. Selected bond distances (Å) and angles (°) for Meaai-$C_{16}H_{33}$ (4b).

Bond distances (Å)		Bond angles (°)	
N(2)-C(1)	1.369(4)	C(12)-C(13)-C(14)	116.1(4)
N (2)-C(3)	1.352(50)	C(13)-C(14)-C(15)	117.4(4)
C(2)-C(3)	1.365(5)	C(23)-C(24)-C(25)	120.7(8)
N(1)-C(2)	1.356(4)	C(24)-C(25)-C(26)	119.8(9)
N(3)-C(1)	1.384(5)		
N(4-N(3)	1.262(4)		
N(4)-C(4)			

bonding strength between N and *p*-tolyl-C [C(8)–N(2), 1.421(4) Å] is weaker than that of N and imidazolyl-C [C(1)–N(1), 1.385(4) Å]. This suggests that the intramolecular interaction between azo-N and imidazolyl-C wave functions is stronger than that of *p*-tolyl orbitals. The C(3)-C(4) bond length (1.364(5) Å) of imidazolyl group is somewhat greater than reported data [41] which may be due to the steric demand of 1-$C_{16}H_{33}$ appended to imidazolyl group. The bond angle values at chain of C(15)-C(16)-C(17) (116.3(4)°) C(20)-C(21)-C(22) (120.6 (6)°), C(24)-C(25)-C(26) (123.6(8)°) etc. show gradual increment which may be due to the removal of steric strain from the neighboring groups arranged in a zigzag manner. The imidazolyl-N is hydrogen bonded, N(3)---H(3)-C(3), (N(3)---H(3), 2.51 Å; N(3)---C(3), 3.390(5) Å and ∠ N(3)---H(3)-C(3), 158°; symmetry, -x, -1/2+y, 1/2-z) to imidazolyl-C-H giving 1D chain (**Figure 3**). Packing view in the unit cell shows π---π interaction (3.3477(16) Å; symmetry, 1-x, 2-y, -z) between adjacent imidazolyl groups to form 2D supramolecular structure (**Figure 4**).

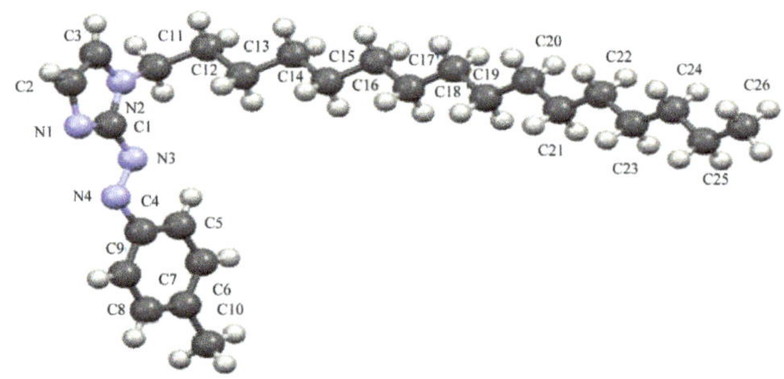

Figure 2. Crystal structure of Meaai-$C_{16}H_{33}$ (4b).

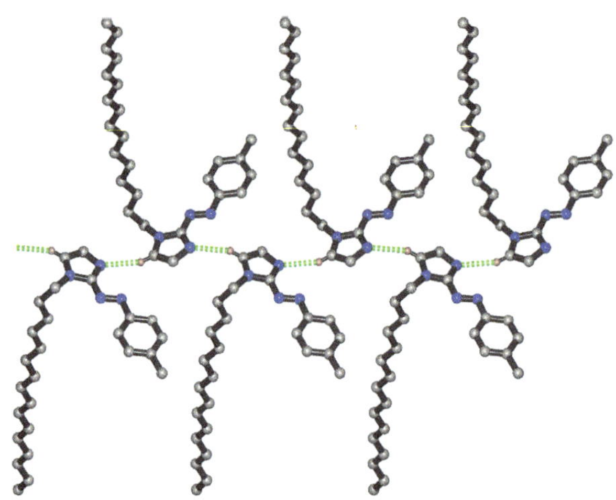

Figure 3. 1D supramolecular chains are formed along crystallographic b-axis through hydrogen bonding interactions of Meaai-$C_{16}H_{33}$ (4b).

3.4. Photochromism

The irradiation of UV light at λ_{max} to a MeOH solution of the ligands has shown the absorption spectral change (**Figure 5**). The intense peak at λ_{max} decreases, which is accompanied by a slight increase at the tail portion of the spectrum until a photostationary state (PSS-I) is reached. Subsequent irradiation at the newly appeared longer wavelength peak reverses the course of the reaction and the original spectrum is recovered up to a point, which is another photostationary state (PSS-II) under irradiation at the longer wavelength peak. The quantum yields of the E-to-Z isomerisation were determined following previously reported procedure [26,27] and the results are set out in **Table 5**.

The absorption spectra of the E-isomer of ligands have changed with isosbestic points upon excitation (**Figure 5**) to the Z-isomer. The ligands show little sign of degradation upon repeated irradiation (**Figure 5(b)**). The quantum yields were measured for the E-to-Z ($\phi_{E\to Z}$) photoisomerisation of these ligands in MeOH. The $\phi_{E\to Z}$ values are significantly dependent on nature of substituents, and chain length of -C_nH_{2n+1}. The Me substituent, Haai-C_nH_{2n+1} to Meaai-C_nH_{2n+1}, and the alkyl substituent C_nH_{2n+1} (n = 10, 12, 14, 16, 18, 20, 22) at N(1)-position reduce $\phi_{E\to Z}$ values. In general, increase in mass of the molecule reduces the rate of isomerisation, E-to-Z. The rotor volume and mass have significant influence on the

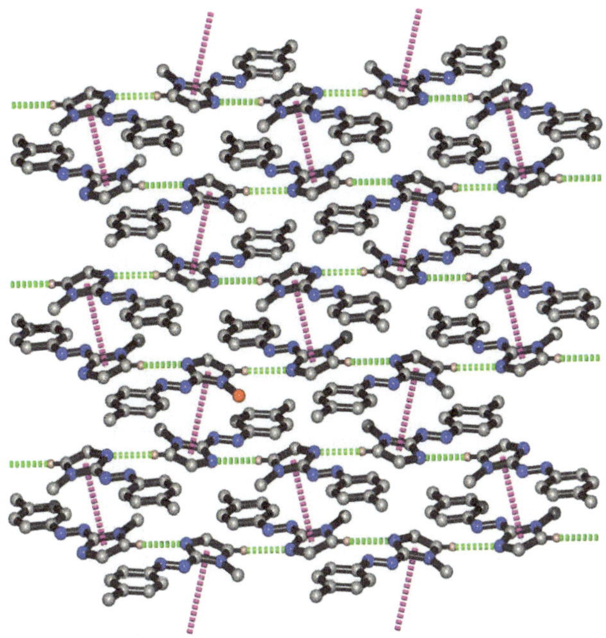

Figure 4. 2D supramolecular structure of Meaai-$C_{16}H_{33}$ (4b) formed by hydrogen bonding (Green) and π---π (magenta) interactions (for clarity the long aliphatic chains started from C15 (Red in color) and hydrogen atoms are not given).

isomerisation rate and quantum yields.

Thermal Z-to-E isomerisation of the ligands was followed by UV-Vis spectroscopy in MeOH at 298 - 313 K (**Table 6**). The Eyring plots in the range 298 - 313 K gave a linear graph from which the activation energy was obtained (**Table 6, Figure 6**). With the increase of the alkyl chain length on N atom of the imidazole moiety, the E_as are reduced which means faster Z-to-E thermal isomerisation. The entropy of activation (ΔS^*) are high negative in the organic compounds having large alkyl groups than that of the compounds having short alkyl chain. This is also in support of increase in rotor volume in the organic fragments.

3.5. Liquid Crystal Behaviour and Tansition Enthalpies of Haai-C_nH_{2n+1} (n = 18 (5a) and 22 (7a))

The thermal properties of the compounds were studied using polarized light optical microscopy (POM) and differential scanning calorimetry (DSC). Out of fourteen compounds only two compounds, Haai-$C_{18}H_{37}$ (**5a**) and Haai-$C_{22}H_{45}$ (**7a**) show mesomorphic response. The sample is sandwiched between untreated slide and a coverslip and is under hot stage of POM. Other twelve compounds show common crystalline-isotropic transition and the DSC plots do not show significant changes. The compound, **5a** exhibits nematic (N) texture at 55°C (**Figure 7**) while **7a** shows unknown Smectic (Sm) phase at 62°C (**Figure 8**) and, finally shift to isotropic (I) phase at 58°C (**5a**) and 65°C (**7a**) respectively. The crystal and nematic phases are under thermal equilibrium. During cooling from isotropic phase, **5a** shows strongly fluctuating textures of the N phase and has been transformed to textures in which "Schlieren" regions can be seen (**Figure 7**). Homeotropic and homogenous oriented domains cause this coexistence, which permits the assignment of the smectic phase in compound **7a**. The transition temperature and associated enthalpies are Cr → N 37.4°C (15.9 KJ·mol^{-1}) → Iso 58°C (0.5 KJ·mol^{-1}/heating cycle) → N 44.1°C (12.6 KJ·mol^{-1}/cooling cycle) for **5a (Figure 9)**.

3.6. Computational Study

The Density Functional Theory (DFT) computation has been used to calculate dipole moment of **5a** and **7a**. It

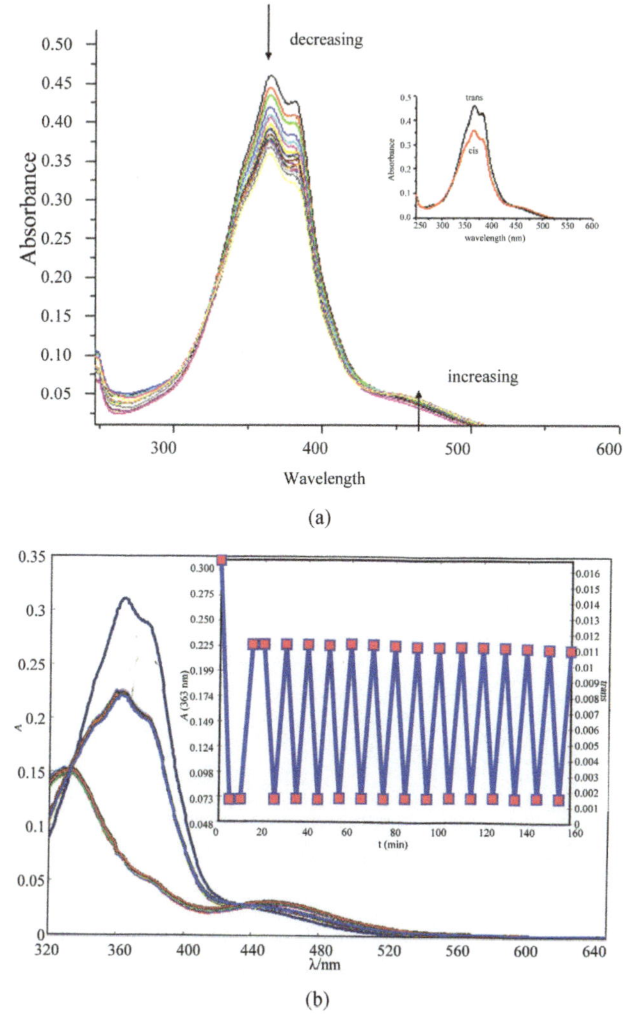

Figure 5. (a) Spectral changes of Haai-$C_{16}H_{33}$ (4a) in MeOH upon repeated irradiation at 365 nm at 3 min interval at 25°C. Inset figure shows the spectra of Z and E isomers of Haai-$C_{16}H_{33}$; (b) Resistance to repeated irradiation in MeOH solution of Haai-$C_{16}H_{33}$ (4a).

Table 5. Results of photochromism, rate of conversion and quantum yields upon UV light irradiation.

Compounds	λ_{π,π^*} (nm)	Isobestic point (nm)	Rate of E→Z conversion × 10^9 (s^{-1})	$\phi_{E\to Z}$ Conversion × 10^9
Haai-$C_{10}H_{21}$ (**1a**)	362	328, 424	34.11	1.61
Meaai-$C_{10}H_{21}$ (**1b**)	363	326, 425	33.12	1.43
Haai-$C_{12}H_{25}$ (**2a**)	363	330, 427	32.76	1.54
Meaai-$C_{12}H_{25}$ (**2b**)	364	334, 425	32.14	1.38
Haai-$C_{14}H_{29}$ (**3a**)	361	326, 426	28.31	1.47
Meaai-$C_{14}H_{29}$ (**3b**)	361	328, 425	26.14	1.35
Haai-$C_{16}H_{33}$ (**4a**)	360	323, 432	25.81	1.37
Meaai-$C_{16}H_{33}$ (**4b**)	362	318, 436	24.80	1.29
Haai-$C_{18}H_{37}$ (**5a**)	362	315, 435	23.24	1.27
Meaai-$C_{18}H_{37}$ (**5b**)	363	312, 432	22.71	1.24
Haai-$C_{20}H_{41}$ (**6a**)	361	318, 429	22.13	1.20
Meaai-$C_{20}H_{21}$ (**6b**)	362	313, 445	21.14	1.19
Haai-$C_{22}H_{45}$ (**7a**)	360	319, 433	20.72	1.17
Meaai-$C_{22}H_{45}$ (**7b**)	363	311, 446	20.17	1.11

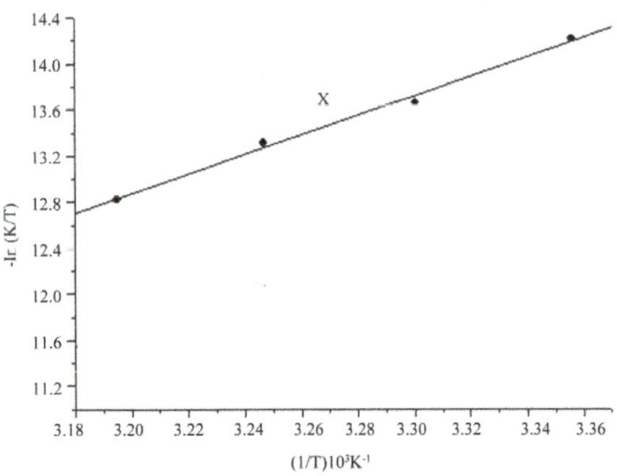

Figure 6. The Eyring plot of rate constants of Z-to-E thermal isomerisation of Haai-$C_{16}H_{33}$ (4a).

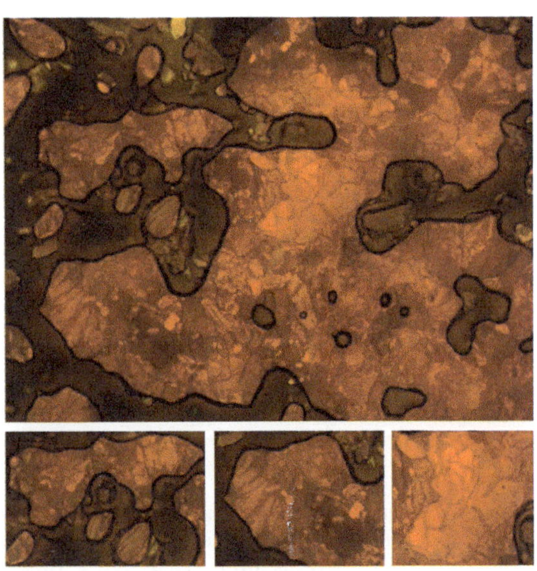

Figure 7. Fluctuating texture with schlieren texture of N phase at 51.2°C of Haai-$C_{18}H_{37}$ (5a).

may determine the presence of dipolar interaction within the molecules (**Figure 10**). The dipole moment of **5a** is 2.997 Debye (X = −1.3726, Y − −2.4777, Z = 0.9814). After geometry optimization of the compound **5a** by addition of atoms step by step, the optimized structure suggests that the molecule possess rod shape. The first optimization was done by the MM2 method, and finally DFT was performed. The maxima of dipole moment are in the negative Y-direction (−2.4777 D), so the direction of net dipole moment should be along the molecular axis (axial). Therefore, the direction of net dipole moment is axial (along molecular axis). Near-neighbor antiparallel correlation of molecules within the layer of N phase will occur in reality if the direction of dipole is axial or along the molecular axis. The direction of rod shaped compound, **5a** in adjacent layers should be antiparallel, so that the layer polarization alternates from layer to layer; leading to a macroscopic apolar antiferroelectric structure which is applicable to N phases and sometime in case of reentrant Smectic phase with switching behavior. Within the layers the molecules are in most cases additionally non-tilted parallel to the polar direction in the case of the Nematic phase. DFT study suggests that the near-neighbour antiparallel correlation exists and in broad-on side position with respect to the direction of dipole within the layers. Based on the direction of the dipole moment, calculated from DFT we have constructed a model of supramolecular arrangement in Nematic phase which is depicted in **Figure 11**.

4. Conclusion

Seven pairs of 1-alkyl-2-(arylazo)imidazoles, Raai-C_nH_{2n+1} (n = 10, 12, 14, 16, 18, 20, 22) with long alkyl chain at imidazolyl-N(1) centre are spectroscopically

Table 6. Rate and activation parameters for Z (cis) → *E* (trans) thermal isomerisation.

Compounds	Temp (K)	Rate of thermal Z→E conversion × 10^3 (s^{-1})	E_a kJ mol^{-1}	ΔH^* kJ·mol^{-1}	ΔS^* J·mol^{-1}·K^{-1}	ΔG^* °kJ·mol^{-1}
Haai-$C_{10}H_{21}$ (**1a**)	298	0.2292	71.67	69.48	−81.94	92.14
	303	0.3898				
	308	0.5496				
	313	0.8867				
Meaai-$C_{10}H_{21}$ (**1b**)	298	0.2345	70.19	70.13	−83.47	90.12
	303	0.3976				
	308	0.5634				
	313	0.8979				
Haai-$C_{12}H_{25}$ (**2a**)	298	0.2367	69.86	68.97	−82.94	91.75
	303	0.3997				
	308	0.5645				
	313	0.8997				
Meaai-$C_{12}H_{25}$ (**2b**)	298	0.2654	69.06	68.89	−84.47	89.97
	303	0.4235				
	308	0.6167				
	313	0.9546				
Haai-$C_{14}H_{29}$ (**3a**)	298	0.2567	67.38	67.78	−84.49	90.34
	303	0.4135				
	308	0.5896				
	313	0.9123				
Meaai-$C_{14}H_{29}$ (**3b**)	298	0.2657	67.78	69.59	−86.67	86.67
	303	0.4767				
	308	0.6578				
	313	0.9545				
Haai-$C_{16}H_{33}$ (**4a**)	298	0.2789	65.34	67.73	−85.69	89.37
	303	0.4567				
	308	0.6189				
	313	1.0989				
Meaai-$C_{16}H_{33}$ (**4b**)	298	0.2577	73.40	70.85	−76.45	94.02
	303	0.4247				
	308	0.6276				
	313	0.9571				
Haai-$C_{18}H_{37}$ (**5a**)	298	0.3467	62.21	63.32	−87.79	87.79
	303	0.5567				
	308	0.7567				
	313	1.4567				
Meaai-$C_{18}H_{37}$ (**5b**)	298	0.4534	62.08	70.56	−89.97	85.54
	303	0.6989				
	308	0.8998				
	313	1.9675				
Haai-$C_{20}H_{41}$ (**6a**)	298	0.4569	60.45	61.89	−91.9	85.67
	303	0.6789				
	308	0.8765				
	313	1.9876				
Meaai-$C_{20}H_{21}$ (**6b**)	298	0.6578	60.09	69.99	−92.45	82.23
	303	0.7568				
	308	0.9678				
	313	2.1434				
Haai-$C_{22}H_{45}$ (**7a**)	298	0.4817	58.64	60.89	−92.38	83.45
	303	0.7541				
	308	0.9574				
	313	2.4270				
Meaai-$C_{22}H_{45}$ (**7b**)	298	0.7140	58.97	78.27	−43.90	91.56
	303	0.9247				
	308	1.2613				
	313	3.729				

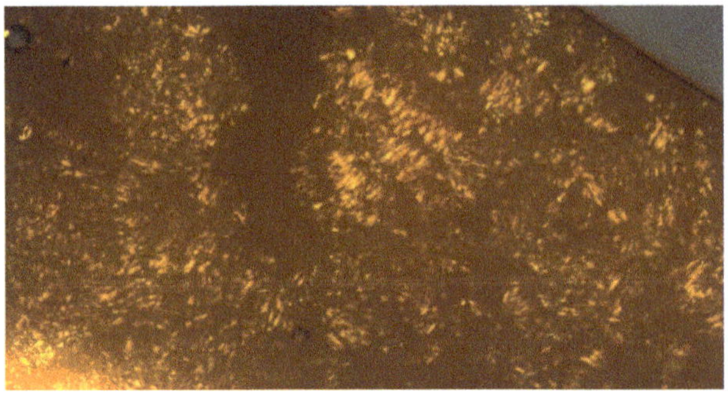

Figure 8. Unknown texture of Sm phase at 62˚C of Haai-$C_{22}H_{43}$.

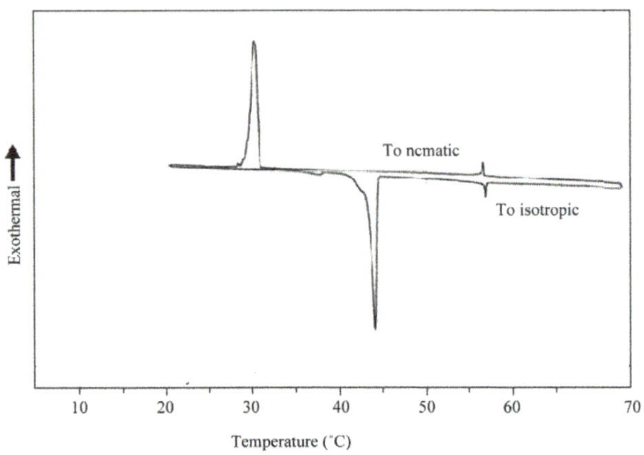

Figrue 9. DSC of Haai-$C_{18}H_{37}$ (5a) at the rate of 2˚C·min^{-1}.

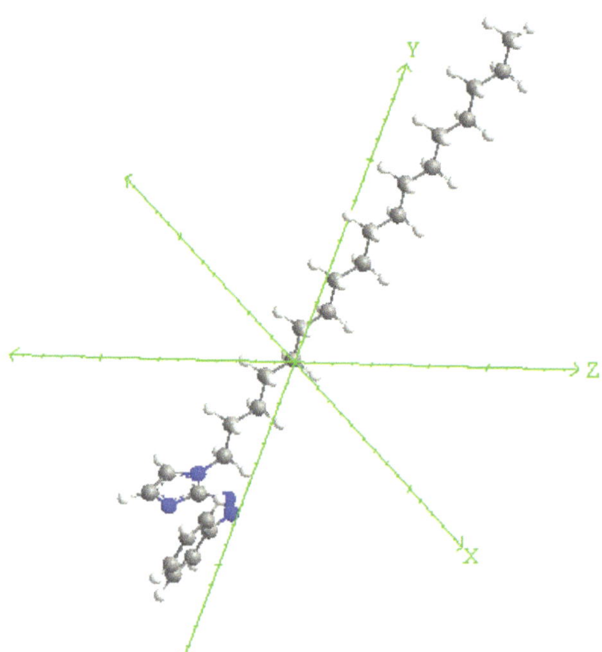

Figure 10. Optimized structure of compound 5a using DFT (B3LYP) level of theory.

characterised. As one of the compounds, 1-hexadcayl-2-(*p*-tolylazo)imidazole, Meaai-$C_{16}H_{33}$ is characterised by

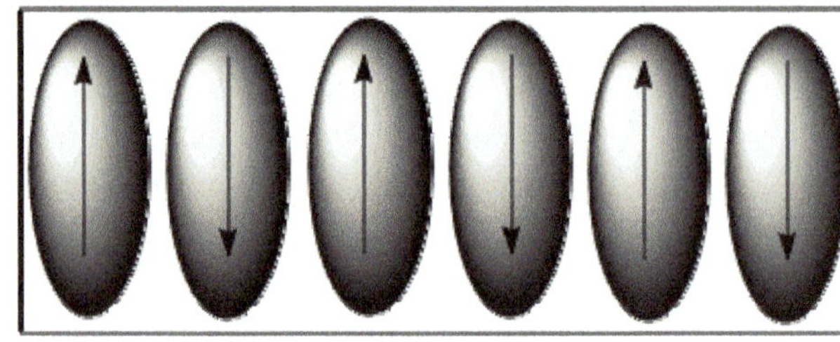

Figure 11. Schematic representation of supramolecular arrangement of dipoles in Nematic phase.

single crystal X-ray diffraction study. Upon UV light irradiation, the compounds show photochromism through geometrical isomerisation, E (trans)-to-Z (cis) forms. The reverse process is followed by thermal route and potential energy barrier (E_a) is calculated. The presence of long chain alkyl group shows long-range orientational order in between solid (crystalline, Cr) and isotropic (I) liquid phases. In some cases Cr and Nematic (N) texture shows thermal equilibrium in heating/cooling cycle. The Schlieren texture of N phase observed in compound **5a** and **7a** exhibits unknown smectic pahse with symmetrical lines. This is a preliminary report of our work; 1, 3-dialkyl-2-(arylazo)imida-zolium salts of different anions and the metal complexes (transition and nontransition) are waiting in this series.

5. Acknowledgements

The authors express their sincere thanks to Department of Science and Technology, New Delhi for financial support. One of the authors (AN) thanks UGC for the award of Fellowship.

REFERENCES

[1] R. B. Wesley and B. L. Feringa, "Light Switching of Molecules on Surfaces," *Annual Reviews*: *Physical Chemistry*, Vol. 60, 2009, pp. 407-428.

[2] N. L. Abbott, C. B. Gorman and G. M. Whitesides, "Active Control of Wetting Using Applied Electrical Potentials and Self-Assembled Monolayers," *Langmuir*, Vol. 11, No. 1, 1995, pp. 16-18.

[3] L. Mu, Y. Liu, S. Y. Cai and J. L. Kong, "A Smart Surface in a Microfluidic Chip for Controlled Protein Separation," *Chemistry—A European Journal*, Vol. 13, No. 18, 2007, pp. 5113-5120.

[4] E. Katz, A. N. Shipway and I. Willner, "Electronic and Optical Transduction of Photoisomerization Processes at Molecular- and Biomolecular-Functionalized Surfaces," In: Z. Sekkat and W. Knoll, Eds., *Photoreactive Organic Films*, Academic, San Diego, 2002, pp. 220-268.

[5] M. Ire, "Photochromism: Memories and Switches—Introduction," *Chemical Reviews*, Vol. 100, No. 5, 2000, pp. 1683-1684.

[6] S. Kawata and Y. Kawata, "Three-Dimensional Optical Data Storage Using Photochromic Materials," *Chemical Reviews*, Vol. 100, No. 5, 2000, pp. 1777-1788.

[7] H. Dürr and H. Bouas-Laurent, "Photochromism: Molecules and Systems," Elsevier, Amsterdam, 2003.

[8] B. L. Feringa, "Molecular Switches," Wiley-VCH, Weinheim, 2001.

[9] J. C. Crano and R. J. Guglielmetti, "Organic Photochromic and Thermochromic Compounds," Kluwer Academic/ Plenum Publishers, New York, 1999.

[10] A. V. El'tsov, "Organic Photochromes," Consultants Bureau, New York, 1990.

[11] S. Furumi, H. Fudouzi and T. Sawada, "Dynamic Photoswitching of Micropatterned Lasing in Colloidal Crystals by the Photochromic Reaction," *Journal of Materials Chemistry*, Vol. 22, No. 40, 2012, pp. 21519-21528.

[12] S. Pu, T. Wang, G. Liu, W. Liu and S. Cui, "A New Photoinduced Fluorescent Switch Based on a Photochromic Diarylethene with a Rhodamine Fluorophore," *Dyes and Pigments*, Vol. 94, No. 3, 2012, pp. 416-422.

[13] M. Irie, "Special Issue Photochromism: Memories and Switches," *Chemical Reviews*, Vol. 100, No. 5, 2000, pp. 1683-1716.

[14] S. Kobatake, S. Takami, H. Muto, T. Ishikawa and M. Irie, "Rapid and Reversible Shape Changes of Molecular Crystals on Photoirradiation," *Nature*, Vol. 446, No. 7137, 2007, pp. 778-781.

[15] T. L. Andrew, H. Y. Tsai and R. Menon, "Confining Light to Deep Subwavelength Dimensions to Enable Optical Nanopatterning," *Science*, Vol. 324, No. 5929, 2009, pp. 917-921.

[16] H. Bouas-Laurent and H. Dürr, "Organic Photochromism," *Pure and Applied Chemistry*, Vol. 73, No. 4, 2001, pp. 639-665.

[17] C. V. Yelamaggad, I. Shashikala, U. S. Hiremath, D. S. S. Rao and S. K. Prasad, "Liquid Crystal Dimers Possessing Chiral Rod-Like Anisometric Segments: Synthesis, Characterization and Electro-Optic Behavior," *Liquid Crystals*, Vol. 34, No. 2, 2007, pp. 153-167.

[18] P. M. Maitlis, A. Giroud-Godquin and M. Angecu, "Metallomesogens: Metal Complexes in Organized Fluid Phases," *Angewandte Chemie International Edition in English*, Vol. 30, No. 4, 1991, pp. 375-402

[19] V. Balzani, "Molecular Devices and Machines," Wiley-VCH, Weinheim, 2008.

[20] S. Gerlich, L. Hackermüller, K. Hornberger, A. Stibor, H. Ulbricht, M. Gring, F. Goldfarb, T. Savas, M. Müri, M. Mayor and M. Arndt, "A Kapitza-Dirac-Talbot-Lau Interferometer for Highly Polarizable Molecules," *Nature Physics*, Vol. 3, No. 10, 2007, pp. 711-715.

[21] H. Hartmann and I. Zug, "On the Coupling of Aryldiazonium Salts with *N,N*-Disubstituted 2-Aminothiophenes and Some of Their Carbocyclic and Heterocyclic Analogues," *Journal of the Chemical Society, Perkin Transactions*, Vol. 1, No. 24, 2000, pp. 4316-4320.

[22] M. M. M. Raposo, A. M. R. C. Sousa, A. M. C. Fonseca and G. Kirsch, "Thienylpyrrole Azo Dyes: Synthesis, Solvatochromic and Electrochemical Properties," *Tetrahedron*, Vol. 61, No. 34, 2005, pp. 8249-8256.

[23] B. A. Trofimov, E. Y. Schmidt, A. I. Mikhaleva, A. M. Vasiltsov, A. B. Zaitsev, N. S. Smolyanina, A. V. Afonin, I. A. Ushakov, K. B. Petrushenko, O. N. K. Kazheva, O. A. Dyachenko, V. V. Smirnov, A. F. Schmidt, M. V. Markova and L. V. Morozova, "2-Arylazo-1-vinylpyrroles: A Novel Promising Family of Reactive Dyes," *European Journal of Organic Chemistry*, Vol. 2006, No. 17, 2006, pp. 4021-4033.

[24] P. J. Coelho, M. C. R. Castro, A. M. C. Fonseca and M. M. M. Raposo, "Photoswitching in Azo Dyes Bearing Thienylpyrrole and Benzothiazole Heterocyclic Systems," *Dyes and Pigments*, Vol. 92, No. 1, 2011, pp. 745-748.

[25] U. Ray, D. Banerjee, G. Mostafa, T.-H. Lu and C. Sinha, "Copper Coordination Compounds of Chelating Imidazole-Azo-Aryl Ligand. The Molecular Structures of Bis[1-ethyl-2-(*p*-tolylazo)imidazole]-bis-(azido)copper(II) and Bis[1-methyl-2(phenylazo)imidazole]-bis(thiocyanato)copper(II)," *New Journal of Chemistry*, Vol. 28, No. 12, 2004, pp. 1437-1442.

[26] J. Otsuki, K. Suwa, K. Narutaki, C. Sinha, I. Yoshikawa and K. Araki, "Photochromism of 2-(Phenylazo)imidazoles," *The Journal of Physical Chemistry A*, Vol. 109, No. 35, 2005, pp. 8064-8069.

[27] J. Otsuki, K. Suwa, K. K. Sarker and C. Sinha, "Photoisomerization and Thermal Isomerization of Arylazoimidazoles," *The Journal of Physical Chemistry A*, Vol. 111, No. 8, 2007, pp. 1403-1409.

[28] K. K. Sarker, A. D. Jana, G. Mostafa, J.-S. Wu, T.-H. Lu and C. Sinha, "Iodo Bridged Lead(II) Compounds of Azo-imidazoles: Single Crystal X-Ray Structures of [Di-iodo-{1-methyl-2-(p-tolylazo)imidazole}lead(II)]$_n$ and {1-Methyl-3-benzyl-2-(p-tolylazo)imidazolium}$_m$-{tri-iodoplumbate(II)}$_m$," *Inorganica Chimica Acta*, Vol. 359, No. 13, 2006, pp. 4377-4385.

[29] K. K. Sarker, D. Sardar, K. Suwa, J. Otsuki and C. Sinha, "Cadmium(II) Complexes of (Arylazo)imidazoles: Synthesis, Structure, Photochromism, and Density Functional Theory Calculation," *Inorganica Chemistry*, Vol. 46, No. 20, 2007, pp. 8291-8301.

[30] K. K. Sarker, B. G. Chand, K. Suwa, J. Cheng, T.-H. Lu, J. Otsuki and C. Sinha, "Structural Studies and Photochromism of Mercury(II)-Iodo Complexes of Arylazo-imidazoles," *Inorganica Chemistry*, Vol. 46, No. 3, 2007, pp. 670-680.

[31] P. Pratihar, T. K. Mondal, A. K. Patra and C. Sinha, "*trans*-Dichloro-bis-(arylazoimidazole)palladium(II): Synthesis, Structure, Photoisomerization, and DFT Calculation," *Inorganica Chemistry*, Vol. 48, No. 7, 2009, pp. 2760-2769.

[32] D. Demus, J. Goodby, G. W. Gray, H. W. Spiess and V. Vill, "Handbook of Liquid Crystals," Wiley-VCH, Weinheim, 1998.

[33] D. Demus, J. Goodby, G. W. Gray, H. W. Spiess and V. Vill, "Dunmur D and Toriyama K. Handbook of Liquid Crystals, Vol 1, Chapter VII," Wiley-VCH, Weinheim, 1998.

[34] K. Binnemans, "Ionic Liquid Crystals," *Chemical Reviews*, Vol. 105, No. 11, 2005, pp. 4148-4204.

[35] G. M. Sheldrick, "SHELXS 97, Program for the Solution of Crystal Structure," University of Göttingen, Göttingen, 1990.

[36] G. M. Sheldrick, "SHELXL 97, Program for the Refinement of Crystal Structure," University of Göttingen, Göttingen, 1997.

[37] A. L. Spek, "Structure Validation in Chemical Crystallography," *Acta Crystallographica Section D*, Vol. 65, No. 2, 2009, pp. 148-155.

[38] Cambridge Soft Corporation: USA, "Chem3D, Version 10," Cambridge Soft, Cambridge, 2006.

[39] M. J. Frisch, "GAUSSIAN 03 (Revision c.02)," Gaussian, Wallingford, 2004.

[40] M. M. Enriquez, M. Fuciman, A. M. LaFountain, N. L. Wagner, R. R. Birge and A. H. Frank, "The Intramolecular Charge Transfer State in Carbonyl-Containing Polyenes and Carotenoids," *The Journal of Physical Chemistry B*, Vol. 114, No. 38, 2010, pp. 12416-12426.

[41] P. Bhunia, B. Baruri, U. S. Ray and C. Sinha, "Azoimidazole Complexes of Manganese(II)-Thiocyanate. X-Ray Structures of 2-(p-tolylazo)imidazole and Bis-[1-methyl-2-(p-tolylazo)imidazole]-di-thiocyanato-manganese(II)," *Transition Metal Chemistry*, Vol. 31, No. 3, 2006, pp. 310-315.

Supplementary Material

Crystallographic data for the structures have been deposited with the Cambridge Crystallographic Data Center; CCDC No. 854093. Copies of this information may be obtained free of charge from the Director, CCDC, 12 Union Road, Cambridge CB2 1EZ, UK (e-mail, deposit@ccdc.cam.ac.uk, or www, htpp://www.ccdc.cam.ac.uk).

3

Study of Oxidation Kinetics in Air of Zircaloy-4 by *in Situ* X-Ray Diffraction

Noureddine Selmi, Ali Sari
Nuclear Research Center of Birine, Birine, Algeria

ABSTRACT

The zircaloy-4 is an alloy of zirconium, which has a very weak thermal neutron absorption, satisfactory mechanical properties and good corrosion resistance at high temperature. For these reasons, zircaloy-4 is used as a material of cladding fuel rod of nuclear reactors. In this environment, it is submitted to different severe conditions of temperature and pressure. The objective of this work is to study the oxidation kinetics of zircaloy-4 in air by the X-ray diffraction technique. The experiments were realized in a "HTK1200" furnace installed as a sample holder in the diffractometer at different temperatures; 25˚C, 350˚C, 500˚C, 830˚C and 1000˚C. The results show that the monoclinic and the tetragonal phases are formed at 350˚C temperature. The volume fraction of these phases increased with the temperature until 1000˚C where the α phase disappears completely. For simulating the case of loss-of-coolant-accident (LOCA), we have done x-ray diffraction of Zry-4 samples water quenched at 1050˚C with different ageing times at this temperature. At 10 seconds and more, there is an important evolution of monoclinic and tetragonal zirconias, which leads to the degradation of zircaloy-4 properties.

Keywords: Zircaloy-4; High Temperature Oxidation; Zirconium Oxides; *In Situ* X-Ray Diffraction

1. Introduction

The fuel rod cladding represents the first barrier of the radioactive products confinement. The use of zirconium alloys for the cladding fuel rod of the pressurized water of the nuclear reactors (P.W.R.) is related to their very weak absorption of the thermal neutrons, good water corrosion resistance, good mechanical properties at high temperature [1], compatibility with the fuel and sufficient resistance to the irradiation effects (embrittlement, swelling ...) (**Figure 1**). This material is submitted to the extreme conditions of the use, and various phenomena which degrade their characteristics. Among these phenomena there is the oxidation by water.

Zircaloy-4 oxidation in air proceeds in a similar way to that described in the pressurized steam. In this work, we studied the oxidation of the zircaloy-4 in air, and we have chosen various temperatures; 25˚C, 350˚C, 500˚C, 830˚C and 1000˚C.

1.1. Allotropic Phases of the Zr

At atmospheric pressure, and temperature inferior to 866˚C, elemental Zr appears as the hcp α-phase. At temperature between 866˚C and 1850˚C, Zr appears as the bcc β-phase. At room temperature and at pressure of about 2 GPa and more, the hcp α-phase is transformed to the hexagonal ω-phase [2].

1.2. Zr-O System

The phase diagram of the binary system Zr-O presented in **Figure 2** [3] shows that a great quantity of oxygen can dissolve in interstitial position (until 30% at 500˚C) on one hand and on the other hand, there exists one stable zirconium oxide with atmospheric pressure, the ZrO_2 zirconia.

At atmospheric pressure, and at a temperature that reaches 1170˚C, the crystalline structure of ZrO_2 is monoclinic, between 1170˚C and 2370˚C, the monoclinic structure becomes tetragonal structure, and cubic structure from 2370˚C to 2706˚C [4,5].

1.3. Oxidation Mechanism

The oxidation of zirconium and its alloys progresses by a diffusive stage of anionic type, *i.e.* the growth of the products of the oxidation of metal is carried out by transport of oxygen of the gas environment towards the interface metal/oxide (intern interface) according to the

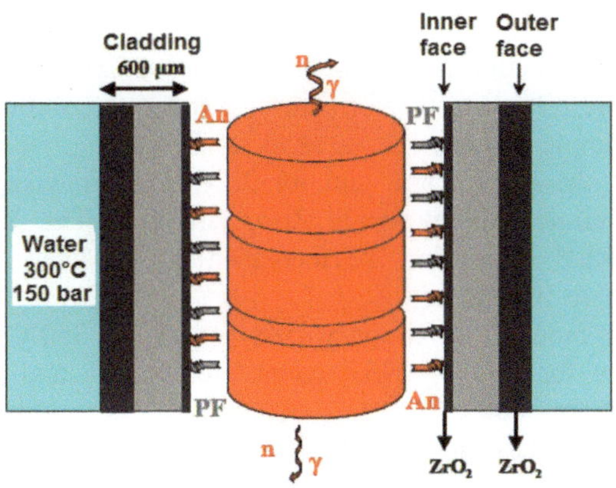

Figure 1. Schematical section of fuel rod in reactor. An and PF represent respectively the actinides and the fission products.

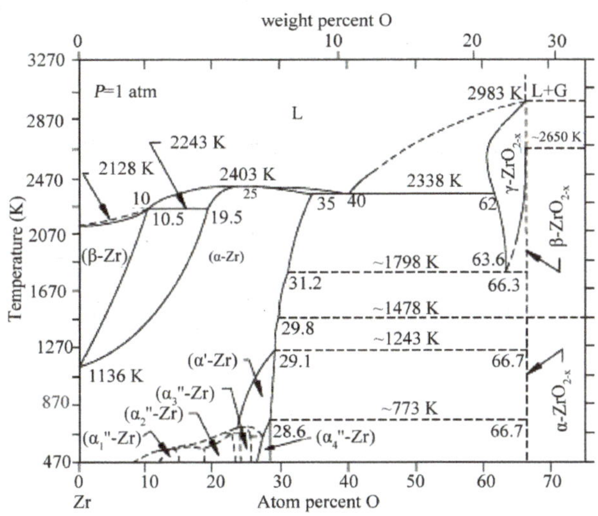

Figure 2. Equilibrium phase diagram for the Zr-O system [2].

following global reaction [6,7]:

$$Zr + (1 - x/2) O_2 \rightarrow ZrO_{2-x}$$

Only a part of the oxygen which arrives at the intern interface enters in this reaction; the other part is dissolved in subjacent metal [8].

Parise gave a relation between the values of oxygen diffusion coefficient in α zirconium between 300˚C and 600˚C [9]:

$$D = 0594.10^{-14} \cdot \exp[-197237/RT]$$

where D is in $m^2 \cdot s^{-1}$, R = 8.314 $J \cdot mol^{-1} \cdot K^{-1}$ and T is in Kelvin.

This same type of relation was obtained by Debuigne for temperatures between 650˚C and 850˚C [8]:

$$D = 444.10^{-4} \cdot \exp[-214{,}016 / RT] m^2 \cdot s^{-1}$$

Concerning the oxygen diffusion coefficients in zirconia resulting from the oxidation of metal, Parise gives the following relation between 150˚C and 500˚C:

$$D = 679.10^{-8} \cdot \exp[-131524 / RT]\ m^2 \cdot s^{-1}$$

and between 400˚C and 850˚C, Debuigne gives:

$$D = 105.10^{-7} \cdot \exp[-122474 / RT]\ m^2 \cdot s^{-1}$$

2. Experimental Procedure

We have studied the microstructure of Zry-4 as received (α phase). To follow the oxidation kinetics, we used, *in-situ* x-ray diffraction for different temperatures (25˚C, 350˚C, 500˚C, 830˚C and 1000˚C), *in-situ* XRD for temperatures of 350˚C and different times of ageing: 00 min, 10 min, 20 min, 40 min and 60 min.

During a postulated loss-of-coolant-accident (LOCA), the cladding tube is submitted in steam environment to a fast increase in the temperature (up to 800˚C - 1200˚C), which induces an accelerated oxidation phenomenon until quenching [1]. To simulate this case, we have done X-ray diffraction with Zry-4 samples water quenched at 1050˚C with different times at this temperature (05 sec, 10 sec, 30 sec, 01 min, 02 min, 05 min, 10 min and 20 min).

The X-ray diffraction is realized on a diffractometer equipped with a vertical goniometric, provided with x-rays tube with copper anticathode ($K\alpha_{Cu}$: 1,5406 Å), the sample stages are spinner and "HTK-1200" furnace with a temperature that reaches 1200˚C.

The X-rays diffraction analysis parameters are the following:

Scan angle (2θ): [15 to 125] and [25 to 65].

Step size (2θ): 0,02.

Counting time (s): 1.

The studied material is the zirconium alloy of nuclear usage; the zircaloy-4. The X-Ray Fluorescence technique (XRF) defines the elementary chemical composition of this alloy (**Table 1**).

3. Results and Discussion

3.1. X-Ray Diffraction

The XRD for Zry-4 sample as received is shown in **Figure 3**.

Crystallographic structure of this material as received is the hexagonal close-packed structure (hcp-Zrα), this material has a polycrystalline metallographic structure with equiaxes grains and a microhardness Hv(0.5) ≈ 180 (kg/mm^2).

3.2. *In Situ* X-Ray Diffraction (Oxidation Kenitics)

The Zry-4 sample has a diameter of 10 mm and a thick-

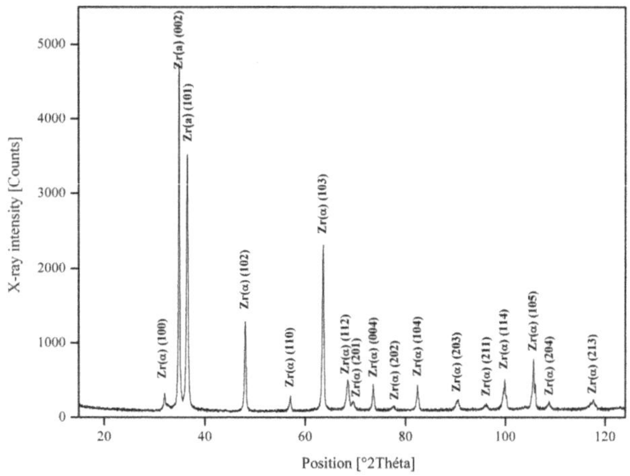

Figure 3. X-ray diffraction pattern of Zry-4 as received.

Table 1. Chemical composition of Zircaloy-4 (%).

Element	Concentration
Zr	97.243
Sn	2.423
Fe	0.220
Cr	0.105
Hf	traces

ness of 0.9 mm, placed in "HTK-1200" furnace of the diffractometer. The different diffraction temperatures in air are: 25˚C, 350˚C, 500˚C, 830˚C and 1000˚C.

When the temperature of furnace is reached, the diffraction is immediately launched and the results are presented in the form of diffractograms.

The results of *in-situ* XRD diffractograms prove that:

- For 25˚C, there is only Zr-α phase, the principal peak of this phase is in $2\theta = 34.83˚$.
- The formation of the monoclinic and tetragonal phases start at 350°C, principal peaks of monoclinic and tetragonal phases are respectively at $2\theta = 28.24˚$ and 30.28˚
- From 350˚C to 1000˚C, there is an important evolution in the two monoclinic and tetragonal zirconias, **Figure 4**.
- When the temperature is equal or up to 830˚C, there is a formation of Zr-β phase and disappearance of Zr-α phase, the principal peak of Zr-β phase is at $2\theta =$ 35.8˚, **Figure 4**.

These results give the oxidation kenitic's of Zry-4, **Figure 5**.

For the room temperature 25˚C, There is only Zr-α phase, the principal peak intensity is 10,787 counts, this intensity increases to 11,747 counts for 350˚C, and decreases rapidly to 8985 counts for 500˚C until 1197 counts for 800˚C and zero intensity for 1000˚C, which confirms the reduction in the volume fraction of this phase.

The reduction in the volume fraction of Zr-α phase is directly related to the formation of the ZrO_2 zirconia on the sample surface.

On the other hand, and for 25˚C, zero intensity for monoclinic and tetragonal ZrO_2 zirconias. The X-rays diffracted intensities are increased progressively from 350˚C to 1000˚C.

For monoclinic zirconia: 388 counts for 350˚C, 697 counts for 500˚C, 2957 counts for 830˚C and 2070 counts for 1000˚C.

For tetragonal zirconia: 371 counts for 350˚C, 438 counts for 500˚C, 827 counts for 830˚C and 629 counts for 1000˚C.

Therefore, we have an increase in the volume fraction of zirconias with the increase in the temperature because of the oxygen diffusion.

3.3. Ageing in HTK-1200 Furnace

We have done *in-situ* XRD at 350˚C and at different times of ageing: 00 min, 10 min, 20 min, 40 min and 60 min (**Figure 6**).

We put the sample during a fixed time of ageing in 350˚C before starting the diffraction again. The holding time enables us to study the influence of the time on the formed phases.

During a typical Loss-Of-Coolant-Accident (LOCA) transient, the fuel cladding tubes are subject to high temperature oxidation that reaches 1000˚C and finally quenched because of the reflooding of the core. The global oxygen content and its distribution affect strongly the proprieties of the cladding which microstructure can be described in terms of zirconia phase.

We have studied this case with x-ray diffraction technique on Zry-4 samples water quenched at 1050˚C with different ageing times (05 sec, 10 sec, 30 sec, 01 min, 02 min, 05 min, 10 min and 20 min) at this temperature, the results are given in **Figure 7**.

For The 1050˚C water quenching after 05 seconds, there is no formation of oxide layer, but at 10 seconds to 20 minutes; there is a clearly evolution of monoclinic and tetragonal zirconias on one hand, and on the other hand, there is a disappearance of the Zr-α phase (**Figure 7**).

3.4. Microhardness

The indentation method allows the hardness of a material to be measured. The Vickers hardness of water quenching Zry-4 with 500 g weight is augmented to 207 (kg/mm^2) at 05 seconds of ageing time to 1120 (kg/mm^2) at 300 seconds of ageing time (**Figure 8**).

4. Conclusions

The oxidation of zircaloy-4 changes completely the

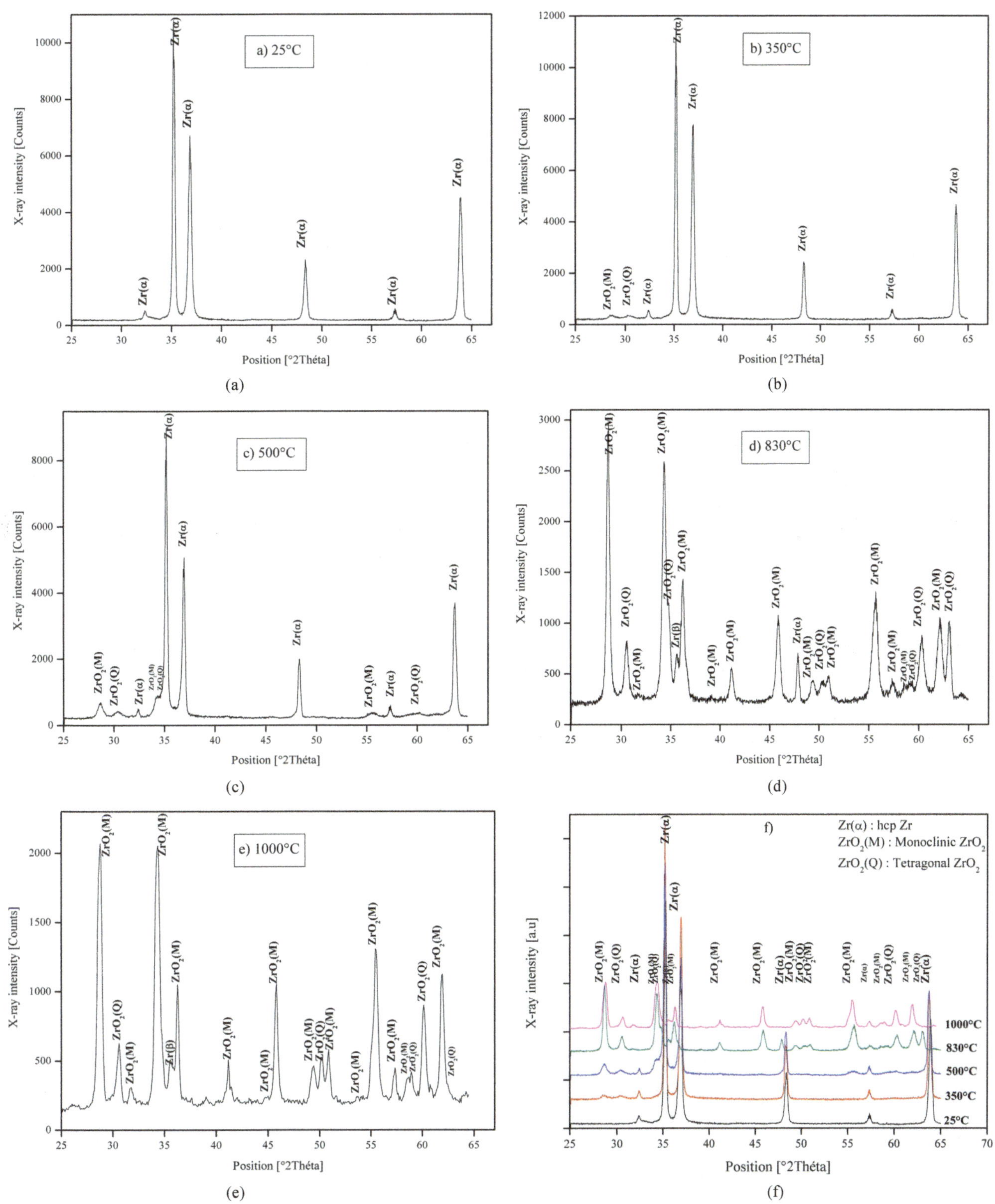

Figure 4. *In-situ* X-Ray diffraction patterns of Zry-4 for different temperatures: (a) 25°C; (b) 350°C; (c) 500°C; (d) 830°C; (e) 1000°C; and (f) Combined graph.

properties of this material. In the presence of oxygen, and in temperature between 25°C and 350°C, a great quantity of oxygen can dissolve in interstitial positions, the zircaloy-4 maintains its properties. Between 350°C and 1050°C, and with the use of x-ray diffraction technique, we have an evolution of the two phases, monoclinic and

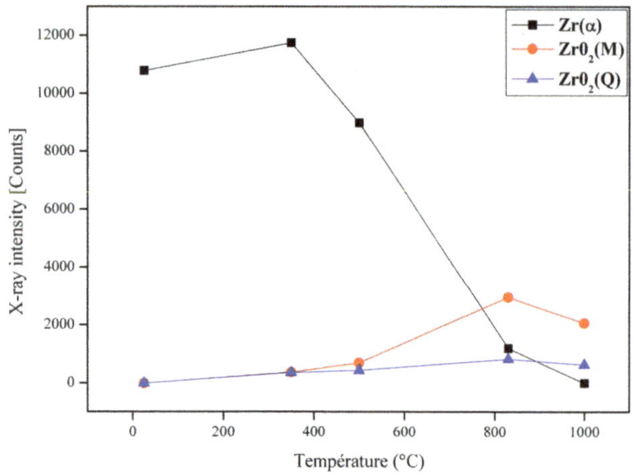

Figure 5. Evolution of th e different phases in function of temperature.

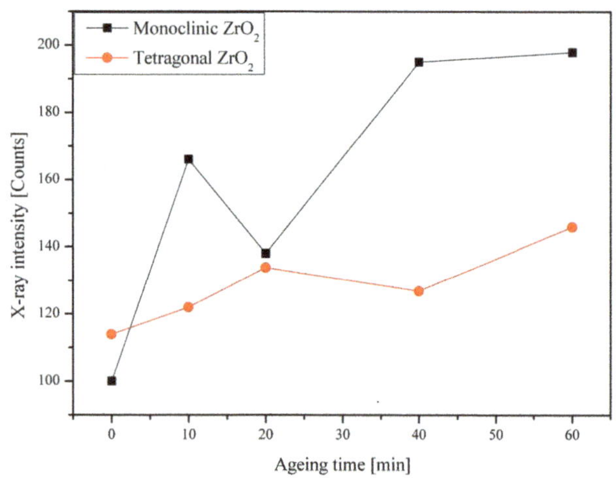

Figure 6. Evolution of the two formed zirconias (monoclinic and tetragonal) in function of ageing time at temperature of 350°C in the HTK furnace.

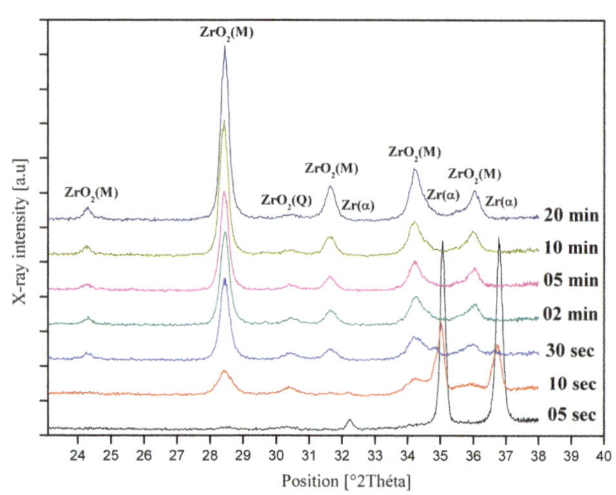

Figure 7. X-ray diffraction pattern of Zry-4 water quenched at 1050°C with different ageing times at this temperature.

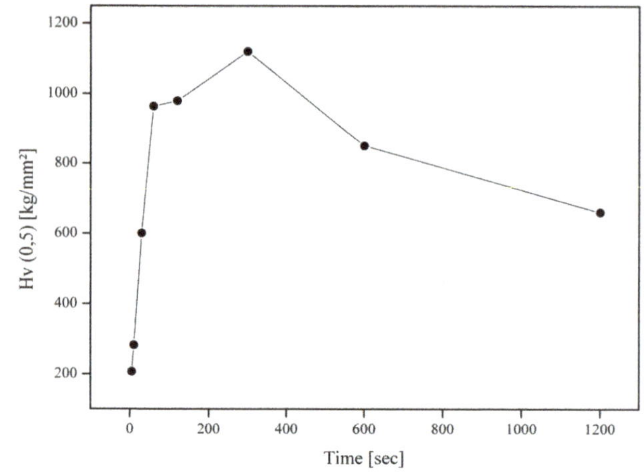

Figure 8. Microhardness (mHv (0.5)) for 1050°C quenched samples with different ageing times.

tetragonal zirconia where the zircaloys-4 loses its properties.

The formation of monoclinic and tetragonal zirconia leads to the increase in hardness and the reduction in thermal conductivity which leads to the degradation of zircaloy-4 mechanical and thermal properties. The microhardness of these oxides (ZrO_2) is very high, which increases the probability of cracking of this material, and consequently, the contamination of the primary circuit.

REFERENCES

[1] A. Stern, J.-C. Brachet, V. Maillot, D. Hamon, F. Barcelo, S. Poissonnet, A. Pineau, J.-P. Mardon and A. Lesbros, "Investigations of the Microstructure and Mechanical Properties of Prior-Structure as a Function of the Oxygen Content in Two Zirconium Alloys," *Journal of ASTM International*, Vol. 5, No. 4, 2008, Article ID: JAI101119.

[2] S. Banerjee and P. Mukhopadyay, "Phase Transformation: Examples from Titanium and Zirconium Alloys," Pergamon Press, Oxford, 2007, pp. 6-11, 42.

[3] D. Yacoub, "Experimental Study of Zirconium Alloys Oxidation at High Water Steam Pressure and Modelling of Mechanisms," Ph.D. Thesis, Superior National School of Mines, Saint-Etienne, 2007.

[4] N. Bererd, "Irradiation Effects on the Zirconium Oxidation and the Uranium Diffusion in the Zirconia," Ph.D. Thesis, Claude Bernard University-Lyon I, Chevarier Alain (Dir.), 2003.

[5] W. Qin, C. Nam, H. L. Li and J. A. Szpunar, "Tetragonal Phase Stability in ZrO_2 Film Formed on Zirconium Alloys and Its Effects on Corrosion Resistance," *Acta Materialia*, Vol. 55, No. 5, 2007, pp. 1695-1701.

[6] N. Pétigny, "Comparison of Tow Zirconium Alloys Oxidation by *in-Situ* and *ex-Situ* X-Ray Diffraction: Texture, Phase," Ph.D. Thesis, Bourgogne University, Dijon, 1999.

[7] P. Putoud, "Alloys Oxidation of the Zircaloy-4 Type with Different Contents of tin at 400°C in the Steam. Analyzes

of Phases in Presence and Their Evolution," Ph.D. Thesis, Compiegne Technology University, Compiègne, 1995.

[8] J. Debuigne, "Contribution in the Study of the Zirconium Oxidation and the Oxygen Diffusion in the Oxide and the Metal," Ph.D. Thesis, Paris University, Paris, 1966.

[9] M. Parise, "Mechanisms of Zirconium Alloys Corrosion: Study of the Oxidation initial Kinetics and the Mechanical Behaviour of the Metal-Oxide System," Ph.D. Thesis, Superior National School of Mines, Paris, 1996.

4

Electroosmotic Water Vapor Transport across Novel, Smart, Functionalized Conducting Polymer Microporous Membranes in Active Mode at Very High Rates, with Concomitant Chemical Warfare (CW) Agent Blocking

Prasanna Chandrasekhar[1*], Petar Pirgov[1], Brian J. Zay[1], David Lawrence[1], Sean Morefield[2], Tilghman L. Rittenhouse[3], Salvatore G. Clementi[3], Quoc Truong[4], Russell R. Greene[5]

[1]Ashwin-Ushas Corporation, Marlboro, NJ, USA

[2]US Army Engineer Research and Development Center, Construction Engineering Research Laboratory (ERDC-CERL), Champaign, IL, USA

[3]Chemical and Biological Technologies Directorate, US Defense Threat Reduction Agency (DTRA), Fort Belvoir, VA, USA

[4]Warrior Science, Technology, and Applied Research Directorate, US Army Natick Soldier Research, Development, and Engineering (RD&E) Center, Natick, MA, USA

[5]Hazardous Materials Research Center (HMRC), Battelle Columbus Laboratories, Battelle Memorial Institute, Columbus, OH, USA

ABSTRACT

Electroosmotic water vapor transport (WVT) across very thin, flexible, functionalized conducting polymer (CP) microporous (μP) membranes at a very high rate is reported. Both *passive* and *active* (6 VDC applied) WVT are reported, the latter *for the first time ever*. WVT occurs with concomitant, effective blocking of chemical warfare (CW) agents, again demonstrated for the first time ever. Initial *active* liquid||membrane||liquid interface studies demonstrated WVT rates of $>1.7 \times 10^{-5}$ $g \cdot mm^{-2} \cdot s^{-1}$, $>3 \times$ the highest prior reported values of 5×10^{-6} $g \cdot mm^{-2} \cdot s^{-1}$. Subsequent vapor||membrane|| vapor interface studies using industry-standard methods (including ASTM E96B Upright Cup ("WVT"), ASTM F2298 ("Dynamic Moisture Permeation Cell") and ASTM F1868 ("Sweating Guard Hotplate") were done at independent, external labs for independent corroboration. These yielded, e.g., WVT values of 2564.4 $g \cdot m^2 \cdot d^{-1}$ (*passive*) and 3706.7 $g \cdot m^2 \cdot d^{-1}$ (*active*), to be compared with the highest (*passive*) value ever reported previously, 984.8 $g \cdot m^2 \cdot d^{-1}$ for a μP-Nylon membrane. More than 15 different membrane configurations, porosities and types were studied, including membranes with CP + organophosphate hydrolase (OPH), an enzyme reactive to CW agents. Efficient blocking of the actual CW agents GB, HD, VX is also reported, using the TOP-8-2-501 standard. Membranes also passed all Industry-standard durability tests, e.g. ASTM D2261 (Tearing), ASTM D5034 (Breaking), ASTM D3886 (Abrasion), ASTM F392 (Gelbo Flex). Incorporation into smart soldiers' garments was demonstrated; power consumption was <1 $W \cdot m^{-2}$. Mechanisms of enhanced WVT and CW agent blocking in the membranes are discussed.

Keywords: Electroosmotic; Transport; Smart; Microporous; Membrane; Conducting; Polymer; Functionalized

1. Introduction

1.1. Electrokinetic Effects and *Passive* and *Active* Electroosmosis

The four *electrokinetic* effects-*electrophoresis* and the related *sedimentation potential*; and *electroosmosis* and the related *streaming potential*-emanate from the existence of electrical charges on the surfaces of particles or walls of channels (such as capillaries or pores) [1-12] are introduced. Of these effects, *electroosmosis* refers to the movement of a liquid or condensed vapor relative to a stationary charged surface (such as that found in a capillary, membrane micropore, or microfluidic channel). The electroosmotic effect is proportional to the *zeta potential*, which in turn is directly related to the charge on the surface in question [5,6].

Electroosmosis may occur not only in a purely liquid

*Corresponding author.

phase, but also in combined vapor/liquid phases. In the latter case, vapors may condense on walls of a capillary or membrane micropore, be transported, and then re-evaporate [1-9,13-19]. This in effect leads to the transportation of vapor across the capillary or micropore in question [13-19]. Electroosmosis is one of the two driving forces behind the permeation of water vapor through "breathable" microporous membranes (such as the well known Gore-Tex® PTFE (poly(tetrafluoroethylene)) membranes used in recreational outerwear). The other driving force is the combined concentration and temperature gradients across the membranes when worn in garments; this is also believed to have an electrochemical basis [13-16]. These gradients assist in driving water vapor, e.g. as generated by sweating, from the inside of a garment (which is warmer and more humid) to the external environment, while simultaneously keeping out liquid water, making the garment "breathable" [13-16]. Quite evidently, the electroosmotic effect in such membranes is *passive*, *i.e.* it emanates from the natural charge or potential difference present across the micropores, rather than due to *active* application of this potential or charge difference. Additionally, of course, such microporous garments do not block against noxious or poisonous vapors permeating from the environment into the interior of the garment.

1.2. Prior Electroosmotic Studies

Active (*i.e.* with voltage or current applied) electroosmotic studies on microporous membranes have been reported in the past [7-12,20-23], but exclusively on [*liquid*||membrane||liquid] interfaces; they use generally aqueous electrolyte solutions of identical composition. In these studies, a voltage or current is typically applied across two electrodes placed on either side of the membrane, and the resulting electroosmotic flow measured. For example, Bowen and Clark [7-10] and subsequently, Mullet *et al.* [12], described variations of a simple "dipped cell" apparatus in which the electroosmotic effect was measured as liquid overflow registered by a capillary or micropipette, recorded by a sensitive balance. The microporous membranes they studied were of rigid alumina and titania. They calculated the electroosmotic flow based on the detailed mathematical treatments set forth by Bowen [7], and Kozak and Davis [11]; these in turn were based on earlier treatments of Smoluchowski [1-2], Henry [3], and Levine and Neale as elaborated by Saville [4] and corroborated later by Hernandez *et al.* [8] and Basu and Sharma [9]. Electroosmotic flow rates measured were extremely low. More recently, Vajandar *et al.* [20] reported electroosmotic "pumping" using silica-coated alumina microporous membranes with pore-sizes of 0.02 to 0.3 μm, for possible applications in biomedical lab-on-a-chip devices, and micropumps for microelectronic cooling; the electroosmotic flow rates they achieved were only marginally better than those of Bowen and Clark and Mullet *et al.* [10,12]. Miller and Martin [21] studied electroosmotic flow in very thin (60 μm) carbon nanotube membranes; once again, flow rates were very low, and there were other issues, such as stability of the membranes. Li *et al.* [22] reported results of electroosmosis and iontophoresis studies with polycarbonate microporous membranes and transdermal electroosmosis and iontophoresis in human epidermal membrane, at applied voltages up to +/− 1.0 V; their results were consistent with a model in which the human epidermal membrane has net-negatively-charged pores. Lee and Hong [23] measured electroosmotic flow rates across microporous membranes, including polysulfone membranes, as a function of pH and ionic strength of the aqueous solutions used; once again, however, the flow rates they reported were extremely low. It may also be noted that all these prior experimental studies comprised, in effect, *active*, electroosmotic transport of liquid across membranes, not transport of vapor across membranes, as we report here.

In other prior work more relevant to a [vapor||membrane||vapor] interface, McCullough *et al.* [17,18] carried out a very detailed and thorough comparative study of *passive* WVT (*i.e.* vapor transport) rates and evaporative resistance of a large number of commercially available microporous membranes; they used a number of standardized tests such as the ASTM tests that we also utilized. Indeed, their work is still used as an "industry-standard" benchmark to compare performance of other membranes and materials, as it will be used in the present work.

1.3. Work Presented in the Present Study

In the present communication, we report *passive* as well as *active* electroosmotic water vapor transport (WVT) across very thin, flexible, functionalized microporous conducting polymer (CP) membranes, the latter *for the first time ever*, to the best of our knowledge. Both the *active* nature of the electroosmotic water vapor transport and its magnitude *two to five times* the highest (*passive*) water vapor transport (WVT) rates previously reported for any microporous membranes [17-19]—represent first, to the best of our knowledge. We show further that it is *the CP that is postulated to be the element that imparts the critical, increased WVT rate*, through the existence of an intrinsic potential difference across the CP membrane. Furthermore, our membrane assemblies are also shown to *simultaneously block chemical warfare* (*CW*) *agents* very effectively, and to be otherwise highly environmenttally durable.

As a *first* segment of our present studies, in an effort to use the methodology of the prior electroosmotic studies reported in the literature cited above [10-12,20-23] for comparative reasons, we studied a very large number of membranes in a [liquid||membrane||liquid] interface. Both galvanostatic (constant current) *and* potentiostatic (constant voltage) studies were carried out. Values for electroosmotic liquid transport up to 50 times than those of the prior studies were obtained in our work. Typical values of liquid transport of 1.4×10^{-5} to 2.1×10^{-4} $g \cdot mm^2 \cdot s^{-1}$ (grams per square millimeter per second, an industry-standard unit of measure) were obtained in our studies, to be compared with values of 4.0×10^{-6} to 5.0×10^{-6} $g \cdot mm^2 \cdot s^{-1}$ reported in the best prior studies, those of Mullet *et al.* [12].

As a *second* segment of our present studies, direct [vapor||membrane||vapor] interface studies of electroosmotic WVT were carried out, again in both potentiostatic and galvanostatic mode. Very large and rapid relative humidity (RH) changes were observed across the membrane interface in a very short time.

As a *third* segment of our studies, a large number of "industry-standard" tests [24-26], *carried out primarily at external testing labs for independent corroboration*, were used to verify the WVT, evaporation resistance, environmental durability, and other relevant properties of our membrane assemblies. For WVT, these tests included, e.g., the ASTM E96 B (WVT), ASTM F2298 (Dynamic Moisture Permeation Cell, DMPC) and ASTM F1868 (Sweating Guard Hotplate, SGHP) tests; these yield results generally in units of $g \cdot m^2 \cdot d^{-1}$, *i.e.* grams of water vapor permeating per square meter per day, another industry-standard unit of measure. Tests were performed in large part on "final-form" fabrics, *i.e.* membranes incorporated into fabrics. A very large number of membrane types, porosities and membrane assembly configurations were studied.

As a *fourth* segment of our studies, CW-agent blocking was clearly demonstrated with CW agents at a facility qualified for performing such tests (Battelle Columbus Labs).

As a *fifth* segment of our studies, a variety of standardized environmental durability and related tests were performed. These ranged from ASTM D 2261 ("Tearing Strength of Fabrics") to ASTM F392/F392M-11 ("Gelbo Flex Test"). Our materials passed all these tests without problems.

A *final* segment of our studies encompassed miscellaneous studies related to *practical* application as a wearable garment. They included, e.g., the use of "smart" controllers to limit the current densities in our *active* membranes to prevent burnout as well as possible fire hazards while simultaneously ensuring adequate WVT; they also included preliminary selection of Li ion batteries. The controllers were also designed to use inputs from humidity and temperature sensors inside and outside the garment, so as to stop application of the *active* applied potential when not needed. They thus contributed to a truly "smart" membrane material.

Also presented in the current work is an *analysis of the likely mechanisms of WVT* (both *passive* and *active*) *as well as of CW agent blocking* in our unique membrane materials, in light of prior experimental studies as well as prior theoretical treatments.

As a final introductory note, it is noted that although this work was initiated in 2003 and much of it completed several years ago, it could not be presented until now for various reasons.

1.4. Objectives of the Present Work

This work was initiated to address a specific need to create a fabric that is lighter, and could be used to create a more comfortable garment when worn in a hot environment.

2. Experimental

2.1. Materials, Electrodes and Conducting Polymer (CP) Microporous Membrane Electrode Substrates

Microporous membrane substrates, comprising poly (sulfone), poly(ether sulfone), PTFE (poly(tetrafluoroethylene)) and poly(ethylene terephthalate), of pore size from 0.05 to 2.0 μm, were procured from GE Osmonics, Pall Gelman Labs and other vendors. Au was deposited on *both sides* of these via e-beam thermal evaporative deposition, to a thickness of ca. 500 nm. This Au/microporous membrane/Au ("Au|μP|Au") comprised the electrode substrate, on which the CP was electropolymerized, as described earlier in extensive detail [27-32, 35-38]. Monomers (aniline, diphenyl amine), dopants (K salt of poly(anetholosulfonate) and its analogue with pendant sulfate groups), CW agent simulants (dimethyl methyl phosphonate (DMMP), diisopropyl methyl phosphonate (DIMP), diethyl (methylthiolmethyl)phosphornate (DEMTMP), others) and all other chemicals and solvents, all of ACS reagent or higher grade, were procured from Sigma-Aldrich and Alfa Aesar. The electrochemical polymerization of the CP was carried out from a mixed monomer solution as described in extensive detail in our earlier communications [27-32,35-38]. The thickness of the CP/dopant layer, monitored coulometrically, lay between 0.4 and 1.6 μm. Although the microporous membrane was deposited on both sides with Au, in nearly all configurations used (see below), the CP was deposited only on one side of the membrane, the other side left as bare Au. The electrode having the CP was then generally designated as the "working" electrode and the other electrode as the "counter" electrode, although this terminology is primarily for convenience. The identity of the CP used did not change in our studies. The CP/Au/membrane electrodes were checked for electrochemical activity via standard cyclic voltammetry (see below); IR reflectance spectroscopy of the CP/Au/membrane surface, was also used, if deemed necessary, for further quality control of the CP electrode substrates (see below).

The gel electrolyte ("GelEl"), which incorporated the dopant, comprised poly(vinyl alcohol), average M_w 31,000 to 50,000, poly(ethylene glycol), average M_n ca. 900, in 95%/5% weight ratio. In some cases, as noted in the Discussion, the GelEl incorporated a low, proprietary concentration of *OPH* (organophosphate hydrolase), an enzyme reactive to the phosphonate group and capable of effectively neutralizing CW agents and simulants while not affecting the vapor and ion transport properties of its host material. These components were dissolved in deionized water having the dopant at 0.6 M concentration, with nearly all the water evaporated off subsequently. This yielded a transparent, flexible, rubbery material. Its preparation has been described in great detail elsewhere [27-32,35-38].

For the few tests carried out with OPH-impregnated electrode substrates, the procedure for impregnation of the OPH enzyme into the CP matrix, based in part on an adaptation of the most efficient procedure reported in earlier literature studies [41,42], was as follows: A saturated (>0.5 mg/mL) solution of OPH enzyme (Libradyne, Inc.) was prepared in deionized water. A previously deposited CP/Au/microporous membrane substrate was then placed, CP-side up, in a petridish filled with this OPH solution, and left overnight in a refrigerator (5˚C). The membrane was then removed, rinsed copiously with deionized water, and air dried at room temperature. The activity and concentration of the OPH within the CP/membrane was determined using Ellman's reagent and Malathion (analyte) according to published procedures [41-42], with Ellman's reagent absorbance monitored at 412 nm and the characteristic OPH IR reflectance peak, monitored at 1650 cm^{-1} [41-42].

Where appropriate, UV-Vis-NIR spectra were taken using a Perkin-Elmer (P-E) Model Lambda 12 spectrometer controlled by P-E software on a PC; when reflectance rather than transmission spectra were required, these were measured using appropriate specular and diffuse reflectance accessories and standards provided by the vendor. Again where appropriate, FTIR spectroscopy, was carried out using a P-E Model Spectrum One FTIR spectrometer, again controlled by P-E software on a PC; these FTIR spectra were primarily reflectance spectra, carried out using specular and, rarely, diffuse, reflectance accessories and standards supplied by the vendor.

Electrochemical measurements, e.g. cyclic voltammetry for quality control of the CP electrode substrates, were carried out using an Ametek-Princeton Applied Research Model 263A potentiostat/galvanostat with 2 A option; potentiostatic electro polymerizations were also occasionally carried out using a Bioanalytical Systems Model PWR-3 potentiostat.

2.2. Membrane Assembly (Fabric) Configurations

The CP microporous membranes as prepared above were then incorporated into composite assemblies with different components or configurations, as represented in the schematics in **Figure 1**. For convenience and reference throughout our work, each type was given a designation (e.g. "Type IIa"), and a shorthand notation, e.g. [inside|Au|μP-PSulf|Au|GelEl|μP-PTFE|outside] for Type IIa, as described at length in Section 2.1 above. For application of the GelEl, prepared as described above, it was heated to ca. 80˚C to render it fluid, then applied, using a syringe, as a mass on top of the center of the assembly,

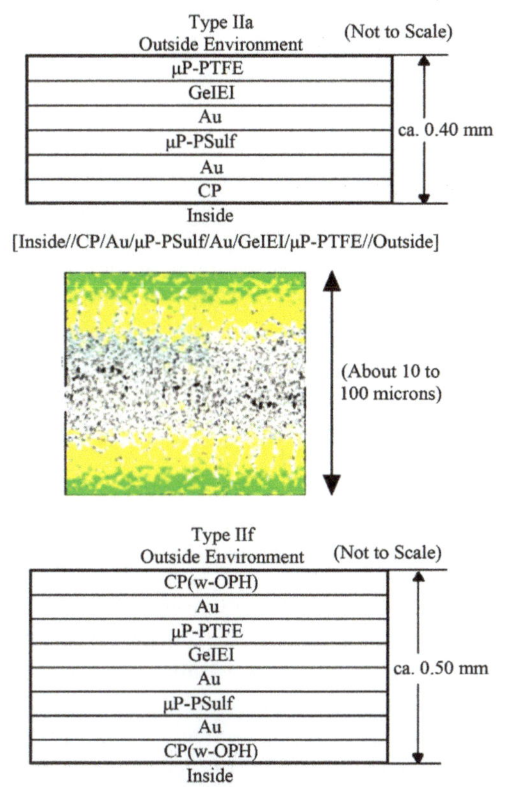

Figure 1. (Top, Bottom): Schematic representation of just two of the several membrane configurations studied in this work. (For the shorthand notation and abbreviations used, see text.) (Center): Schematic of the CP/Au/membrane structure (green = CP, yellow = Au, speckled white = microporous membrane); shown here for one of the double-sided electrodes. It shows that the CP, Au, etc. layers are not solid layers, but rather, permeate well into the micropores. This "permeative" feature has an important bearing on the electroosmotic effect, as discussed in the text.

then spread to even thickness using a doctor blade technique. The entire assembly was then clamped between glass plates and the GelEl left to cure overnight. What resulted was a flexible, fabric-like material.

2.3. Fabrication of Complete, *Active* "Cloth"

To arrive at complete, *active* fabrics usable to build a garment, additional components were added to the above membrane assemblies and complete fabrics assembled, as shown in **Figure 2** (the photos therein show the assembly for a Type IIa configuration). As a first step, individual "pixels", of approximate size 10 cm × 10 cm, were prepared. Next, they were assembled into arrays, with like electrodes (counter, working) connected in parallel, using very thin, flexible ribbon cable and electrically contacted with a urethane-base (*i.e.* flexible) Ag adhesive. The various layers of the assembly were bonded together and onto Nylon Ripstop camouflage cloth, available commercially from many vendors (e.g. Rockywoods, www.rockywoods.com), as shown in **Figure 2**; this was done using environmentally durable, neutral, pressure-sensitive adhesive (PSA) of a type used commercially in extreme-weather clothing, e.g. Therm-O-Web # 3738 adhesive (www.thermowebonline.com), formed into a small-footprint, mesh-type matrix, again as shown in **Figure 2**. (*Ripstop* fabrics are fabrics woven using a special reinforcing interweave technique that makes them resistant to tearing and ripping.)

The resulting material was a very lightweight, flexible, drapable "cloth". This complete cloth was then used for all testing, such as the WVT testing, as described below. In actual, final use, the inner fabric layers will not constitute Nylon Ripstop cloth, but rather a much thinner, highly porous material.

2.4. Liquid-Liquid Interface ([liquid||Membrane||Liquid]) Tests

As a first series of tests, liquid-liquid interface tests were carried out using the apparatus depicted schematically in **Figure 3**, emulating the apparatus and methodology used in the electro-osmotic measurements of Mullet *et al.* [12], Kozak *et al.* [11] and Bowen and Clark [10]. These tests thus comprised a first set of electro-osmotic studies primarily to serve as a comparison of the performance of our membranes with those of prior studies. Electro-osmotic transport was measured as a capillary overflow, as depicted in **Figure 3**, to be measured with a microgram balance; in practice, large milligram quantities of liquid overflow were observed. The membrane area tested was typically 17.4 cm^2 (membrane diameter 4.7 cm) Electrolytes tested included Na_2SO_4 and NaCl, typically of concentration 0.01 to 0.001 M. Reference membranes tested included Whatman Anodisc 47 (poresize 0.2 μm), ex-

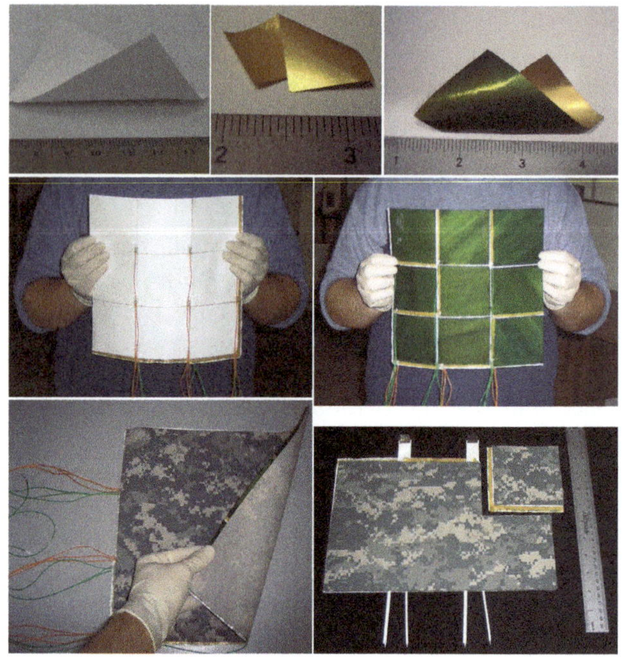

Figure 2. Photos of the components of functional fabric in various stages of fabrication (cm scale, where shown): (Top row, L to R): 1) Microporous poly(sulfone) membrane only (white). 2) Same, coated with Au on both sides. 3) Same, now with CP (green) deposited on one side. (Middle row, L to R). 4) 9-pixel section of garment with "Type IIa" membrane, PTFE side. 5) Same, CP side. (Bottom row, L to R). 6) Further assembled garment, now with Ripstop Nylon fabric bonded on both sides. 7) Same with more refined, ribbon-cable connectors, the final, wearable garment (individual pixel at top right; cm-scale shown for clarity).

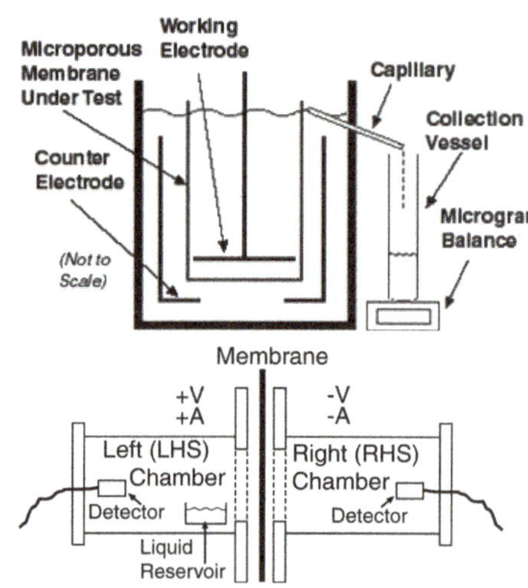

Figure 3. (Top): Schematic of apparatus used for [liquid|| membrane||liquid] studies. It emulated that used in prior studies [7-12], and so allowed for a direct comparison of our membranes' performance with that in the prior work. (Bottom): Schematic of simple apparatus used in [vapor|| membrane||vapor] studies. Water vapor was driven from the LHS to the RHS chamber. The detector used gave a real-time reading and record of humidity and temperature every second.

Actly emulating that used by Mullet *et al.* [12] This yielded a very useful, direct comparison of our data with earlier, published data. The electro-osmotic flow rate, per unit area of membrane per second, was then expressed in $g \cdot mm^{-2} \cdot s^{-1}$. The studies were carried out both potentiostatically and galvanostatically. The voltages applied in the potentiostatic mode ranged from 0 V to (−)10 V. The currents applied in the galvanostatic mode were such that the current densities were very close to those reported in the studies of Mullet *et al.* [12], ca. 30 $\mu A \cdot cm^{-2}$; the current densities observed for the potentiostatic mode were also in this region. These studies served as a very useful benchmark for comparison of our membranes with those reported in the literature.

2.5. Vapor-Vapor Interface ([Vapor||Membrane||Vapor]) Tests

As a second series of tests, [vapor||membrane||vapor] interface tests were carried out using the apparatus depicted schematically in **Figure 3**. These were a step up from the liquid-liquid tests and, to our knowledge (and as noted above), have never been reported before. Here, the difference in RH in the RHS (right hand side) compartment, effected upon the application of the voltage (potentiostatic mode) or current (galvanostatic mode) as a function of time, was monitored. The humidity/temperature sensor modules used, placed as depicted in **Figure 3** in each compartment, were the "TAI-8540 module" manufactured by TAI Inc., and incorporating the Honeywell Model HIH-3610 humidity sensor. This module measured both humidity and temperature, with the following parameters: Humidity range 0% to 100% RH (relative humidity), accuracy +/-2% RH; temperature accuracy +/−1˚C. Digital output from the module could be directly read by a PC interfaced with USB or RJ11 connectors.

2.6. Industry-Standard Water Vapor Transport (WVT) Tests

It is noted at the outset that these tests [24-26] also of course naturally comprised [vapor||membrane||vapor] interface tests. They were carried out at external laboratories, primarily for purposes of independent corroboration. They are listed and described in detail in Section 3.4 below.

2.7. GC/MS Studies of CW Agent Simulant Permeation through Membranes

These studies were carried out at Ashwin-Ushas and M-Scan Laboratories (now SGS, West Chester, PA, USA) in order to continually screen the progressively improving membrane assembly configurations for simulant permeation, to supplement the actual, agent permeation tests. The latter could only be carried out in a limited manner at a qualified facility (Battelle Columbus labs, see below) An adaptation of **TOP 8-2-501** procedure was used (US Army Test and Evaluation Command, Test Operations Procedure # 8-2-501, dated 17 January 2004, "Permeation and Penetration of Air-Permeable, Semipermeable, and Impermeable Materials with Chemical Agents or Simulants" [45]. DIMP was the primary simulant tested. The sample collection procedure was nearly identical to that used at Battelle Columbus Labs for the agent tests, and is now briefly described.

A special permeation test cell (Part# PTC 600, www.astmpermeationtestcell.com) was used, with the membrane assembly under test (*i.e.* the sample) clamped therein and with air flow through the membrane controlled via an AirChek 2000 flow apparatus calibrated to 0.3 lpm (liters/min) using a field rotameter (all SKC Inc., www.skcinc.com). To collect the DIMP vapors a solid sorbent tube (SST) from Camsco (www.camsco.com part # T13030) charged with Chromosorb 106 (60/80 mesh, 235 mg, 15 mm bed length) was used. All connecting tubes and couplings were made of Teflon® (PTFE). 10 mg of DIMP was added to the "challenge" (*i.e.* exposure) side of the sample using a micropipette. The "sampling" side of the membrane was first purged with dry N_2. An SST was then placed in line and air was commenced to be drawn through the system. Over a 24 h period, new SSTs were inserted at 0, 2, 6 and 16 h, with permeation samples collected corresponding to 0 - 2, 2 - 6, 6 - 16 and 16 - 24 h. Each SST was sealed in a separate enclosure after use, until the end of the experiment. The analyte (DIMP) was then extracted from the sorbent in the SSTs using chloroform containing 25 ppm DEMTMP (which was used as an internal standard). Each of the 4 extracts (0, 2, 6 and 16 h) of DIMP/chloroform were stored in sealed vials and sent for GC/MS analysis, as described below. The GC instrument used was a P-E Clarus 500 GC with autosampler, split/splitless injector and FID detector. The mass spectrometer was a P-E Turbomass Quadrupole mass spec with heated electron ionization (EI) source. The GC parameters were as follows: Column Restek RTX-1701 (30 m length); temperature program-start @ 80˚C, hold 5 min, ramp at 20˚C/min to 250˚C, hold 2.5 min, total run time 16 min; carrier gas He, 1.0 mL/min; injector temperature 270˚C; injector split flow-splitless for 0.5 min (FID), split 20 mL/min (MS), injection volume 1.0 μL. The mass spectrometer parameters were as follows: Ion mode-EI+, full scan and single ion recording, full scan range *m/z* 10-250, SIR ion monitored *m/z* 97, SIR dwell time 0.1 s, total run time 16 min, solvent delay 4 min, MS source temperature 200˚C, transfer line temperature 200˚C, electron energy 70 V, calibration gas heptacosa. The FID parameters were as follows: H_2 flow 45 mL/min, air flow 450 mL/min, FID temperature

275˚C, time constant 200 ms. A Total Ion Current (TIC) chromatogram with a di-isopropyl methyl phosphonate (DIMP) standard yielded a main peak at 8.97 min, whose Electron Ionization (EI) mass spectrum matched NIST data for DIMP. Diethyl methylthiomethyl phosphonate (DEMTMP) was used as an internal standard, yielding a TIC peak at 11.51 min, confirmed through its EI mass spectrum as being DEMTMP. Using an enhancement of this method with Single Ion Recording and monitoring of the *m/z* 97 peak, the detection limit of the procedure was established to be 20 ppb DIMP.

2.8. Agent Permeation Tests

These were carried out at the Battelle Hazardous Materials Research Center (HMRC), part of Battelle Columbus Labs, Columbus, OH (USA), one of the few facilities in the US qualified to carry out such tests. The tests were carried out per the **TOP 8-2-501** procedure [24-26] already described above with the static flow (AVLAG, Aerosol Vapor Liquid Assessment Group) method, using standard AVLAG cells. Membrane materials were tested for 24 hours using four sampling intervals at 0 - 2, 2 - 6, 6 - 16 and 16 - 24 hours (as for the simulants, described above). Bubblers filled with ethylene glycol diacetate (EGDA) were utilized to capture any HD and VX vapor permeation while Chromosorb 106 solid sorbent tubes (SSTs) were used to capture any GB vapor permeation. The SSTs were extracted with 10 mL of Chloroform. The chloroform extract and the EGDA were analyzed by gas chromatograph (GC) to quantify the agent permeation through the materials. *Passive* test samples were assembled into standard AVLAG test cells and loaded in the particular orientation desired (e.g. CP-coated side towards the "interior"). The membrane samples were tested using three repetitions for each agent GB, HD, and VX. All cells were leak checked prior to being agent challenged.

Two slightly differing procedures were used for the *passive* (no voltage applied) and *active* (voltage applied, + 6 VDC) tests, as follows:

Passive Tests Only: Prior to CW agent application, all samples were preconditioned at the required environmental conditions for twenty minutes. Upon completion of the preconditioning period, swatches were challenged with the required agents and reinstalled into the test fixture. After all cells were reinstalled, airflow was initiated and the trial started for the 24 hours. Values below the detection limit of 0.5 g/cm^2 were reported as 0.25 g/cm^2 which were used for any subsequent calculations. (This is a standard practice for manipulation of swatch permeation data.) Spike controls were utilized during the spiking process for all agent challenges. While based on only three replicates, the data were consistent within each subset of material and agent combination.

Active Tests Only: *Active* test samples were also assembled into standard AVLAG cells with the exception that Teflon® plates were used in place of the standard aluminum plates to isolate the voltage applied to the swatch. Each sample was oriented in the cell so that the leads attaching to the electrodes were directed toward the front of the cell. The membranes were tested using two repetitions for each agent GB, HD, and VX. All cells were leak checked prior to being agent challenged. After all cells were assembled and loaded into the test fixture, the sample leads were attached to their respective jumper cable and voltage was applied in 1 V increments, cycled through all samples until 6 V was reached. The voltage remained on for the remainder of the test. All samples were preconditioned at the required environmental conditions for twenty minutes. Upon completion of the preconditioning period, cells were challenged with the required agents and reinstalled into the test fixture.

The additional test parameters may be summarized as follows: Temperature 32˚C ± 1˚C; relative humidity (RH) 80% ± 5%; CW agent challenge: 10 mg ± 1.0 mg per swatch; air flow: static flow, constant at 0.3 lpm ± 0.01 lpm; sample area 10 cm^2; solvent volume 10 mL.

2.9. Industry-Standard Environmental Durability Tests

The industry-standard tests [24-26] performed on the membranes under this category included:

1) ASTM D 2261, "Tearing Strength of Fabrics".

2) ASTM D 5034 (Grab), "Breaking Strength of Textile Fabrics".

3) ASTM D 3886, "Stoll Flat Abrasion Resistance".

4) ASTM F392/F392M-11, "Standard Test Method for Flex Durability of Flexible Barrier Materials", also called the "Gelbo Flex Test".

The first three tests were carried out at Intertek Labs, Springfield, NJ (USA). The 4th was carried out at Ashwin-Ushas using a custom test setup, as shown in **Figure 4**. Details of these tests are available in the open literature [24-26].

2.10. Other Miscellaneous Tests

"*Seaming*" *tests*: Our final-form fabric, contains "pixels" of our membrane assemblies, of size approximately 10 cm × 10 cm. Important issues in final use are therefore whether the seams between the pixels: 1) Are impermeable to CW agents. 2) Impede WVT in any way. 3) Create any problems in flexibility and "drapability" of a full garment. Now the seams between these pixels are ca. 2 to 4 mm in width. They are hermetically sealed using an environmentally benign, highly flexible, temperature-stable, translucent, rubber-like polyurethane adhesive (Cytec Conathane EN series, available from various

Figure 4. Custom Gelbo flex apparatus, in actual use. Various views are shown, including completely flexed with sample.

vendors, e.g. Ellsworth, at http://www.ellsworth.com) specially formulated for textiles. In order to test these seams, a series of "seaming tests" were performed on the final-form garments. The tests included WVT (ASTM E96 B, Open Cup) and DIMP permeation through the seam part of garment. They also included a physical and visual test of the drapability of the garment. As noted in the Discussion, the garments passed all these tests successfully.

3. Results and Discussions

3.1. Membrane Assembly Construction and Configuration

The fabrication of the membrane assemblies, including deposition of the conducting polymer (CP) onto the Au-coated microporous membranes, is described at length in the Experimental Section. As noted therein, these CP-functionalized microporous membrane assemblies are very similar to those used in our prior work for completely unrelated applications, such as variable emittance skins for spacecraft thermal control, military IR camouflage, and electrochemical voltammetric sensing [27-40].

To very briefly describe membrane assembly here then, the base membrane was a microporous membrane, typically a poly(sulfone) of poresize 0.1 to 0.4 μm, with Au deposited either on one side only or on both sides. In the latter case, this resulted in a single, 2-sided, 2-electrode membrane; CP could then be deposited (electrochemically polymerized) on either side [27-32,34]. Gel electrolyte ("GelEl") [33-35], which permeated into the micropores, also served to function as a good adhesive, holding the various membrane layers well together. A large number of membrane configurations (designated "Type IIa", "Type IIe", etc.) were tested. Two of these are shown in **Figure 1**. Shorthand notation could be used to distinguish the various configurations. Thus, e.g., **Type IIa 1)**), shown in **Figure 1**, was, ii n shorthand notation, denoted as [inside|Au|μP-PSulf|Au|GelEl||μP-PTFE|outside], where μP = microporous, PSulf = poly(sulfone), GelEl = gel electrolyte, a double bar (||) ndicates an interface between phases, and "inside" and "outside" denote, respectively, the inside of a worn garment and the exterior environment. The other membrane configurations studied included: **2) Type IIe**: [inside|CP|Au|μP-PSulf|Au|GelEl|| μP-PTFE|GelEl||μP-PTFE|outside]. **3) Type IIf**: [inside| CP-w-OPH|Au|μP-PSulf|Au|GelEl||μP-PTFE|Au|CP-w-O PH|outside], where "w-OPH" signifies "impregnated with organophosphate hydrolase", OPH being an enzyme that hydrolyses organophosphates. **4) Type II**: Type II ([inside||CP|Au|μP-PSulf|Au|GelEl||outside]. Membrane porosities, from 0.05 μm to as high as 5.0 μm were studied, with optimal porosities identified; e.g., for Type IIa, these were 0.01 μm (for poly(sulfone)) and 1.0 μm (for PTFE)). Additionally, a "+V", "–V", "+A" or "–A" at either end of the configuration indicated the application of a + or (–) voltage or current.

As shown in the color figure in **Figure 1** (center), the Au and CP were not solid layers, as represented in the schematic figures, but rather, permeated into the pores of the membrane. Additionally, the GelEl, while again shown as a single, solid layer in the schematics, actually also permeated throughout all the microporous membranes. This "permeative" feature of the CP/Au/membrane interface structure [27-40] has an important bearing in understanding the electroosmotic effect therein, as discussed further below. **Figure 2** shows components of the final, wearable soldier garment, in various stages of fabrication, as described.

The gel electrolyte (GelEl) [35-38] used in our membrane assemblies had a primary poly(vinyl alcohol) (PVA) base, with several other critical ingredients (see Experimental Section). The GelEl cured overnight to form a very thin, translucent, rubbery, highly flexible material [35-38]. Several features of this GelEl made it unique and vital for membrane assembly function. Firstly, it allowed permeation of water vapor, at a very high rate. This was verified with independent [vapor||GelEl||vapor] studies, as described below. Secondly, it served as the primary mechanism of blocking, as shown in the results below. Thirdly, it provided an adhesive function, prior to cure, bonding the two membrane layers on either side of it very well.

The CP used in our membrane assemblies was a copolymer of aniline and diphenyl amine with a polymeric dopant also used, surprisingly, in our earlier, unrelated, IR electrochromics and spacecraft thermal control work [27-40]. The CP was always placed on the garment "inside" or interior, (*i.e.* closest to the skin when worn as a garment); this ensured that the electroosmotic flow was in the correct direction (from the "inside" to the "outside"). A typical membrane configuration such as Type IIa displayed an open circuit potential (OCP), between the "outside" facing and "inside" facing electrodes, of ca. + 0.2 V to + 0.3 V (with the + polarity observed at the CP). The significance of this in relation to the mechanisms of electroosmotic WVT in our membranes is discussed in more detail further below.

Type IIa, Type IIe and Type IIf membrane assemblies (described above) comprised the primary, "workhorse" configurations used in nearly all our work and the further testing reported here. In Type IIf, the function of the layer of CP-OPH (organophosphate hydrolase) was to serve as an additional barrier to immediately intercept and neutralize or block CW agents: OPH has been established in much earlier work as capable of blocking, through binding or reaction, moieties with phosphonate groups, such as the agents targeted in the present work [41-42]. In the final analysis, *Type IIa and Type IIe were found to be the best performing membranes in terms of WVT rates in combination with agent blocking.*

In the *active, potentiostatic* (constant voltage) mode of operation, typical voltages applied across the primary membrane (*i.e.* the one on the "inside" side of the assembly) were in the range of 1 to 6 V DC, with a maximum voltage of 10 V tested. The polarity of the potential applied at the *exterior* electrode, in 2-electrode mode, was always negative (−). Now ordinarily, such extreme voltages applied on the aniline-diphenyl-amine copolymer CP used would cause irreversible degradation of the CP. This would, e.g., be the case in our variable emittance or IR electrochromic devices containing the same CP/Au/ microporous-membrane electrodes [27-40], which use ionic liquid electrolytes. However, in the case of the present electroosmotic membrane assemblies, the GelEl used had a very low conductivity, in the region of 0.005 mS/cm (to be compared with ca. 12 to 18 mS/cm for the ionic liquid electrolytes used in our variable emittance devices and even higher for the electrolyte used in our IR electrochromic devices.) Due to this poor GelEl conductivity, there was a large resistance barrier (*i.e.* conductivity drop) between the working and counter electrodes of the electroosmotic membranes. Thus, even with large potential differences, of up to 10 V, applied across the electrodes in these membranes, the potential actually "sensed" by the CP was much smaller. This was confirmed by a relatively simple experiment: In our IR electrochromic devices, with the use of an aqueous-based gel electrolyte with high conductivity, H_2 evolution was observed at the CP electrode at about (−)1.5 V applied (in 2-electrode mode). However, H_2 evolution did not occur at the CP electrode in the electroosmotic membranes until (−)15.0 V was applied, a factor of 10.

In the *active, galvanostatic* mode of operation, typical current densities applied across the primary membrane in our studies were in the region of ca. 8 $mA \cdot cm^{-2}$. Even for the highest currents densities applied, the observed potential difference never exceeded ca. 2.5 V (cf. **Figure 5(e)**); consequently, again, no degradation of CP was observed.

3.2. [Liquid||Membrane||Liquid] Interface Studies

As noted briefly above, as a first segment of our studies, it was attempted to emulate the methodology of prior electroosmotic studies [7-12,20-23] on [liquid||membrane|| liquid] configurations, so as to directly compare our membranes' performance with the prior, published work. For this purpose, an apparatus exactly emulating the work of Mullet *et al.* [12] and Bowen *et al.* [7-10] and Kozak and Davis [11] was used. This is depicted schematically in **Figure 3**. As electrolyte, an aqueous solution of 0.01 M NaCl was used, approximating that of the earlier work. A constant voltage was applied across the two electrodes depicted therein, and the resulting electroosmotic flow, was collected using the microcapillary as shown and measured using a sensitive balance. The area of the membrane at the interface, subject to electroosmotic flow, was 1215 mm^2. The current densities used in our potentiostatic studies, ca. 33 to 42 $\mu A \cdot mm^{-2}$, were very close to those used in the work of Mullet *et al.* [12] and Bowen *et al.* [7-10] (ca. 41 $\mu A \cdot mm^{-2}$).

Several different membrane assembly configurations were tested. For quite evident reasons, the GelEl was not used in these [liquid||membrane||liquid] interface studies (among other things, slow dissolution of this into the aqueous solution used would occur). A typical configuration tested, represented using the shorthand notation described above, was [(Pt Cathode, (−)V)||CP|Au|μP-Sulf| Au||Pt Anode(+)V], where the microporous poly(sulfone) membrane (μP-PSulf) had a porosity of 0.2 μm. Voltages applied at the cathode, in 2-electrode mode, ranged from (−)1.0 to (−)9.8 V.

The *electroosmotic flow rate*, ***V***, in these studies could be calculated based on the detailed mathematical treatment of Bowen [7] and Kozak and Davis [11], Smoluchowski [1-2], Henry [3] and Saville [4] as also used in the experimental studies of Bowen and Clark [10] and Mullet *et al.* [12], as follows:

$$V = \frac{I\varepsilon\zeta}{\eta\lambda_P} f(kr,\zeta) \quad (1)$$

where: V—electroosmotic flow rate in volume units; I—applied current; ε—electrolyte permittivity; ζ—zeta potential; η—electrolyte viscosity; λ_p—electrolyte conductivity in the pores; k—Debye radius; r—average pore radius; $f(kr, \zeta) \approx 1$ for the conditions of all our experiments.

Now in our experiments, a volume of liquid was collected after a particular period of time and its mass, ***m***, was measured. Also measured was $I(t)$, the current as a function of time, at constant potential (E):

$$\boldsymbol{m} = \rho\int Vdt = \int\frac{I\varepsilon\zeta}{\eta\lambda_P} f(kr,\zeta)\,dt = \rho\frac{\varepsilon\zeta}{\eta\lambda_P} f(kr,\zeta)\int Idt \quad (2)$$

where: ***m*** is the mass (measured in the experiment); ρ is the electrolyte density; t is the time.

Introducing m/A (mass of electroosmotic flow per unit area of the membrane), and since we are primarily interested in the relative flow rates for various, different membranes, we were able to compare different experiments:

$$m/A \sim \int jdt \sim \frac{1}{A}\int Idt \quad (3)$$

where: j is the current density.

Table 1 summarizes representative results from these [liquid||membrane||liquid] studies; it also lists typical results from the studies of Mullet *et al.* [12], at nearly identical current densities and electrolyte concentrations, for comparison. It is observed that the electroosmotic liquid transport values for our membranes *are more than an order of magnitude higher than the highest values reported previously*, which are those of Mullet *et al.* [12]. We note again that the current densities used in Mullet *et al.*'s and others' prior work [7-11,20-23] were nearly identical to those in our work, all being in the region of ca. 30 $\mu A \cdot mm^{-2}$, thus allowing for a truly direct comparison. It is also to be noted that our tests with the same membranes used in the prior work, e.g. the microporous alumina membranes used by Mullet *et al.* [12], (see Experimental Section) yielded results nearly identical to that of Mullet *et al.* [12] and the other prior work, further corroborating our methodology.

3.3. [Vapor||Membrane||Vapor] Interface Studies

As a further step up from prior studies, a second segment of our work studied electroosmotic WVT across [vapor||membrane||vapor] interfaces, in both potentiostatic and galvanostatic mode. As noted above, to our knowledge this is the first such study of *active* (*i.e.* with voltage or current applied) WVT across such [vapor||membrane||vapor] interfaces.

Figure 3, above, has shown a schematic of the dual-chamber [vapor||membrane||vapor] apparatus used in our work (see Experimental Section for more detail). A reservoir of ca. 1 mL of aqueous 0.01 M NaCl was placed in the left hand side (LHS) chamber, to simulate a humid environment; this would be the type of warm, sweating, humid environment that would be expected inside clothing. The LHS then corresponded to the "inside". Water vapor was driven from the LHS to the RHS chamber. A constant, negative voltage or current was applied at the RHS of the membrane, *i.e.* the one facing the RHS chamber, as indicated, and the change in relative humidity (RH) as a function of time monitored using the humidity sensors. The detectors used gave a real-time reading and record of humidity and temperature every second. (The interval could be adjusted from 0.05 s onwards, but a 1 s interval was found adequate.)

Typical *potentiostatic* results are summarized in the raw data in **Figures 5(a)-(c)**. The different data therein correspond to two different membrane configurations tested, respectively: **1)** [(+V)||CP|Au|μP-PSulf|Au||(−V)] (where μP-PSulf = microporous polysulfone membrane, poresize 0.2 μm); and **2)** [(+V)||CP|Au|μP-PSulf|Au|CP||(−V)]. It is to be noted that the second configuration has a layer of CP on both sides. As these data show, a very rapid increase in RH was observed in the RHS chamber, interfaced to the (−) polarity electrode, indicating a very rapid transfer of water vapor into this chamber from the LHS chamber. For example, typical changes in RH (*i.e.* ΔRH) for the (−) 3 V experiment (with a membrane area of ca. 12 cm^2) were 28% in 240 s (*i.e.* 4 min), rising to 40% at 600 s (10 min). Furthermore, it was observed that, as expected, the rate of change of RH was higher for higher applied voltages. Finally, it was observed that the rate of change of RH was slower when a layer of CP was present on the exterior face as well as on the interior face of the membrane assembly. Also depicted in **Figure 5** (part (c)) is a chronoamperogram, showing that the steady-state current is very small.

Importantly, it is to be noted that the 0 V data in **Figure 5** correspond of course to *passive* WVT, analogous to that observed, e.g., in commercial Gore-Tex® PTFE membranes as reported by McCullough *et al.* [17,18], but at a much higher rate. That is to say, as noted briefly above, *our CP membrane materials possessed an inherent, passive WVT capability which itself was much higher than that of "conventional" microporous membranes.*

Typical *galvanostatic* results are shown in **Figure 5(d), (e)**, with current densities indicated in the figure legend.

Table 1. Representative results from [liquid||membrane||liquid] studies. Also given, for comparison, are: results with alumina membranes, as used in prior work of Mullet *et al.* [12] and others, but using our apparatus; and the original, published results from this prior work.

Membrane Configuration (μP-PSulf = poly(sulfone, 0.2 μm poresize; 0.01 M NaCl electrolyte, *current density ca.* 35 μA/mm²; 39 μA/mm²for reference Mullet et al. study cited)	Time (Δt) [s]	m [g]	$\frac{1}{\Delta t}\int jdt$ [A·mm^{-2}]	*Electro-osmotic flow, g per unit area per second* $\frac{m}{A\Delta t}$ [g·mm^{-2}·s^{-1}]
[(+V)‖CP|Au|μP-PSulf|Au‖(−V)]	3745	4.263	4.145×10^{-4}	**1.434×10^{-5}**
[(+V)‖Au|μP-PSulf|Au|CP‖(−V)]	2887	4.034	4.765×10^{-4}	**1.760×10^{-5}**
[(+V)‖CP|Au|μP-PSulf|Au|CP‖(−V)]	5834	4.663	4.172×10^{-4}	**1.007×10^{-5}**
Alumina membrane (Whatman Anodisc, 47 mm dia., 0.2 μm poresize), *exactly emulating Mullet et al. study* [12], but with our apparatus	510	1.201	2.9×10^{-4}	**1.3×10^{-6}**
Actual *Mullet et al. study* [12], [Alumina/titania/silica Membrane (64/27/9 w/w%), 0.9 μm poresize), 0.01 M NaCl electrolyte, current density ca. 39 μA/mm²]				**4.0×10^{-6}**

Also shown therein are chronopotentiograms, *i.e.* the behavior of the potential during the constant current application. It is seen here that after the initial spike, the potential quickly stabilizes to a relatively small value (magnitude <2.5 V) and within a very narrow range, thereafter showing little change. *These data clearly show that there is no "potential overload" observed.*

In other [vapor||membrane||vapor] results which varied the concentration of the electrolyte used, it was observed that dilution of the electrolyte concentration caused progressively higher WVT rates, in line with theory and prior experiment [1-12]. For example, in an experiment analogous to that represented in **Figures 5** and **6** above, at a time duration of 1000 s, the values of ΔRH observed for 0.1 M, 0.01 M and 0.001 M aqueous NaCl, were, respectively, 20.1%, 23.3% and 29.3%. These data are not shown graphically for space reasons.

The WVT rate *of the GelEl alone* was verified, for reference, by performing a [vapor||membrane||vapor] experiment as described above, but on a [vapor||Au|GelEl-impregnated-membrane|Au||vapor] configuration, *i.e. without* any *CP*. (The membrane used was microporous poly(sulfone), poresize 0.2 μm). In a *passive* mode, *i.e.* with no voltage or current applied across the interface, the ΔRH observed was ca. 20% higher than the 0 V results shown **Figure 5** above. These results confirmed that the rubbery GelEl actually acted as an efficient transporter of water vapor in the *passive* mode, with the driver in this case being the concentration gradient of the water vapor (and the resulting electrochemical potential difference-see discussion below) on either side of it.

3.4. [Vapor||Membrane||Vapor] Interface Studies, Cont.: Water Vapor Transport (WVT) and Related Studies Using Industry-Standard Tests

Three industry-standard WVT tests [24-26] were carried out (see detail in Experimental Section) at external laboratories (for purposes of independent corroboration). They included:

1) ASTM E96 B, Upright Cup (American Society for Testing and Materials, standard# ASTM E96, "Standard Test Methods for Water Vapor Transmission of Materials").

2) DMPC, ASTM F2298: Dynamic Moisture Permeation Cell test ("Standard Test Methods for Water Vapor Diffusion Resistance and Air Flow Resistance of Clothing Materials Using the Dynamic Moisture Permeation Cell").

3) SGHP, ASTM F1868 (Sweating Guard Hotplate test). This is a "skin model" test which simulates heat and mass transfer processes occurring next to human skin and produces accurate, repeatable measurements of the thermal resistance and vapor permeability of textiles.

It is important to note that the above three tests monitor different performance parameters that *cannot be compared directly across the different tests*, but *can* be compared within the tests between different materials and against certain "reference" or "standard" materials. It is thus customary in the industry to simultaneously test reference materials (standards), for which extensive published data exist [17,18]. Such reference or standard materials include Gore-Tex® XCR microporous PTFE membranes or an eVent® microporous Nylon membranes (both of poresize 0.2 μm). The first two tests above yield results in units of g·m^2·d^{-1}, i.e. grams of water vapor transported (WVT) per square meter of area per 24 h; however, again, since they are measuring different performance parameters, the range of numbers obtained from the two tests results is different [17-18].

A recent publication of McCullough *et al.* [17-18] gives a most recent and very comprehensive evaluation of the WVT properties of a very wide variety of fabrics, using a variety of WVT tests, including all three tests used in our work; in many ways, this work is used as a

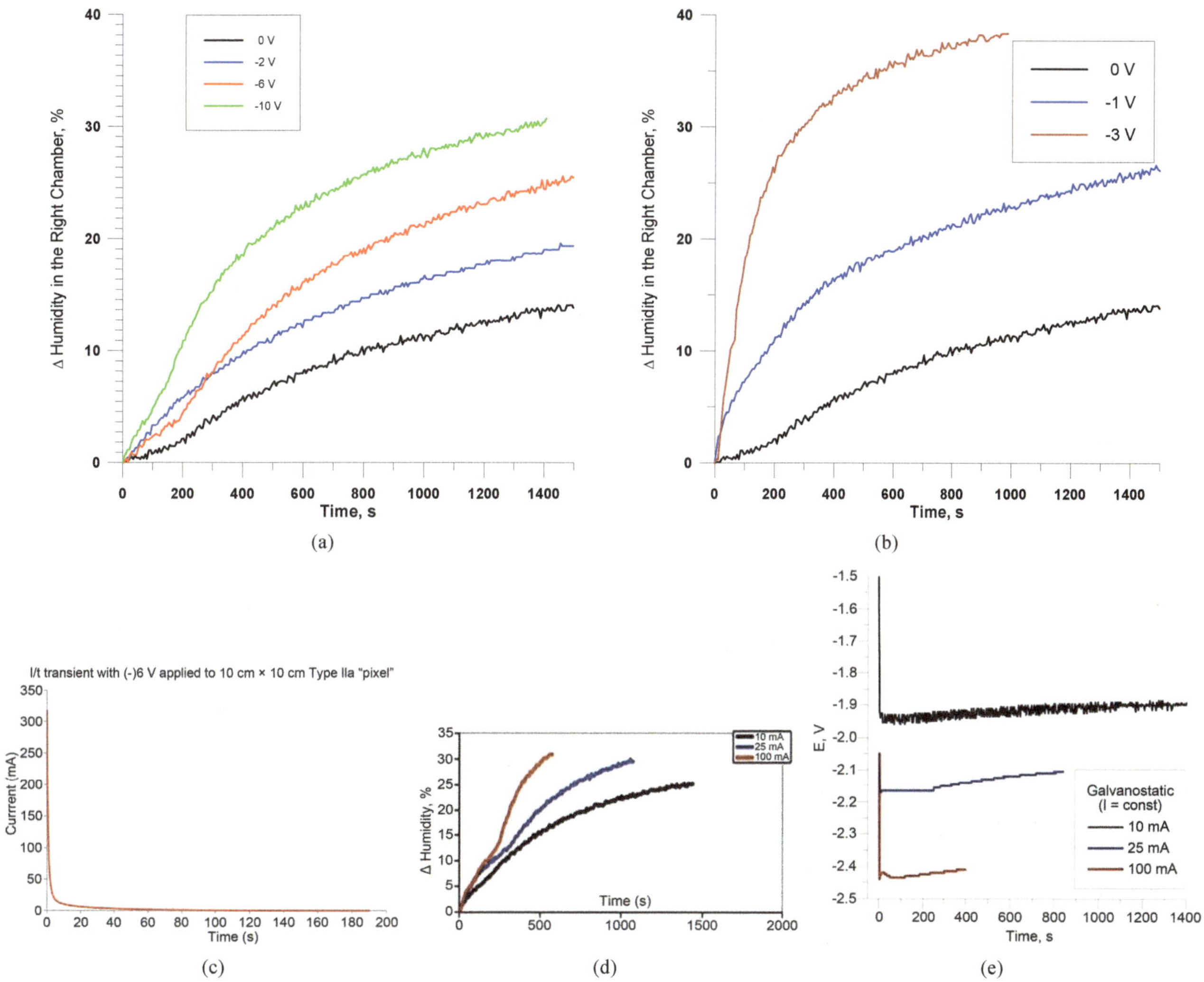

Figure 5. (a-c): Typical *potentiostatic* results on [vapor||membrane||vapor] interfaces. (a) For configuration [(+V)||CP|Au|µP-PSulf|Au||(−V)] (where µP-PSulf = microporous polysulfone membrane, poresize 0.2 µm and "+V" indicates polarity of voltage applied, etc.); and (b) For configuration [(+V)||CP|Au|µP-PSulf|Au|CP||(−V)], which has a layer of CP on both sides. (c) Typical current-time transient for (−)6V application on a single "pixel", showing minimal steady-state current. (d, e): Typical *galvanostatic* results on [vapor||membrane||vapor] interfaces: (d) For the configuration [(+A)[Au||µP-PSulf||Au|CP](−A)], at different currents as indicated, corresponding to current densities of 0.8 mA·cm^{-2} (for 10 mA), 2.0 mA·cm^{-2} (for 25 mA) and 8.0 mA·cm^{-2} (for 100 mA). (e): Chronopotentiograms corresponding to these three experiments, which clearly show that there is no "potential overload".

seminal reference work in the industry [13-18]. Thus, McCullough *et al.* [17-18] report a WVT value for the Gore-Tex® XCR microporous PTFE membrane is **864.4 g·m^2·d^{-1}** for the ASTM E96 B Upright Cup test, but it is **3193.3 g·m^2·d^{-1}** for the same membrane using the DMPC (ASTM F2298) test; these obviously quite different values are both accepted as industry standards. It is to be noted that these values are *passive* values, *i.e.* with no voltage of any kind applied. It is also very important to note that *Gore-Tex, Nylon and other microporous fabrics do not of course provide any protection against agents* (in contrast to our membranes, which do).

The third test (the SGHP test) yields parameters measuring the evaporative resistance, in units of m^{2}PaW^{-1}. One parameter, R_{et}, measures the total resistance to evaporative heat transfer of the material, including the surface air layer and all barriers. A second parameter, R_{ef}, measures the resistance to evaporative heat transfer of the material alone; it generally provides a better comparison across materials of different types, and is the one used in our results presented here. A lower value of R_{ef} indicates a *higher* WVT. The reference (standard) used in the industry is again that for Gore-Tex XCR®, **4.9 m^{2}PaW^{-1}**. **Table 2** summarizes the results from these three tests, for both our membranes and the industry-standard membranes.

A more extensive set of DMPC WVT measurements were carried out following the initial results (shown in

Table 2. Summary results from studies of WVT using the ASTM E96 B Upright Cup method and the Dynamic Moisture Permeation Cell (DMPC, ASTM F2298) method. ((Abbreviations: μP-PSulf = microporous poly(sulfone), poresize 0.2 μm; μP-PTFE = microporous PTFE, poresize 2.0 μm; GelEl = gel electrolyte; CP = conducting polymer).

Membrane Configuration	Method	Voltage Applied	WVT [g-m^{2}d^{-1}]	% Increase with Voltage Applied
[inside‖CP|Au|μP-PSulf|Au|GelEl‖outside]	ASTM E96 B	None	2564.4	
[(+V)inside‖CP|Au|μP-PSulf|Au|GelEl‖outside(−V)] (**Type II**)	ASTM E96 B	(−)6 V	3706.7	44.5%
[inside‖CP|Au|μP-PSulf|Au|GelEl-μP-PTFE‖outside] (**Type IIa**)	ASTM E96 B	None	2008.0	
[(+V)inside‖CP|Au|μP-PSulf|Au|GelEl|μP-PTFE‖outside(−V)] (**Type IIa**)	ASTM E96 B	(−)6 V	3154.8	57.1%
Gore-Tex® XCR (PTFE, 0.2 μm poresize) [17-18]	ASTM E96 B	None	864.4	
eVent® (**Nylon**, 0.2 μm poresize) [17-18] [a]	ASTM E96 B	None	984.8	
[inside‖CP|Au|μP-PSulf|Au|GelEl‖outside] (**Type II**)	DMPC ASTM F2298	None	7760.0	
[inside‖CP|Au|μP-PSulf|Au|GelElμP-PTFE‖outside] (**Type IIa**)	DMPC ASTM F2298	None	4330.0	
Gore-Tex® XCR (**PTFE**, 0.2 μm poresize), *our measurement*	DMPC ASTM F2298	None	2990.0	
Gore-Tex® XCR (**PTFE**, 0.2 μm poresize), literature[17-18] [a]	DMPC ASTM F2298	None	3193.3	

[a] Fabric showing the highest WVT using this method in prior published data.

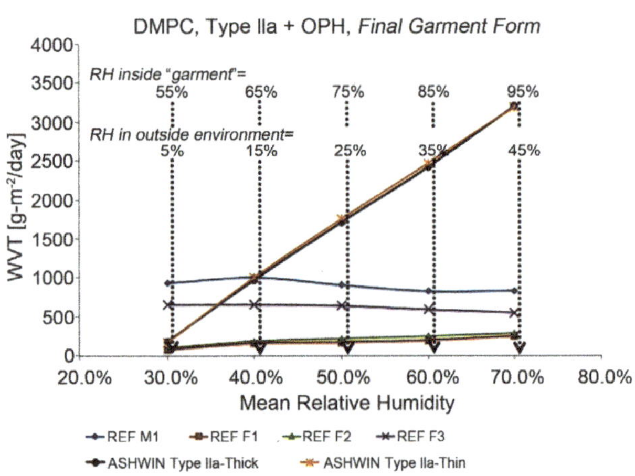

Figure 6. More extensive WVT tests using the DMPC (ASTM F2298) method. All except the references/control materials were carried out on membrane assemblies incurporated into a final, fabric material. (See text for detailed discussion of these results).

Table 2). These were done on the Type IIa membrane configuration, but with OPH (the enzyme organophosphate hydrolase [41-42], see above) incorporated into the CP; the resulting configuration was [inside‖CP(OPH)|Au|μP-PSulf|Au|GelEl|μP-PTFE‖outside]. Most importantly, these measurements (except for the ones on the reference materials) were performed with our membrane assembly incorporated into a *final-form, wearable garment*, including additional outside Nylon-Ripstop camouflage cloth and inside cloth liner (see **Figure 2**). These DMPC tests were done at the US Army Natick Soldier RD&E Center.

The results from these are summarized in **Figure 6**. A set of five measurements were done, each maintaining a difference in relative humidity (ΔRH) of 50%. In the first set, the "inside" RH was maintained at 55% whilst the "outside" was maintained at 5%, corresponding to very dry, desert conditions. In each set, the RH was progressively increased by 10% on both ends, so that the fifth said had an "inside" RH of 95% and an "outside" RH of 45%, corresponding to a sub-tropical environment. All experiments were thermostatted at 25°C. Additionally, a number of reference materials were tested: 1) "REF M1", an expanded microporous PTFE membrane somewhat similar to Gore-Tex® XCR. 2) "REF F2", a 3-layer fabric with enhanced WVT. 3) "REF F1", a microporous-PTFE-membrane-based 3-layer laminate fabric. 4) "REF F3", an additional type of microporous-PTFE-membrane-based 3-layer laminate fabric.

The DMPC results in **Figure 6** show, quite dramatically, that while the reference fabrics (including multi-layer, multi-component fabrics with specially enhanced WVT, denoted REF F2, FEF F1 and REF F3) show a *constant* WVT rate with increased humidity inside the garment, our final-garment materials, incorporating our Type IIa membrane, show *sharply increasing* WVT with increased internal humidity; this is of course exactly what one would want to increase comfort. And of course, most importantly, in spite of expelling water vapor at a far higher rate than the reference fabrics, our materials also protect against agents, whereas the reference fabrics do not.

With regard to the ASTM F1868 (Sweating Guard Hot Plate) test, just one set of measurements were carried out, for the Type IIa membrane configuration with no voltage applied. This gave a normalized R_{ef} value of 1.01

m^2PaW^{-1}, with a sample of Gore-Tex XCR® measured in the same batch of tests yielding a normalized R_{ef} value of 4.9 m^2PaW^{-1}, corresponding closely to the literature value [17-18]. This indicates a WVT capacity for our Type IIa membrane configuration in *passive* mode (no voltage applied) and under sweating human skin conditions, of 4.9/1.01 = 485% of that of the Gore-Tex® standard, while our membrane also provides protection against CW agents which the Gore-Tex® membrane of course doesn't.

To briefly reiterate the salient findings from the above three types of tests at independent labs then, very high WVT rates were observed for our materials. For example, in a direct comparison, the best performing prior art membrane tested (in *passive* mode only) by McCullough *et al.* [17-18], a microporous Nylon, yielded a WVT rate, using the ASTM E96 B Upright Cup method, of **984.8 $g \cdot m^2 \cdot d^{-1}$**. This may be compared with **2564.4 $g \cdot m^2 \cdot d^{-1}$** and **3706.7 $g \cdot m^2 \cdot d^{-1}$** for one of our membrane assemblies (for *passive* and *active* WVT, respectively), using the same ASTM method. And most importantly, of course, none of the materials studied by McCullough *et al.* or others [7-23] are able to block CW agents.

3.5. Permeation/Blocking Studies with Chemical Warfare (CW) Agent *Simulants* Using Industry-Standard Tests

For a proper frame of reference in appreciating the results on simulants, it is pertinent to comparatively view the structures of the simulants and the agents that they are models for. **Figure 7** provides this, along with structures of common pesticides. The simulants studied included D(M/I/E)MP = di(methyl/isopropyl/ethyl) methyl phosphonate; DEMTMP = diethyl(methylthiomethyl) phosphonate; Cl-Eth = bis(2-Cl-ethyl amine:HCl), all shown in **Figure 7**, and methyl salicylate (simulant for HD) and 3-hepten-2-one (simulant for G agents). The methodology used for these studies, based on the **TOP-8-2-501** standard, has been described at length in the Experimental Section. DIMP was the primary simulant studied; other simulants were studied briefly to confirm applicability of the methodology. GC-MS formed the primary method of detection and quantitation. Results for DIMP permeation are summarized in **Table 3.**

These results show, firstly, that the values for total permeation (cumulative, *i.e.* over the 24 h period) for the simulant DIMP, as measured by GC/MS using our methodology, closely correspond with those for the live agent GB, as measured at Battelle; this is true for both the *passive* (no voltage applied) and *active* (−6 V applied) modes, thus validating our methodology. Secondly, they show that the introduction of an additional layer of GelEl, as in Type IIe vs. Type IIa, *drastically reduces the DIMP*

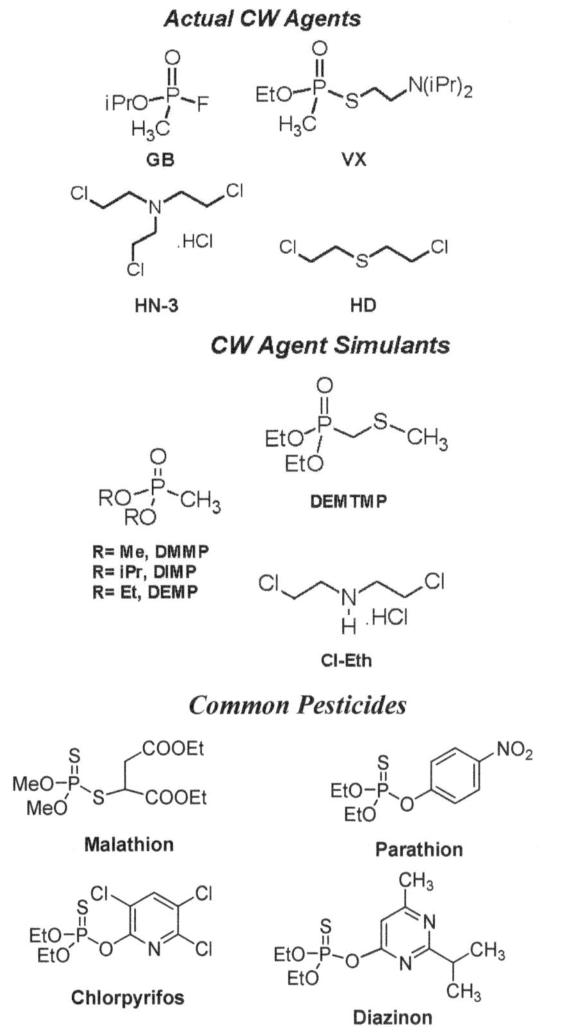

Figure 7. Structures of chemical warfare (CW) agents and simulants/models for these. Also shown, for info only, are structures of common pesticides; the stark similarity of their structures with those of the other categories is evident. (Abbreviations for simulants: D(M/I/E)MP = di(methyl/isopropyl/ethyl) methyl phosphonate; DEMTMP = diethyl(methylthiomethyl) phosphonate; Cl-Eth= bis(2-Cl-ethyl amine: HCl); Mal = Malathion, a common commercial pesticide).

permeation (total permeation drops from 236.23 μg/cm^2 to 54.04 μg/cm^2). Thirdly, they show that, for Type IIf, *i.e.* membranes with the enzyme OPH impregnated into the CP, there is a *very drastic reduction in the DIMP permeation when a voltage is applied* (7.90 μg/cm^2) as compared to when no voltage is applied (79.8 μg/cm^2), indicating electrochemical activation of the OPH enzyme, as observed by others as well [41-42].

Results for permeation of other simulants were very similar to those for DIMP above and are not given in detail for space reasons. However, we mention that particularly noteworthy was the nearly complete blockage of Malathion (a commercially available pesticide that is a model for the CW agent GB), when present at any con-

Table 3. Representative results from studies of DIMP permeation, using the GC/MS method described above, for several membrane configurations, as indicated.

Sample#	Membrane Configuration	Voltage Applied	DIMP permeation, [μg·cm^{-2}]			Live agent (GB) permeation, *total* (Battelle), [μg·cm^{-2}] [*a*]
			2 - 6 hrs	6 - 24 hrs	*Total*	
cc_143, cc_165	[inside‖CP|Au|μP-PSulf|Au|GelEl|μP-PTFE‖-outside] (*Type IIa*)	None (open circuit)	23.16	213.07	236.23	242.89
cc_164	[inside‖CP|Au|μP-PSulf|Au|GelEl|μP-PTFE‖-outside] (*Type IIa*)	−6.0 V	14.86	115.99	130.85	117.25
cc_158	[inside‖CP|Au|μP-PSulf|Au|GelEl|μP-PTFE|-GelEl|μP-PTFE-‖outside] (*Type IIe*)	None (open circuit)	5.02	49.02	54.04	
Q09_p13b	[inside‖CP+OPH|Au|μP-PSulf|Au|GelEl|μP-PTFE|Au|CP+OPH-‖outside] (*Type IIf*)	None (open circuit)			79.8	
Q09_p13c	[inside‖CP+OPH|Au|μP-PSulf|Au|GelEl|μP-PTFE|Au|CP+OPH-‖outside] (*Type IIf*)	−6.0 V			7.90	

[*a*] As measured at Battelle.

centration above 0.065 mg·cm^{-2}, in Type IIf membranes; this concentration may be compared with the US Army AEGL-2 (Acute Exposure Guideline Level-2, "disabling") criterion for GB of 1 mg·cm^{-2}. (It may be recalled that in Type IIf, the CP is impregnated with the enzyme OPH [41-42]).

It is also to be noted that, as per prior published data on gas permeation through microporous PTFE membranes [13-18] and as expected, *unmodified* microporous PTFE membranes (GoreTex®, poresize 0.45 μm), were found to be *extremely* permeable to DIMP in our studies, with permeability within the first 2 hours of tests being about 10× that of the membranes listed in the above table. That is to say, the "plain" microporous membranes would provide no protection whatsoever against agents, as expected.

3.6. Permeation/Blocking Studies with Agents

The methodology for these studies is described at length in the Experimental Section. The structures of the agents tested, known by their military appellations (GB, HD, VX), are shown in **Figure 7**, where they may also be compared with their corresponding simulants/models. Representative results are given in **Table 4**. This table also gives, for reference, values of the US Army guidelines for exposure, known as the Acute Exposure Guideline Levels (AEGL), of which AEGL-1, AEGL-2 and AEGL-3 represent, respectively, the "non-disabling", "disabling" and "lethal" exposure limits.

It is seen from these results (**Table 4**) that *our membranes effectively block agents*. GB is the most volatile of the three agents tested; for this agent only, some permeation is seen in Type II membranes. In later work, this permeation was effectively blocked through modification of this membrane configuration, in the Type IIe and Type IIf configurations; these introduced an additional membrane/GelEl layer on the exterior that effectively blocked GB. This was seen in tests with the simulant DIMP (see above) with these modified membrane configurations; tests with live GB agent for these new configurations (Type IIe, Type IIf) are planned for early 2014. More interestingly, the membranes exhibited very slightly poorer blocking in *active* mode (*i.e.* with voltage applied) as compared to *passive* mode. It is proposed that the redox of the CP that occurs on application of voltage may cause a physical change in the conformation of the CP chain and thus a small expansion of the poresize, facilitating increased CW agent permeation. This is also in line with the increased *water* vapor permeation seen in the *active* mode as compared to the *passive* mode (results above). (The structure of the CP/Au/membrane interface, as described above, is again noted in this respect: Rather than clearly defined, solid layers, the Au and the CP actually permeate well into the fibers of the microporous membrane).

3.7. Environmental Durability and Related Studies Using Industry-Standard Tests

The industry-standard tests [24-26] performed on the membrane assemblies under this category included: 1) ASTM D 2261, "Tearing Strength of Fabrics". 2) ASTM D 5034 (Grab), "Breaking Strength of Textile Fabrics". 3) ASTM D 3886, "Stoll Flat Abrasion Resistance". 4) ASTM F392/F392M-11, "Standard Test Method for Flex Durability of Flexible Barrier Materials", also called the "Gelbo Flex Test".

Results of these tests are given here in summary form in **Table 3** above. They may also be briefly summarized

Table 4. Agent permeation results, in *passive* (no voltage applied) and *active* (−6 VDC applied) modes. The membrane configurations tested were, as indicated, Type IIa ([inside||CP|Au|µP-PSulf|Au|GelEl|µP-PTFE||outside]) and Type II ([inside||CP|Au|µP-PSulf|Au|GelEl||outside]. (*ND = not detected; h = hours.*) A value of 0.250 µg/cm^2 essentially indicates ND, since that was the detection limit of the instrument.

Agent	Sample#	Membrane Confign.	Mode	Permeation, ([µg/cm^2], except Controls g/m^2)					AEGL-2 (disabling) Level, *for reference* [µg/cm^2] [a]
				0 - 2 h	2 - 6 h	6 - 16 h	16 - 24 h	Cum. (0 - 24 h)	
HD	M1-P-S5	Type II	*passive*	0.250	1.851	9.132	3.930	15.163	50.0
HD	M1-P-S6	Type II	*passive*	0.250	1.871	8.670	4.728	15.519	50.0
HD	M2-P-S5	Type IIa	*passive*	0.250	0.250	1.505	0.950	2.955	50.0
HD	M2-P-S6	Type IIa	*passive*	0.250	0.250	1.717	0.831	3.040	50.0
GB	M2-P-S1	Type IIa	*passive*	92.60	0.250	0.250	0.250	93.35	100.0
GB	M2-P-S2	Type IIa	*passive*	43.50	0.250	0.250	0.250	44.25	100.0
VX	M1-P-S7	Type II	*passive*	0.250	0.250	0.250	0.250	1.00	50.0
VX	M1-P-S9	Type II	*passive*	0.250	0.250	0.708	0.250	1.458	50.0
VX	M2-P-S7	Type IIa	*passive*	0.250	0.250	0.250	0.250	1.00	50.0
VX	M2-P-S8	Type IIa	*passive*	0.250	0.250	0.250	0.250	1.00	50.0
HD	Spike Control	Control, 10 ± 1.0 g/m^2	*passive*					10.25 g/m^2	
GB	Spike Control	Control, 10 ± 1.0 g/m^2	*passive*					10.66 g/m^2	
VX	Spike Control	Control, 10 ± 1.0 g/m^2	*passive*					10.24 g/m^2	
HD	M1-A-S3	Type II	*active*	ND	ND	47.44	18.91	66.35	100.0
HD	M1-A-S4	Type II	*active*	ND	ND	50.38	32.0	82.98	100.0
HD	M2-A-S3	Type IIa	*active*	ND	0.76	8.63	5.49	14.88	100.0
HD	M2-A-S4	Type IIa	*active*	0.25	ND	ND	1.49	1.74	100.0
VX	M1-A-S5	Type II	*active*	0.25	ND	ND	ND	0.25	50.0
VX	M1-A-S6	Type II	*active*	ND	ND	ND	0.25	0.25	50.0
VX	M2-A-S5	Type IIa	*active*	0.25	0.25	0.25	3.015	3.765	50.0
VX	M2-A-S6	Type IIa	*active*	0.25	0.25	0.25	0.566	1.316	50.0
HD	Spike Control	Control, 10 ± 1.0 g/m^2	*active*					10.09 g/m^2	
GB	Spike Control	Control, 10 ± 1.0 g/m^2	*active*					10.73 g/m^2	
VX	Spike Control	Control, 10 ± 1.0 g/m^2	*active*					9.82 g/m^2	

[a] US army acute exposure guideline levels (AEGL). AEGL-1: Non-disabling; AEGL-2: Disabling; AEGL-3: Lethal.

here: 1) *Test*: ASTM D 2261, "Tearing Strength of Fabrics". *Result, Type IIa membrane*: 0.067 lb-f/N. *Reference results*: 1 to 2 lb-f/N for typical microporous fabrics such as GoreTex®. 2) *Test*: ASTM D 5034 (Grab), "Breaking Strength of Textile Fabrics". *Result, Type IIa membrane*: 16.4 lb-f. *Reference results*: 10 to 15 lb-f for typical microporous fabrics such as GoreTex®. 3) *Test*: ASTM D 3886, "Stoll Flat Abrasion Resistance". *Result, Type IIa membrane*: No noticeable wear at 1000 cycles. *Reference results*: Some wear is observed at about 500 cycles for typical microporous fabrics such as GoreTex®. 4) *Test*: ASTM F392/F392M-11, "Standard Test Method for Flex Durability of Flexible Barrier Materials", also called the "Gelbo Flex Test". For this test, the specimen fabric (15 cm × 9.5 cm) was subjected to flexing, which consisted of a twisting motion followed by a horizontal motion, undergoing a maximum of 180 degree of twist (see photos of apparatus and actual tests in **Figure 4**). The sample was flexed for 20 cycles at 15 cpm. WVT (ASTM E 96 B) and DIMP permeation were monitored (GC/MS method outlined above), before and after the flexing. The deterioration in WVT was observed to be 12%. The increase in DIMP permeation was not observable (DIMP permeation of pre- and post-Gelbo flex test samples were both below the detection limit of the method used).

3.8. Practical Aspects of Field Use of Final Garment, Including Power Requirements

From the data above, voltage and power requirements may be arrived at for garment use in the field. Our results indicate that a potentiostatic mode of operation (with (−)6 V as the standard voltage applied) is the preferred mode in practical field applications; among other things, it is much easier to apply a constant potential than a constant current from a power source such as a Li ion battery. In this respect, it is also noted that, as indicated by the chronoamperogram in **Figure 5(c)**, after an initial current spike, *the current plateaus to a steady state that is well below the initial spike value*. In the field studies currently underway, a "current-limiting" function has been built into the smart Controller used to coordinate application of voltage to the garment. This cuts off the power (*i.e.* the membrane is brought to open circuit) if the current density exceeds a pre-designated maximum. Another feature of the Controller is input from humidity and temperature sensors both on the inside and the outside of the garment. On the basis of these inputs, the Controller's control algorithm decides whether to shut off the voltage being applied; the humidity on the internal side of the garment is the main determinant of this decision. The overall product is thus a "smart" garment requiring little user input. (The Controller is the subject of a separate patent application in preparation.)

With a view to such practical field use, current-limiting studies were carried out in a laboratory setting. In a typical such study, a current-limiting DC power supply was used and set at a maximum current output of 50 mA @ (−)6.0 V applied, for a freshly fabricated membrane assembly test sample of dimension 10 cm × 10 cm; this yielded a maximum possible current density of 0.5 $mA\cdot cm^{-2}$ or 5 $A\cdot m^{-2}$. The test sample was allowed to sit in the lab, in ambient air, for ca. 20 h. It was then hooked up to the current-limiting power supply and the (−)6.0 V applied to it. The chronoamperogram observed showed a maximum current (spike) of 1.6 mA at 1.0 s, which rapidly decayed to <0.5 mA @ 120 s, subsequently maintaining this steady-state level over the 1 h tested. This steady-state current computes to a current density of 0.005 $mA\cdot cm^{-2}$ or 0.05 $A\cdot m^{-2}$, and the power computes to **0.3 $W\cdot m^{-2}$** (6 V × 0.05 $A\cdot m^{-2}$). On repeat of this procedure several times, it was observed that the "spike" in the initial current essentially disappeared, yielding only the steady state current, which in turn then fell to ca. half its original value, *i.e.* ca 0.25 mA. Several such experiments confirmed that freshly fabricated membrane assemblies would need only an initial such "break-in", on initial voltage application, and would then stabilize to a very low power consumption value, **0.15 $W\cdot m^{-2}$** or less.

The above data combined with the extensive [vapor|| membrane||vapor] data presented earlier (including both the lab experiments and the industry-standard tests) allow for the estimation of power requirements (from the voltages applied and current densities measured). These lie in the region of **0.05 $W\cdot m^{-2}$ to 0.50 $W\cdot m^{-2}$** (*i.e. per square meter*). To start with, these are extremely low power consumption values. More significantly, these are for *constant, continuous* application of a voltage (or current) to the membrane garment. Now in actual use with the smart Controller described above, a voltage (or current) would not be continuously applied to the garment. Rather, the Controller would monitor the humidity on the inside of the garment with a sensor and turn the garment "ON" only when the threshold humidity is crossed. It is also noted that the field use garments comprise "pixels" of the membranes of dimension 10 cm × 10 cm each, electrically connected in parallel (cf. **Figure 2**). The seams between these pixels, of width ca. 2 to 4 mm, are sealed with a flexible, agent-blocking polyurethane adhesive. These seams were tested for agent simulant permeation and were found to be impermeable before and after several stress tests, including the Gelbo Flex test (see results in Supplementary Information).

An issue of concern for field use, and eventual decontamination in the event of agent exposure, is the retention of residual CW agent within the fabric. To evaluate this, full garments incorporating Type IIe and Type IIa membranes were surface-saturated with agent simulant (DIMP). The evaporative loss of DIMP at 25°C and atmospheric pressure was monitored using weight-loss, GC/MS and other methods. It was found that ca. 33% of the original DIMP was linearly lost every hour, with complete loss of the DIMP within 3.25 h. This demonstrated the membranes would not retain any residual agent within just a few hours of exposure. Other issues of concern for use include the potential denaturing of the enzyme OPH in any membranes incorporating it, especially if a prolonged shelf life is involved. For this reason, and because the membranes with OPH did not perform markedly better than membranes without it except with the agent simulant Malathion, it appears that membranes that incorporate OPH should not be used within the fabric. As of this writing, the fabric samples incorporating our best membrane configurations are being evaluated for consideration in a future garment.

3.9. Potential Mechanism of Large WVT Rates Observed and of Agent Blocking

3.9.1. WVT

Our results, as presented above, and as confirmed with a multitude of tests at independent, external laboratories (also presented above), clearly establish: 1) That our membranes clearly demonstrate *passive* WVT rates that are multiples of the best rates reported previously. 2) That they display *active* WVT, demonstrated for the first time ever, at rates which are further increased from the *passive* rates. 3) And that they simultaneously and efficiently block agents. These results are clearly established. What is not established however is a possible mechanism for the observed WVT in our unique membranes. We now briefly discuss some potential mechanisms that may be postulated, in light of prior, published theoretical and experimental studies [1-23].

As noted above, the CP in our membrane assemblies was placed on the "inside" or interior, (*i.e.* closest to the skin when worn as a garment). Now a typical membrane configuration, such as Type IIa, displayed an open circuit potential (OCP, as measured with a potentiostat) between the "outside" facing and "inside" facing electrodes, of ca. +200 mV to +300 mV (with the + polarity observed at the CP, *i.e.* on the "inside" of the garment). These values may be compared with typical *zeta potentials*, observed in electroosmotic studies with capillaries, membranes and other naturally charged surfaces, of about +50 mV [1-10]. That is to say, *as compared to all prior studies on microporous and other membrane and capillary materials, the intrinsic electroosmotic potential observed in our CP-functionalized membranes is ca. four to six times in magnitude*. Now it is well established from prior studies [1-13] that *this potential difference is the primary driver of the electroosmotic flow*. Furthermore, an additional, externally applied voltage may be used to further supplement this intrinsic potential difference, as in the case of the *active* mode in our membranes.

Thus, one potential, clear explanation for the very large WVT rates observed in the case of our functionalized CP microporous membranes, even in a *passive* mode (*i.e.* with no voltage applied), is that *the effective, intrinsic electroosmotic potentials in our materials are much larger than those observed in conventional capillaries or microporous (or other) membranes*. Now when a *further* voltage is applied across our membranes (e.g. the 6 V typically applied in our *active*-mode studies, with the polarity always (+) at the CP on the "inside"), the resultant electric field further enhances the intrinsic electroosmotic potential of our membranes. However, this enhancement may not be significant, due primarily to the large (electrical) resistance of the gel electrolyte; this causes the actual voltage sensed at the CP electrode surface to be significantly less than the 6 V applied; we have confirmed this in OCP (open circuit potential) measurements, as cited earlier. As a result, the observed enhancement of WVT rates on application of a voltage is not large (ca. 44% was the maximum enhancement observed in our work).

It may be argued that the theoretical treatment and experimental observations of electroosmotic flow [1-12, 20-23] intrinsically stipulate that flow occurs only in the presence of an electrical field, *i.e.* in *active* mode. However, in this respect, it may be noted that electroosmotic flow is always "*active*", because there is always a potential difference, corresponding approximately to the zeta potential, across a membrane or between the ends of a capillary. Indeed, the observed potential difference between the ends of the micropores of our membranes (ca. 200 - 300 mV, see data above and below) can be considered to be equivalent to an actively applied electrical field. Additionally, an analysis of the intrinsic electroosmotic potentials and their gradients present in capillaries and micropores, again based on prior theoretical and experimental work, shows that these inevitably vary significantly, by as much as 75 mV, from one end of any capillary or micropore to another [1-12]. Thus, it may be argued, the term "*passive* electroosmotic flow" is inherently incorrect, because the electroosmotic flow is in fact always *active*. Furthermore, when a species, e.g. water in vapor form, is introduced, that changes the electrochemical potential of one end of the capillary or micropore as compared to the other end, the electrical field gradient is further increased. Thus, what one may ascribe as a *water* vapor *concentration difference* between one end of a micropore or capillary and the other end may also be considered as an *electroosmotic potential difference*. During WVT then, the water vapor condenses on the walls of the capillary/micropore at one end and re-evaporates at the other end. In other studies in our lab, this intrinsic potential difference and resulting electrical field gradient have been documented, in measurements of potential difference (ΔV) across interfaces using a laboratory potentiostat. Some typical results are as follows: 1) (RH = relative humidity, all measurements at 25°C) Interface#1 [45% RH||Au|μP-PSulf|Au||45% RH], ΔV across Au electrodes = below noise of potentiostat, *i.e.* ~0 mV. 2) Interface#2 [90% RH||Au|μP-PSulf|Au|| 45% RH], ΔV across Au electrodes = 95 mV. 3) Interface#3 [90% RH||CP|Au|μP-PSulf|Au||45% RH], ΔV across Au electrodes = 295 mV. These measurements were highly reproducible and reversible. They clearly show that: 1) There is no observable ΔV when the RH at both ends of the membrane is the same. 2) When the RH at both ends is different (90%/45%), the presence of the CP layer causes a dramatic increase in the ΔV (from 95 mV to 295 mV). Finally, it should also be noted that the applicability of electroosmotic flow to combined liquid /vapor phases was clearly demonstrated by our [vapor||μPmembrane||vapor] experiments cited above, which showed a very large WVT upon application of a voltage across the interface.

The above discussion not with standing, regardless of the mechanism of WVT, which may properly be further elucidated in more detailed studies, the fact of the ***very high rate WVT*** (*in both passive and active modes*), as corroborated by several very different methods and at several independent laboratories, ***is clearly established*** in our work, as is the fact that the WVT is greatly enhanced with the inclusion of the CP in the membrane.

3.9.2. CW Agent Blocking

With regard to the effective blocking of the CW agents (and simulants) by our membrane assemblies, this may be ascribed primarily to the GelEl. The unique, patented [35-38] composition of the GelEl allows for water vapor penetration and transport within it, while serving as an impermeable barrier to organic molecules (in our case, those possessing phosphonate and structurally related moieties). This is confirmed by the observation, reported above, that our membrane assemblies, when they do *not* have GelEl, are significantly permeable to agents and simulants.

4. Conclusions

In this communication, we have demonstrated *passive* as well as *active* electroosmotic WVT, across unique, functionalized CP microporous membranes, the latter for the first time to our knowledge. This occurs together with simultaneous blocking of CW agents and simulants, again for the first time to our knowledge. The membrane assemblies demonstrate typical *passive* WVT rates more than 2.5 times as high and *active* rates more than 3.5 times as high as the highest previously reported (*passive*) WVT rates, for Nylon microporous membranes; these latter of course do not possess any agent blocking capability as our membranes do. Agent blocking was demonstrated with agents. All WVT and other measurements were independently corroborated at external laboratories. The membrane assemblies passed all relevant environmental durability tests. They were incorporated into "smart" final-form fabrics. Likely mechanisms of the significantly enhanced electroosmotic transport as well as the CW agents blocking in these membranes were discussed. Many issues relating to field use, such as power consumption and "smart" control were resolved. Garments incorporating these unique membranes are currently under test. The ultimate objective of this work is to create lightweight, comfortable protective garments.

5. Acknowledgements

Partial support of this work by the U.S. Army Engineering Research Development Command-Engineering Research Lab (ERDC-CERL) and U.S. Defense Threat Reduction Agency (DTRA) through Contract #W9132T07C0012 to Ashwin-Ushas Corporation is gratefully acknowledged.

REFERENCES

[1] M. Smoluchowski, " Elektroosomosis," In: L. Graetz, Ed., *Handbuch der Elektrizitaet und des Magnetismus*, Barth Verlag, Leipzig, 1921, p. 366.

[2] M. Smoluchowski, "Elektroosmosis," *Zeitschrift für Physikalische Chemie*, Vol. 93, 1918, p. 129.

[3] D. C. Henry, "The Cataphoresis of Suspended Particles. Part I. The Equation of Cataphoresis," *Proceedings of the Royal Society A*, Vol. 133, No. 133, 1931, p. 106.

[4] D. A. Saville, "Electrokinetic Effects with Small Particles," *Annual Review of Fluid Mechanics*, Vol. 9, 1977, p. 321.

[5] Y. Xu, "Tutorial: Capillary Electrophoresis," *The Chemical Educator*, Vol. 1, No. 2, 1996, pp. 1-14.

[6] Zetasizer Nano Series User Manual, "Zeta Potential Theory," Chapter 16, Malvern Instruments Ltd., Malvern.

[7] B. D. Bowen, "Theory of Electrokinetic Measurements in Sandwich Cells," *Journal of Colloid and Interface Science*, Vol. 98, No. 1, 1984, pp. 236-244.

[8] A. Hernandez, R. Lopez, J. I. Calvo and P. Pradanos, "A Network Microcapillary Model for Electrokinetic Phenomena through Microporous Membranes," *Colloids and Surfaces A: Physicochemical and Engineering Aspects*, Vol. 145, No. 1-3, 1998, p. 11.

[9] S. Basu and M. M. Sharma, "An Improved Space-Charge Model for Flow through Charged Microporous Membranes," *Journal of Membrane Science*, Vol. 124, No. 1, 1997, pp. 77-91.

[10] B. D. Bowen and R. A. Clark, "Electro-Osomosis at Microporous Membranes and the Determination of Zeta-Potential," *Journal of Colloid and Interface Science*, Vol. 97, No. 2, 1984. pp. 401-409.

[11] M. Kozak and E. J. Davis, "Electrokinetic Phenomena in Fibrous Porous Media," *Journal of Colloid and Interface Science*, Vol. 112, No. 2, 1986, pp. 403-410.

[12] M. Mullet, P. Fievet, J. C. Reggiani and J. Pagetti, "Surface Electrochemical Properties of Mixed Oxide Ceramic Membranes: Zeta-Potential and Surface Charge Density," *Journal of Membrane Science*, Vol. 123, No. 2, 1997, pp. 244-265

[13] B. Halford, "Raincoats," *Chemical & Engineering News*, Vol. 84, No. 24, 2006, p. 39.

[14] L. K. Wolf, "Shielding Soldiers with Fabric," *Chemical & Engineering News*, Vol. 90, No. 15, 2012, pp. 36-38.

[15] W. L. Gore & Associates. http://www.gore.com

[16] DuPont breathable fabric. http://www2.dupont.com/Active_Layer/en_US/assets/downloads/K-06012-1%20Active%20Layer.pdf http://www2.dupont.com/Active_Layer/en_US/assets/do

wnloads/APA60015NC000.pdf

[17] E. A. McCullough, M. Kwon and H. Shim, "A Comparison of Standard Methods for Measuring Water Vapour Permeability of Fabrics," *Measurement Science and Technology*, Vol. 14, No. 8, 2003, pp. 1402-1408.

[18] E. A. McCullough, J. Huang and C. S. Kim, "An Explanation and Comparison of Sweating Hot Plate Standards," *Journal of ASTM International*, Vol. 1, No. 7, 2004, Article ID: JAI12098.
http://www.astm.org/JOURNALS/JAI/PAGES/JAI12098.htm

[19] J. Walker, H. Schreuder-Gibson, W. Yeomans, F. Hoskin, R. C. Cheng, R. Yin and R. Hill, "Development of Self-Decontaminating Textiles with Microporous Membranes," Report, US Defense Technical Information Center, 2002.

[20] S. K. Vajandar, D. Xu, D. Li, D. Markov, J. Wikswo and W. Hofmeister, "SiO_2-Coated Porous Anodic Alumina Membranes for High Flow Rate Electroosmotic Pumping," *Nanotechnology*, Vol. 18, No. 27, 2007, Article ID: 275705.

[21] S. A. Miller and C. R. Martin, "Controlling the Rate and Direction of Electroosmotic Flow in Template-Prepared Carbon Nanotube Membranes," *Journal of Electroanalytical Chemistry*, Vol. 522, No. 1, 2002, pp. 66-69.

[22] S. K. Li, A.-H. Ghanem and W. I. Higuchi, "Pore Charge Distribution Considerations in Human Epidermal Membrane Electroosmosis," *Journal of Pharmaceutical Sciences*, Vol. 88, No. 10, 1999, pp. 1044-1049.

[23] C. K. Lee and J. Hong, "Characterization of Electric Charges in Microporous Membranes," *Journal of Membrane Science*, Vol. 39, No. 1, 1988, pp. 79-88.

[24] Detailed Descriptions of All the American Society for Testing and Materials (ASTM) Tests quoted in this communication are available on the ASTM website at:
http://www.astm.org/Standard/index.shtml.

[25] Details of Related Tests of the International Standards Organization, ISO, Geneva.
http://www.iso.org/iso/home.html

[26] Test Operations Procedure (TOP) 8-2-501, Permeation and Penetration of Air-Permeable, Semipermeable, and Impermeable Materials with Chemical Agents or Simulants.
http://ammtiac.alionscience.com/ammt/iacdocs.do?115337

[27] R. B. Lindsay, A. G. Pappas and J. M. Baranoski, "Test Results of Air-Permeable Charcoal Impregnated Suits to Challenge by Chemical and Biological Warfare Agents and Simulants," Report #A368614, US Defense Technical Information Center, May 2003.

[28] CW Protective Equipment: An Overview of Respiratory and Body Protection.
http://www.opcw.org/resp/html/equip.htm

[29] http://www.army.mil/factfiles/equipment/nbc/jslist.html
http://tech.military.com/equipment/view/88668/jslist.html

[30] J. M. Sloan, E. Napadensky, D. M. Crawford and Y. A. El-Abd, "Nanostructured Polymer Membranes for Chemical Protective Clothing," Weapons and Materials Research Directorate, Army Research Lab, Aberdeen Proving Ground, MD, 2004.

[31] E. Napadensky and Y. A. El-Abd, "Breathability and Selectivity of Selected Materials for Protective Clothing," Report# A602524, US Defense Technical Information Center, 2004.

[32] E. Klemperer, "Toxic Industrial Chemical Tests of Resistance to Permation by Protective Suits," Report# A063134, US Defense Technical Information Center, 2006.

[33] P. Chandrasekhar, B. J. Zay, G. C. Birur, S. Rawal, A. A. Pierson, L. Kauder and T. Swanson, "Large, Switchable Electrochromism in the Visible Through Far-Infrared in Conducting Polymer Devices," *Advanced Functional Materials*, Vol. 12, No. 2, 2002, pp. 95-103.

[34] P. Chandrasekhar, B. J. Zay, D. A. Ross, T. M. McQueeney, G. C. Birur, T. Swanson, L. Kauder and D. Douglas, "Chromogenic Phenomena in Polymers: Tunable Optical Properties," S. A. Jenekhe and D. J. Kiserow, Eds., *ACS Symposium Series*, American Chemical Society, Oxford University Press, New York, 2004, pp. 66-79.

[35] P. Chandrasekhar, US Patent No. 5,995,273, 1999.

[36] P. Chandrasekhar, US Patent No. 6,033,592, 2000.

[37] P. Chandrasekhar, Canadian Patent No. 2,321,894, 2007.

[38] P. Chandrasekhar, European Patent No. 99908208.4, 2009.

[39] P. Chandrasekhar, "Conducting Polymers: Fundamentals and Applications. A Practical Approach," Springer Verlag, Berlin, 1999.

[40] P. Chandrasekhar, B. J. Zay, T. M. McQueeney, G. C. Birur, V. Sitaram, R. Menon and R. L. Elsenbaumer, Physical, Chemical, Theoretical Aspects of Conducting Polymer Electrochromics in the Visible, IR and Microwave Regions," *Synthetic Metals*, Vol. 155, No. 3, 2005, pp. 623-627.

[41] J. Chen, B. Winther-Jensen, C. Lynam, O. Ngamna, S. Moulton, W. Zhang and G. G. Wallace, "A Simple Means to Immobilize Enzyme into Conducting Polymers via Entrapment," *Electrochemical and Solid-State Lett*ers, Vol. 9, No. 7, 2006, pp. H68-H70.

[42] A. Mulchandani, P. Mulchandani, W. Chen, J. Wang and L. Chen, "Amperometric Thick-Film Strip Electrodes for Monitoring Organophosphate Nerve Agents Based on Immobilized Organophosphorus Hydrolase," *Analytical Chemistry*, Vol. 71, No. 11, 1999, pp. 2246-2249.

[43] Information on the US Army's Joint Service Lightweight Integrated Suit Technology (JSLIST), including the suits (uniforms) currently deployed.
http://www.army.mil/factfiles/equipment/nbc/jslist.html.

[44] E. Wilusz, Q. Truong, D. H. McCullough III., J. Li and S. L. Regen, "Perforated Monolayers for Enhanced Permselectivity in Chemical Biological Barrier Membranes,"

Report# A650334, US Defense Technical Information Center, 2004.
http://www.dtic.mil/dtic/

[45] A. Prasad, "Polyethylene, Low-Density," In: J. E. Mark, Ed., *Polymer Data Handbook*, Oxford University Press, Oxford, 1996, p. 571.

Electrochemical Self-Assembly of Nanoporous Alumina Templates Title

Muna Moussa Abbas
Department of Physics, College of Science, Baghdad University, Baghdad, Iraq

ABSTRACT

Porous alumina was fabricated electrochemically through anodic oxidation of aluminum by means of such a self-organized method. Anodic aluminium oxide (AAO) template with nanopores was grown by two-steps anodization processes from a high purity aluminium foil. The anodization process was carried out in a phosphoric acid electrolyte at ambient temperature with a different duration of anodization. The analysis observation by Atomic Force Microscopy (AFM) showed that nanopore size increased with anodization time. The nanopore sizes of porous alumina were (16.04, 26.19 and 37.39 nm) for (1, 2 and 3 hour) respectively.

Keywords: Anodic Alumina; Porous Alumina; Self-Assembly

1. Introduction

The discovery of nanoporous materials opened a tremendous flurry of worldwide research, because of their promising and potential industrial and technological applications for magnetic, electronic, optoelectronic structures and nanometric devices.

Moreover, the application of ordered nanochannel-arrays as two-dimensional photonic crystals has generated an increasing interest in recent years. Anodic porous alumina film is a promising candidate for starting materials of nanofabrication of various devices [1,2].

Porous alumina is arrays of a large 2D adjustable configuration forms on the surface of aluminum in a self-organized manner when anodized in an acid under specific conditions [3,4].

Since Masuda *et al.* [5] reported the two-steps anodization process to obtain a self-ordered of pores, new areas of applications have emerged in the fields of magnetic storage, solar cells, carbon nanotubes, catalysts, and metal nanowires. This increasing attraction of porous alumina as template is mainly due to it enables mass production without the use of expensive lithographical tools [6].

Anodic aluminium oxide (AAO) templates have received considerable attention in synthesizing nanostructure material due to their particular characteristics such as controllable pore diameter and periodicity, and extremely narrow distribution of pore size. The pore has ideal hexagonal shape. AAO is characterized by a homogeneous morphology of parallel pores which grow perpendicular to the template surface with a narrow distribution of diameter, length and interpore spacing, all of which can be easily controlled by a suitable way of choosing the anodization parameters such as pH of the electrolyte, anodization voltage and duration of anodization [7,8].

Wang and Han [9] showed that AAO preparation after a two-steps anodization possesses controllable pore diameter and periodicity; therefore further pore adjustment can be caused by exposure to phosphoric acid, which is called pore-widening.

Extensive researches have been carried to a better understanding of the physical and chemical properties of porous alumina. The preparing conditions of highly ordered AAO template are highly precise, especially for those with pore diameters under 25 nm. The anodization time is one of the main factors of fabrication for the order of nanopore arrays.

Nielsch *et al.* [10] showed that the domain size does not linearly increase with time but has a maximum size and he suggested that it is an optimum anodization time to obtain the largest domain size. This goes beyond the results of Li *et al.* [11], Oh *et al.* [12] and Zaraska *et al.* [13] they reported that the average domain area is a linear function of anodization time.

Lee *et al.* [14] investigated the anodization of aluminum in malonic acid ($H_4C_3O_4$) solution. They found that the ordering of cells shows an inverse time dependence on the anodization time.

The purpose of this work is to understand and improve the self-assemble of porous alumina under a specific condition with the variation of duration of the anodization.

2. Experimental Details

The AAO templates were fabricated by a two-step anodization process. Specimens of High purity (99.93%, 0.3 mm thick) aluminium foils annealed at 500˚C for 3 hour in an N_2 atmosphere then it was purged by acetone in an ultrasonic cleaner to clean out the possible grease on its surface. To reduce surface roughness, electropolishing is carried out in a mixture consisting of 1:4 volume mixture of ($HClO_4$) and (C_2H_5OH) at 10 V for 3 min at 10˚C with magnetic stirring. The samples were later rinsed in distilled water; this step repeats ones again under the same conditions. After the electrochemically polishing procedure, the specimens were anodized in two electrodes electrochemical cell with a magnetic stirrer was used, Pt foil serves as cathode where the Al foil as the anode, the electrolyte was 10% sulfuric acid , the process conducted with 18 V DC at 5˚C for different duration of anodizing (1, 2 and 3 hour). A two steps anodizing procedure were conducted in which the first anodized layer removed by solution of (1.8 wt chromic acid and 6% wt phosphoric acid) at 60˚C. The samples cleaned again as mentioned above to repeat the anodizing process at the same conditions of the first step, then immersed in 6% of phosphoric acid to open the pores where the phosphoric acid is thought to help connect cylindrical pores in the bottom layer to the top layer of the AAO template.

Atomic Force Microscopy (AFM-AA3000) of Angstrom Advanced Inc., USA) was used to obtain quantitative information on the grain sizes of the AAO films as well as the surface roughness.

3. Results and Discussion

The morphology in the three-dimensional AFM images and the statistical results of the anodized aluminum porous films in a three different times, (1 - 3 hour) are shown in **Figures (1-3)** respectively. The results show distinctly that an ordered honeycomb structure with this method.

It was observed that the pore diameter is 16.04 nm for AAO prepared under 1 hour, increasing the time up to 2 hour produced AAO with a pore of 26.19 nm size in diameter as shown in **Figure 2**. In addition, the pores were possessed irregular structure and separated by a thick pore walls. The pores were continuity formed on the surface; its structure becomes more open by increasing the anodizing time up to 3 hour. The measuring of the pore

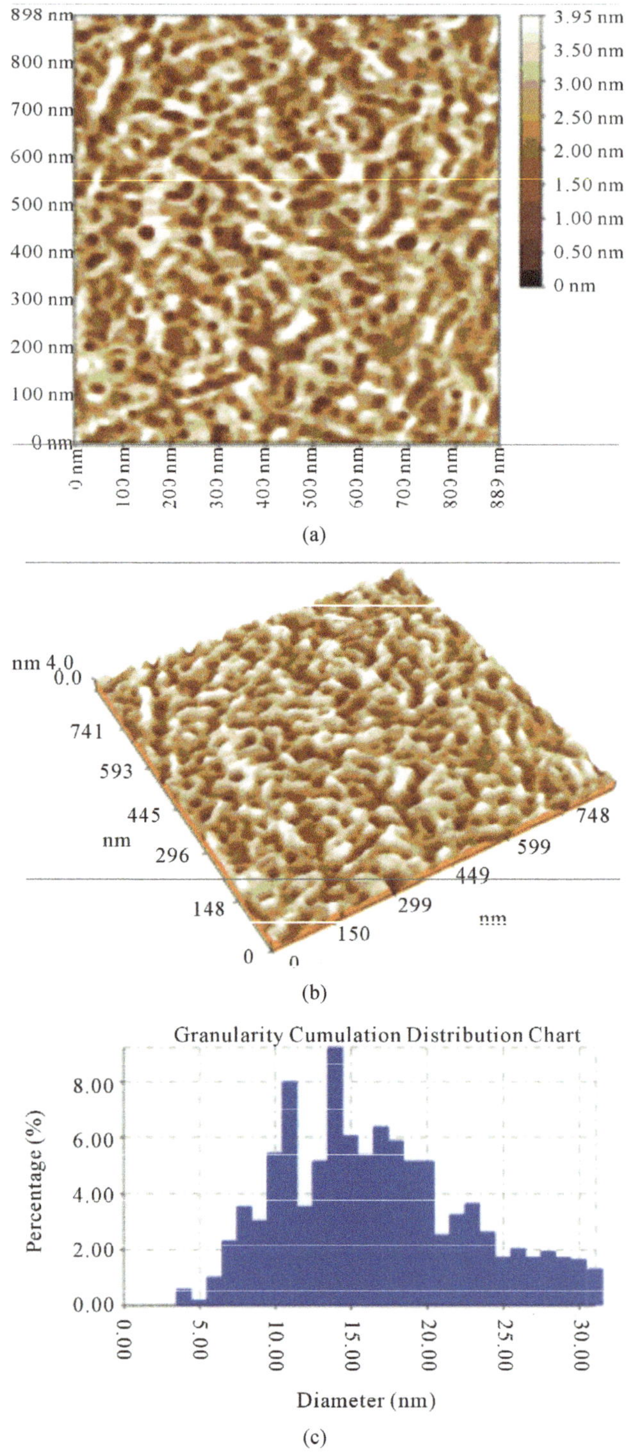

Figure 1. Atomic force microscope topographic views of AAO ionized during 1 hour (a) 2D image (b) 3D and (c) normal granulites distribution report of AAO.

diameter appeared to be 37.39 nm as illustrates in **Figure 3**. The structure of the pore was more circular with a large opening and separated by a thick pore walls.

Meanwhile, the pores were more circular with a large opening and separated by a thick pore walls; it had been

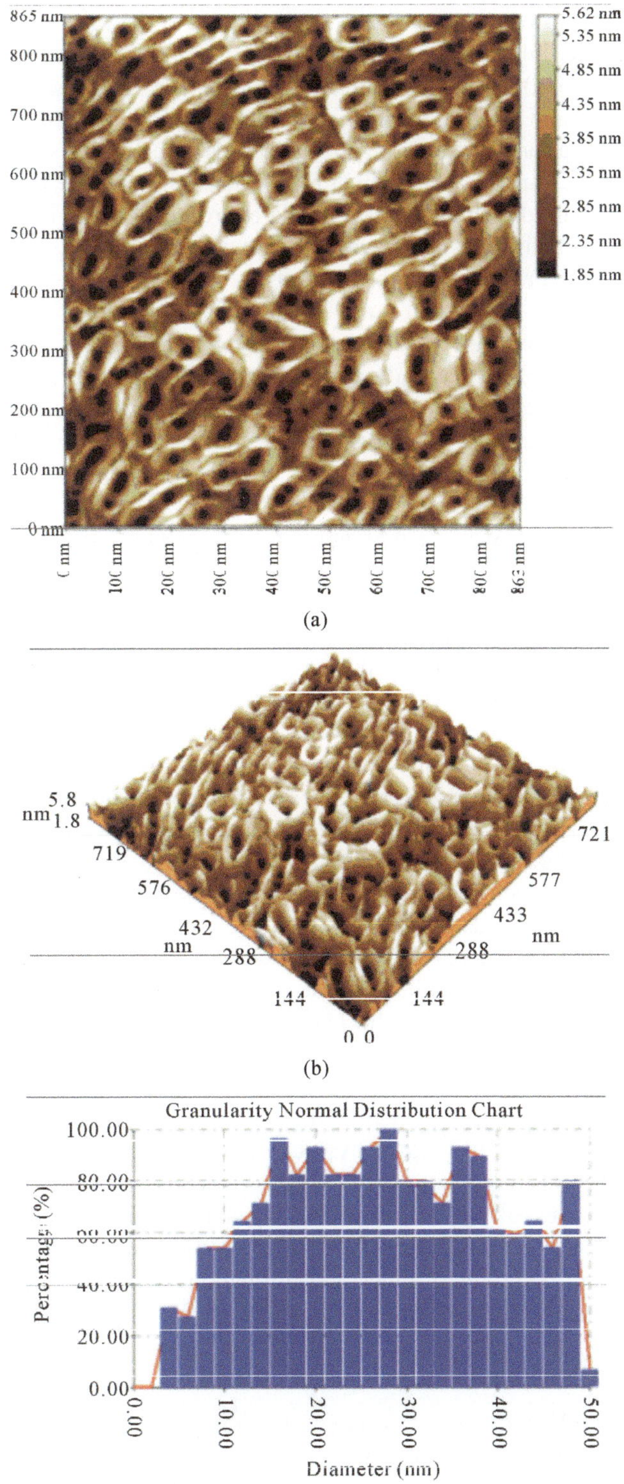

Figure 2. Atomic force microscope topographic views of AAO ionized during 2 hour (a) 2D image (b) 3D and (c) normal granulites distribution report of AAO.

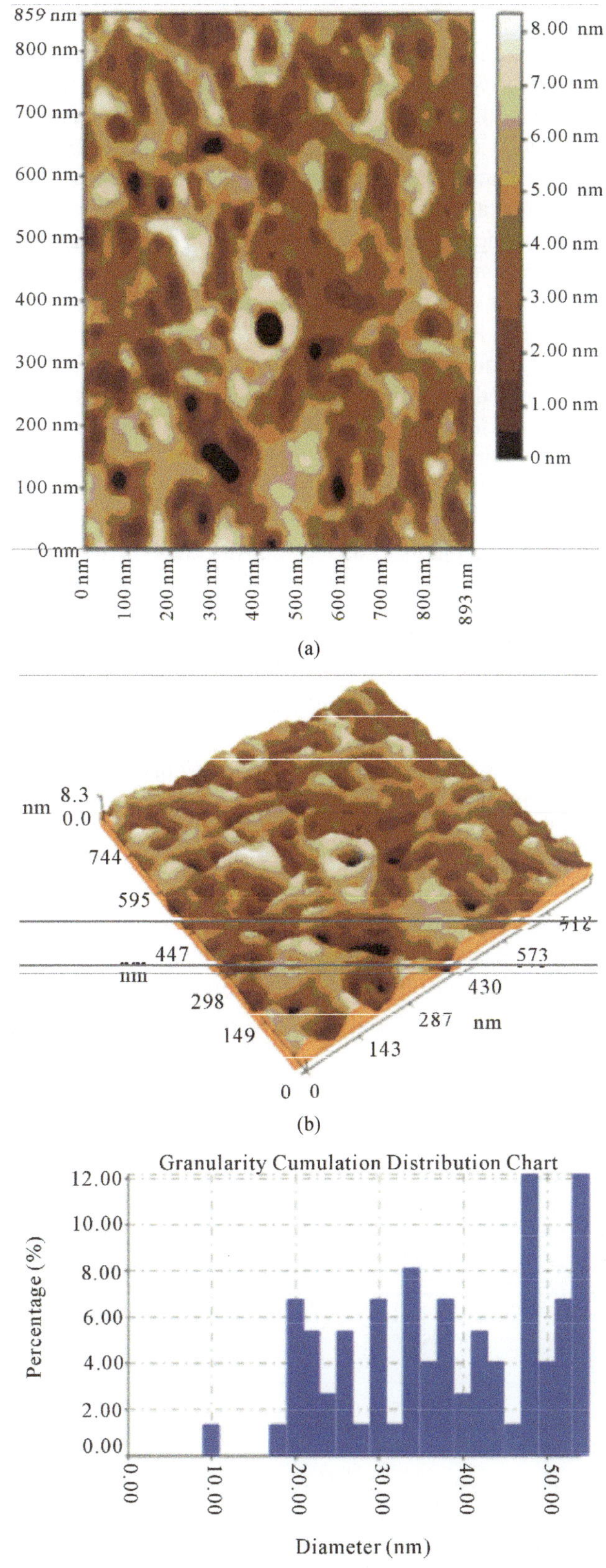

Figure 3. Atomic force microscope topographic views of AAO ionized during 3 hour (a) 2D image (b) 3D and (c) normal granulites distribution report of AAO.

shown that the size distribution become inhomogeneous at longer anodization time. It can be conclude that connection of pores is an intermediate step in porous alumina on the way to the conventional hexagonal pattern.

However, the results showed the average pore diameter linearly increased with anodization time. This increase in the pore diameter can be related as a result of prolonged etching of alumina in the acidic electrolyte used for anodizing. In addition, the complex network of pores containing subpores disappears due to the electrolyte etching action. This confirms that the pore diameter can be controlled by the anodization time. This is in agreement with the results obtained by Zaraska *et al.* and Li *et al.* [11,13].

It was noticed that by increasing the duration of anodization led to the increase of the films roughness due to the increasing of thickness these results are similar to those reported by Wei *et al.* [15] who argue that the thickness of the porous oxide film was affected by the anodizing time. These results can be explained through the film morphology, as the film thickness increases the grain size increases too. The surface roughness increases with increasing grain size and appearance of enlarge pores caused by the extension of the aluminum surface, which led to possess an irregular structure and got separated by a thick pore walls. Similar results have been observed by Alam *et al.* [6].

AFM results for the AAO nanopurs films prepared at a different duration of anodization are summarized in **Table 1**.

It is worth noting that with a significant extension of the anodizing time a considerable improvement of pore arrangement have not been observed. Therefore, further study is needed to find how the surface condition of aluminum foil can modify homogeneity of pores and significant improvement in surface roughness on an anodic alumina surface.

The mechanism of the formation Al_2O_3 film and nanopurs arrays is proposed by many authors [11,16-25] they assumed that nanotubes formed as a result of a competition between:

1) Field-assisted anodic oxidation, defined as the formation of the layer under applied electric field and Chemical filed/assisted dissolution of aluminum oxide (or alumina) at the oxide/electrolyte interface. The dissolution and thinning of the oxide layer is mainly due to the hydration reaction of the formed oxide layer.

2) Alumina formation at the oxide/aluminum interface. The oxidation takes place at the entire metal/oxide interface mainly by the migration of oxygen containing ions (O_2^- or OH^-) from the electrolyte.

Pores are first formed at certain micro-rough region where the current density is concentrated on after the formation of a steady barrier layer and then the pores grow perpendicularly to the surface with equilibrium of field-enhanced oxide dissolution at the oxide electrolyte interface and oxide growth at the metal/oxide interface. The formed alumina is assumed to be Al_2O_3. Therefore, the atomic density of aluminium in alumina is by a factor of 2 lower than in metallic aluminium. This means that the volume of the anodized alumina expands to about twice the original volume. This volume expansion leads to compressive stress during the oxide formation in the oxide/metal interface. The expansion in the vertical direction pushes the pore walls upwards. Meanwhile, the repulsive interaction between the pores drives the structural adjustment to form pore arrays like in a honeycomb. The barrier layer at the bottom of the nanotubes (at the metal/oxide interface) acts as resistance to the flow of ions, which needs to move through the anodic film to keep the oxidation active. Greater resistance can be overcome by increasing the applied anodization voltage, which can cause the increase of the longitudinal growth of the pores due to the lager migration velocity of the reactive cation and anion ions, which can be confirmed by the thickening of the AAO the horizontal growth also leads to the enlargement of the pore diameter and spacing. Meanwhile, the adjusted velocity among the pores cannot keep up with the growth in velocity of the AAO and the features of the order of the structure might decrease. Moreover the degree of the volume expansion of aluminum determines either the formation of self-ordered pores or the formation of disordered pores.

Table 1. Atomic force micproscope results.

AFM results	Duration of anodization (hour)		
	1	2	3
Pore size (nm)	16.04	26.19	37.39
Ra (Roughness Average) nm	0.692	1.01	1.32
Rq (Root Mean Square) nm	0.835	1.17	1.61
Ry (Peak-Peak) nm	3.8	3.78	8.26
Rz (Ten Point Height) nm	3.8	3.77	7.55
Rk (Core Roughness Depth) nm	2.43	3.39	4.39
Rvk (Reduced Valley Depth) nm	1.25	0.0073	1.46

4. Conclusion

Anodic aluminium oxide (AAO) film with the nanoporous was successfully prepared by the use of a two-steps anodized method. This study confirms that the duration of anodization is a conditional parameter that strongly influences the porous diameters and the roughness of the films. It was obtained that the pore size increased (16, 26 and 37 nm) in (1, 2 and 3 hour) respectively.

5. Acknowledgements

I want to thank Dr. Kareem AL-Samuraee for his supporting to carry out this work.

REFERENCES

[1] S. Shingubara, "Fabrication of Nanomaterials Using Porous Alumina Templates," *Journal of Nanoparticle Research*, Vol. 5, No. 1, 2003, pp. 17-30.

[2] N. Taşaltina, S. Öztürka, H. Yüzerb and Z. Z. Öztürk, "Simple Fabrication of Highly Ordered AAO Nanotubes," *Journal of Optoelectronic and Biomedical Materials*, Vol. 1, No. 1, 2009, pp. 79-84.

[3] S. Phok, S. Rajaputra and V. P. Singh, "Copper Indium Diselenide Nanowire Arrays by Electrodeposition in Porous Alumina Templates," *Nanotechnology*, Vol. 18, No. 47, 2007, pp. 1-8.

[4] V. Sadasivan, C. P. Richter, L. Menon and P. F. Williams, "Electrochemical Self-Assembly of Porous Alumina Templates," 2005.

[5] H. Masuda and K. Fukuda, "Ordered Metal Nanohole Arrays Made by a Two-Step Replication of Honeycomb Structures of Anodic Alumina," *Science*, Vol. 268, No. 5216, 1995, pp. 1466-1468.

[6] C. Sousab, D. C. Leitaoa, J. Venturaa, A. M. Pereiraa, M. Amadoa, J. B. Sousaa and J. P. Araújo, "Nanoporous Self-Organized Anodic Alumina Templates," 2007.

[7] P. Chowdhury, K. Raghuvaran, M. Krishnan, H. C. Barshilia and K. S. Rajam, "Effect of Process Parameters on Growth Rate and Diameter of Nanoporous Alumina Templates," *Bulletin of Materials Science*, Vol. 34, No. 3, 2011, pp. 423-427.

[8] K .M. Alam, A. P. Singh, S. C. Bodepudi and S. Pramanik, "Fabrication of Hexagonally Ordered Nanopores in Anodic Alumina: An Alternative Pretreatment," *Surface Science*, Vol. 605, No. 3-4, 2011, pp. 441-449.

[9] X. Wang and G.-R. Han, "Fabrication and Characterization of Anodic Aluminum Oxide Template." *Microelectronic Engineering*, Vol. 66, No. 1-4, 2003, pp. 166-170.

[10] K. Nielsch, J. Choi, K. Schwirn, R. B. Wehrspohn and U. Go1sele, "Self-Ordering Regimes of Porous Alumina: The 10% Porosity Rule," *Nano Letters*, Vol. 2, No. 7, 2002, pp. 677-680.

[11] F. Li, Z. Zang and R. M. Metzger, "On the Growth of Highly Ordered Pores in Anodic Aluminum Oxide," *Chemistry of Materials*, Vol. 10, No. 9, 1998, pp. 2470-2480.

[12] H. J. Oh, J. H. Lee, Y. Jeong, S. J. Suh and C. S. Chi, "A Control of the Nanopore Regularity in Alumina Layer," *Materials Science Forum*, Vol. 475-479, 2005, pp. 3919-3922.

[13] L. Zaraska, G. D. Sulka1 and M. Jaskuła, "Properties of Nanostructures Obtained by Anodization of Aluminum in Phosphoric Acid at Moderate Potentials," *Journal of Physics: Conference Series*, Vol. 146, No. 1, 2009, Article ID: 012020.

[14] W. Lee, K. Nielsch and U. Gosele, "Self-Ordering Behavior of Nanoporous Anodic Aluminum Oxide (AAO) in Malonic Acid Anodization," *Nanotechnology*, Vol. 18, No. 47, 2007, Article ID: 475713.

[15] W. Wei, S. Li, J. Liu and J. Zhang, "Size-Controlled Synthesis of Highly Ordered Nanoporous AAO Membranes," *Nanoscience*, Vol. 11, 2006, pp. 157-160.

[16] Y. Xu, G. E. Thompson and G. C. Wood, "Mechanism of Anodic Film Growth on Aluminium," *Transactions of the Institute of Metal Finishing*, Vol. 63, 1986, pp. 98-103.

[17] O. Sullivan and G. C. Wood, "The Morphology and the Mechanism of Formation of Porous Anodic Films on Aluminium," *Proceedings of the Royal Society A*, Vol. 317, 1970, pp. 511-543.

[18] V. P. Parkhutik and V. I. Shershulsky, "Theoretical Modeling of Porous Oxide Growth on Aluminum," *Journal of Physics D: Applied Physics*, Vol. 25, No. 8, 1992, pp. 1258-1263.

[19] J. Siejka and C. Ortega, "Study of Field-Assisted Pore Formation in Compact Oxide Films on Aluminium," *Journal of the Electrochemical Society*, Vol. 124, No. 6, 1977, pp. 883-891.

[20] G. E. Thompson and G. C. Wood, "Anodic Films on Aluminum," In: J. C. Scully, Ed., *Treatise on Material Science and Technology*, Academic Press, New York, 1983.

[21] K. Shimizu, K. Kobayashi, G. E. Thompson and G. C. Wood, "A Novel Marker for the Determination of Transport Numbers during Anodic Oxide Barrier Growth of Aluminum," *Philosophical Magazine Part B*, Vol. 64, No. 3, 1991, pp. 345-353.

[22] O. Jessensky, F. Múller and U. Gósele, "Self-Organized Formation of Hexagonal Pore Arrays in Anodic Alumina," *Applied Physics Letters*, Vol. 72, No. 10, 1998, pp. 1173-1175.

[23] A. P. Li, F. Müller, A. Birner, K. Nielsch and U. Gösele, "Hexagonal Pore Arrays with a 50 - 420 nm Interpore Distance Formed by Self-Organization in Anodic Alumina," *Journal of Applied Physics*, Vol. 84, No. 11, 1998, pp. 6023-6027.

[24] A. P. Li, F. Müller, A. Birner, K. Nielsch, U. Gösele and J. Vac, "Polycrystalline Nanopore Arrays with Hexagonal Ordering on Aluminum," *Journal of Vacuum Science & Technology A*, Vol. 17, No. 4, 1999, pp. 1428-1432.

[25] G. E. Thompson, R. C. Furneaux, G. C. Wood, J. A. Richardson and J. S. Goode, "Nucleation and Growth of Porous Anodic Films on Aluminum," *Nature*, Vol. 272, No. 5652, 1978, pp. 433-435.

A New Heptamethine Cyanine-Based Near-Infrared Fluorescent Probe for Divalent Copper Ions with High Selectivity

Zhixiang Han[1,3*], Qingqing Yang[1], Lihui Liang[2*], Xiaobing Zhang[3]
[1]School of the Environment and Safety Engineering, Jiangsu University, Zhenjiang, China
[2]Hunan Provincial People's Hospital, Changsha, China
[3]State Key Laboratory of Chemo/Biosensing and Chemometrics, College of Chemistry & Chemical Engineering, Hunan University, Changsha, China

ABSTRACT

A new near-infrared fluorophore 2-(2-Aminoethyl) pyridine-tricarbocyanine (**1**) was rationally designed and synthesized as a fluorescent probe for detection of Cu^{2+} with high selectivity. The response of Probe **1** is based on the fluorescence quenching upon binding to Cu^{2+}. The sensing performance of the proposed Cu^{2+}-sensitive Probe **1** was then investigated. The probe can be applied to the quantification detection of Cu^{2+} with a linear concentration range covering from 4.8×10^{-7} to 1.6×10^{-4} mol/L and a detection limit of 9.3×10^{-8} mol/L. The experimental results showed that the response of **1** to Cu^{2+} was independent of pH in medium condition (pH 6.0 - 8.0), and exhibited excellent selectivity towards Cu^{2+} over other common metal cations.

Keywords: Near-Infrared Dye; Fluorescent Probe; Cu^{2+} Ions

1. Introduction

The design and synthesis of fluorescent probes for selective and sensitive detection of metal ions have attracted wide-spread interests of chemists, biologists, clinical biochemists and environmentalists in recent years [1]. Copper is an essential trace element in both plants and animals, including humans. Among the essential heavy metals, the abundance of copper ranks the third in human body. It participates in many biological processes, such as haemoglobin synthesis (in utilization of Fe and regeneration of Hb), development of connective tissue, normal functions of the central nervous system, and oxidative phosphorylation [2-4]. Nevertheless, copper of high concentration is highly toxic to some organisms such as many bacteria and viruses [5]. Owing to its toxicity for bacteria, elevated concentrations of copper would ham- per the self-purification capability of the sea or rivers, and destroy the biological reprocessing systems in water. Copper is also found to be harmful to human at high concentration and has been suspected to cause the damage of infant liver in recent years. Accordingly, searching for efficient and reproducible analytical methods for the copper assay is of great importance for environment and human health.

Many analytical methods for detection of copper, including atomic absorption spectrometry (AAS) [6], inductively coupled plasma-mass spectroscopy (ICP-MS) [7], inductively coupled plasma-atomic emission spectrometry (ICP-OES) [8], spectrophotometry [9], voltammetry [10] and fluorescence spectroscopy [11], have been developed so far. Among these methods, fluorescence spectroscopy offers significant advantages due to its nondestructive character, high sensitivity and specificity, and the availability of a wide range of indicative dyes. Several fluorophores have been used to design fluorescent probes for divalent copper ions including calcein [12], rhodamine [13], naphthalimide [14,15], pyrene [16,17], tris(2,2'-bipyridine)-ruthenium(II) [18], ben-

*Corresponding author.

zoxazole [19], porphyrin [20,21], spiropyran [22,23], BODIPY [24], and so on. Unfortunately, limitations of the currently available probes for Cu^{2+} include low sensitivity, and/or excitation profiles in the ultraviolet or visible region, which can damage living samples and cause interfering autofluorescence from native cells.

The light in the near-infrared region (NIR) around 650 - 900 nm can penetrate more deeply into tissues, which is of importance to study on living organism imaging. Moreover, it has a further advantage that autofluorescence is not observed upon NIR excitation. Heptamethine cyanine dyes [25], one of the important kinds of NIR dyes, has been widely used in various fields, and been employed as fluorescent labels in the studies of fluorescence imaging with biological mechanisms. And a few probes based on heptamethine cyanine dyes have been employed to detect metal ions or small molecules [26]. However, to the best of our knowledge, only few NIR fluorescent probes based on cyanine dyes have been reported for divalent copper ion assay detection [27]. Searching for new NIR probe for copper detection with high selectivity is still an active field as well as a challenge for the analytical chemistry research.

Herein, we report the synthesis and properties of a novel NIR fluorescent probe 2-(2-Aminoethyl) pyridine-tricarbocyanine (**1**) for the detection of Cu^{2+} with good selectivity and high sensitivity. Tricarbocyanine and 2-(2-aminoethyl) pyridine were selected as the reporter and cheleator, respectively. The probe exhibited stable response towards Cu^{2+} over the concentration range from 4.8×10^{-7} to 1.6×10^{-4} mol/L with a working pH range from pH 6.0 to 8.0.

2. Experimental

2.1. Reagents

Before being used, N, N'-dimethylformamide (DMF) was subjected to simple distillation from K_2CO_3. IR-780 iodide was purchased from Sigma-Aldrich. 2-(2-Amino-ethyl) pyridine was purchased from Alfa Aesar. All other chemicals were of analytical reagent grade, purchased from Shanghai chemical Reagent Corporation (Shanghai, China), and used without further purification. Twice distilled water was used throughout all experiments. Thin layer chromatography (TLC) was carried out using silica gel 60 F254, and column chromatography was conducted over silica gel (100 - 200 mesh), both of which were obtained from the Qingdao Ocean Chemicals (Qingdao, China).

2.2. Synthesis of Compound 1

Synthetic route for Compound **1** was depicted in **Scheme 1**. Briefly, IR-780 iodide (11.3 mg, 0.0167 mmol) and 2-(2-Amino-ethyl) pyridine (20.4 mg, 0.167 mmol) were

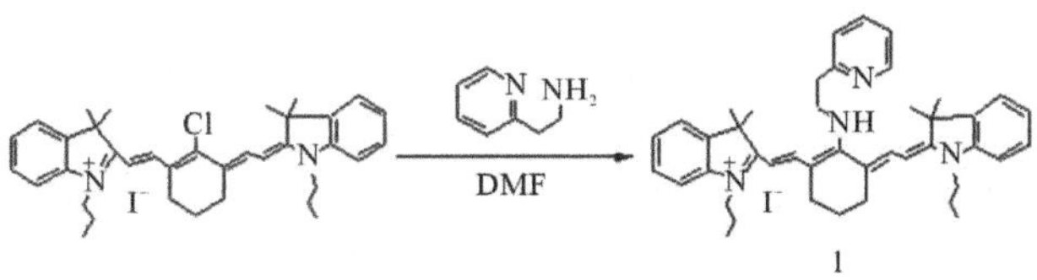

Scheme 1. Synthetic pathway of Compound 1.

dissolved in anhydrous DMF (3 mL) in a 25 mL round bottom flask. The mixture was stirred at 80°C for 4 h under an argon atmosphere. The solvent was removed under reduced pressure, then purified on silica gel chromatography eluted with CH_2Cl_2/ethanol (100:1, V/V) to afford the desired product as a blue solid (7.1 mg, yield 56%). ^1H NMR (400 MHz, $CDCl_3$): δ 0.93 (t, 6H, J = 7.4 Hz), 1.55(s, 12H), 1.63 - 1.73(m, 6H), 2.45 (t, 4H, J = 6.0 Hz), 3.21(t, 2H, J = 6.4 Hz), 3.92 (t, 4H, J = 6.8 Hz), 4.14(d, 2H, J = 6.4 Hz), 5.76 (m, 2H), 7.05(t, 2H, J = 6.8 Hz), 7.15(d, 2H, J = 8.0 Hz), 7.27 - 7.33(m, 4H), 7.43(m, 2H), 7.60(d, 2H, J = 12.8 Hz), 7.75(m, 1H), 8.50(d, 1H, J = 5.2 Hz), 8.68(br, 1H). ESI-MS: $[M-I^-]$ = 625.3, calculated: $[M]^+$ = 625.9.

2.3. Apparatus

^1H NMR spectra were recorded on a INOVE-400 (Varian) spectrometer operating at 400 MHz. All chemical shifts are reported in the standard δ notation of parts per million. LC-MS analyses were performed using an Agilent 1100 HPLC/MSD spectrometer; UV-Vis absorption spectra were recorded with a Shimadzu MultiSpec-1501 spectrophotometer. All fluorescence measurements were carried out on a HITACHI F4500 (Japan) with excitation slit set at 10.0 nm and emission at 20.0 nm. The pH measurements were carried out on Mettler-Toledo Delta 320 pH meter (Shanghai, China).

2.4. Measurement Procedures

4-(2-Hydroxyethyl) piperazine-1-ethanesulfonic acid (HEPES) buffer {0.1 mol/L, pH = 7.4, I = 0.1 ($NaNO_3$)} were prepared by dissolving appropriate HEPES and $NaNO_3$ in water, adjusting to pH = 7.4 by 1.0 mol/L NaOH with the volume to 1000 ml in a volumetric flask.

A 2.0×10^{-5} mol/L stock solution of **1** was prepared by dissolving **1** in CH_3CN. A stock standard solution of Cu^{2+} (0.01 mol/L) was prepared by dissolving an appropriate amount copper in water and adjusting the volume to 100 ml, then further diluted to 1×10^{-3} - 1×10^{-7} mol/L stepwise. The buffered solutions of wide pH range were obtained by adjustment of 0.1 mol/L HEPES solution with HCl or NaOH solution. The complex solution of Cu^{2+}/**1** was prepared by adding 5.0 mL the stock solution of **1** and 1.0 mL mentioned above solution of Cu^{2+} in a 10 mL volumetric flask. Then the mixture was diluted to 10mL with pH 7.4 HEPES buffer solution. In the obtained solution, the concentrations were 1×10^{-5} mol/L of **1** and 1×10^{-3} - 1×10^{-8} mol/L of Cu^{2+}. Blank solution of **1** was prepared under the same conditions without Cu^{2+}. All the solution above were protected from light and kept at 4˚C for further use.

3. Results and Discussion

3.1. Spectral Characteristics

Figure 1 showed the fluorescence spectra of **1** in HEPES buffer solutions with different concentrations of Cu^{2+}, which recorded at excitation wavelength of 640 nm and emission wavelength of 670 - 800 nm. The spectrum of free **1** exhibited very strong fluorescence emission in a buffer solution. Addition of Cu^{2+} to a solution of **1**, fluorescence signal exhibited a remarkable quenching. The fluorescence intensity of **1** was gradually decreased with increasing Cu^{2+} concentration. These results provided a proof for the formation of an inclusion complex of **1** with Cu^{2+}, which constituted the basis for the determination of Cu^{2+} concentration with **1**. It is worthy to note that the fluorescence intensity of Probe **1** can be recovered upon addition of the coordinating reagents ethylenediaminetetraacetic acid (EDTA).

In order to better understand the variation of fluorescence intensity with the concentrations of Cu^{2+}, the absorption spectra of **1** in the absence and presence of Cu^{2+} were recorded (**Figure 2**). In the absorption spectrum of **1**, the results showed a strong absorption band at 643nm in the absence of Cu^{2+}, while the addition of Cu^{2+} ions decreased with no obvious shift in absorbance at 643 nm. From the fluorescence spectra and UV-is spectra, it is indicated that the fluorescence changes of **1** were more likely to be caused by the change of quantum yield rather than spectral shifts. Similar results were reported by Tang *et al.* [28].

In addition, Probe **1** works well and no detectable change in the linear range, detection limit or other analytical performance is found after it has been stored for several weeks in the dark at 4˚C, which implies that the NIR fluorescent probe used is stable.

3.2. Principle of Operation

The complexation equilibrium of **1** (A) with Cu^{2+} (B) with an association constant K can be expressed by the following equation:

$$mB + nA \xleftrightarrow{K} A_nB_m \quad (1)$$

where Cu^{2+} (B) and **1** (A) is established by formation of a complex with a complexing ratio of m:n. According to the modified Stern–Volmer equation [29], the relationships for the changes of fluorescence intensities, the concentration of Cu^{2+} [B] in solution and the concentration of **1** [A] in solution can be expressed as follows:

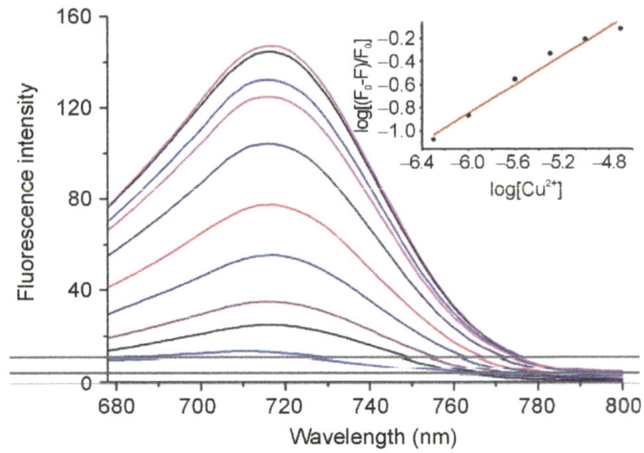

Figure 1. Fluorescence emission spectrum of 1 (10 µmol/L) with different concentration of Cu^{2+} (From top to bottom: 0, 0.5, 1.0, 2.5, 5.0, 10, 20, 50, 100, 200 µmol/L) in CH_3CN/H_2O (1:1, v/v) solution. Inset shows the linear responses with divalent copper ions concentrations.

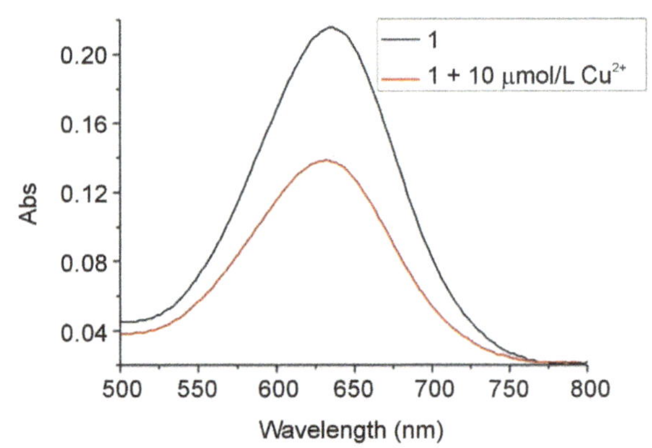

Figure 2. Chages in the UV-vis spectra of 1 (10 µmol/L) (black line) upon the addition of Cu^{2+} (10 µmol/L) (red line).

$$\frac{F_0 - F}{F} = K_q [A]^{n-1} [B]^m \quad (2)$$

Assuming $\Delta F = F_0 - F$, one can obtain:

$$\log\left(\frac{\Delta F}{F}\right) = \log K_q + (n-1)\log[A] + m\log[B] \quad (3)$$

Here F_0 and F denote the fluorescence intensities of **1** in the absence and presence of Cu^{2+}, respectively. K_q is fluorescence quenching constant. The calibration curve was constructed by recording the fluorescence intensity values of **1** in the presence of different Cu^{2+} concentration. In the range between 1.0×10^{-7} and 5.0×10^{-5} mol/L Cu^{2+}, the fluorescence intensity is linearly dependent on the Cu^{2+} concentration. The dependency can be described by the following equation:

$$\log\left(\frac{\Delta F}{F}\right) = 0.9343\log\left[Cu^{2+}\right] + 4.80 \quad (4)$$

It is obvious from Equation (4) that m is the slope of

log(ΔF/F) versus log[Cu^{2+}], which was calculated to be 1 approximately. Quenching constant (Kq) is 6.3 × 10^4. The relative fluorescence intensity α is defined as the ratio of free A $[A]_f$, to the total amount of A $[A]_t$, in the solution. It can be experimentally determined by measuring the fluorescence intensity of **1** in the solution:

$$\frac{[A]_f}{[A]_t} = \frac{F - F_t}{F_b - F_t} \quad (5)$$

Here F_b is the fluorescence intensity of **1** in the blank buffer solution and F_t represents the fluorescence intensity of **1** in the solution when **1** is completely complexed with Cu^{2+}. F is the fluorescence intensity of **1** actually measured when in contact with Cu^{2+} solutions of a given concentration. The relationship between the α and Cu^{2+} concentration [B] can be represented as:

$$\frac{\alpha^n}{1-\alpha} = \frac{1}{nK[A]_t^{n-1}[B]^m} \quad (6)$$

The response of **1** for different concentrations of Cu^{2+} was shown in **Figure 3**. Three curves are calculated using Equation (6) with different K and ratios of Cu^{2+} and **1**. It can be seen that the best curve was 1:1 complex ratio and an appropriate K of 1.09 × 10^5 fits form the experimental data. The curve can serve as the calibration curve for the detection of Cu^{2+} concentration. A practically usable range for quantitative determination covered from 4.8 × 10^{-7} to 1.6 × 10^{-4} mol/L ($0.05 \le \alpha \le 0.95$) [29]. The detection limit was 9.3 × 10^{-8} mol/L (defined as three times standard deviation of blank solution).

3.3. Effect or pH

The effects of pH on the fluorescence intensity of **1** in the presence of Cu^{2+} were carried out at a pH range from 5.0 to 9.0 with fixed the Cu^{2+} concentration at 5 µmol/L (**Figure 4**). In lower pH value, the fluorescence intensity of **1** decreased with decreasing pH value, which might be caused by the protonation of Compound **1** without binding with the metal ion. On the other hand, too high pH would lead to form the precipitation of $Cu(OH)_2$, and reduce its complexation with **1**. In a wide range of pH from 6.0 to 8.0, acidity did not affect the determination of Cu^{2+} with Compound **1**. In other words, the response behavior of Compound **1** is independent of pH in medium condition, which is convenient for practical applications of the proposed probe in determination of Cu^{2+}.

3.4. Selectivity

Under the same conditions, the ability of **1** to recognize Cu^{2+} was further investigated by mixture 100 µmol/L Cu^{2+} with the other background anions and metal ions. The experiments were carried out by recording the changes of the fluorescence intensity before and after adding the interferants into the pH 7.4 HEPES buffer solution. As shown from **Figure 5**, one can see that the proposed probe exhibited a relatively high selectivity for Cu^{2+} ions over a large number of mono-, bi-, and trivalent cations. Fortunately, normal interferents like Hg^{2+} do not interfere, which is better than that of the probe reported in literatures.

3.5. Preliminary Analytical Application

The proposed probe was applied to the determination of copper ions in water samples of Xiang River. The river water samples were simply filtrated and showed that no Cu^{2+} was present in them. All the water samples were spiked with standard Cu^{2+} solutions at different concentration levels, and then analyzed their concentrations with proposed Probe **1**. Results are shown in **Table 1**. One can see that the recovery study of spiked Cu^{2+}

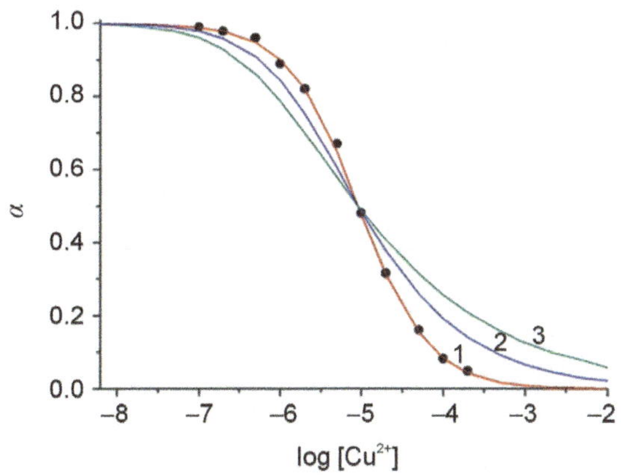

Figure 3. Relative fluorescence intensity α of 1 as a function of log[Cu^{2+}] in CH_3CN/H_2O (1:1, v/v) solution. The curves fitting the experimental data were calculated from Equation (6). (1) m:n = 1:1, K= 1.09 × 10^5; (2) m:n = 1:2, K= 1.09 × 10^{10}; (3) m:n = 1:3, K= 1.46 × 10^{15}. (•) data points experimentally obtained.

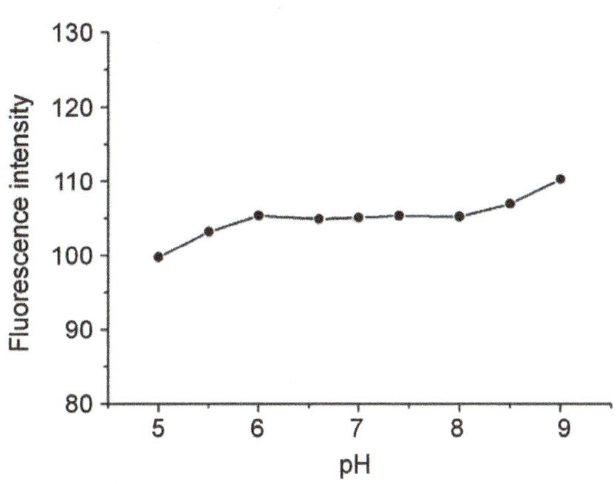

Figure 4. Effect of pH on the emission of 1 (10 µmol/L) with 5 µmol/L Cu^{2+} at 715 nm in CH_3CN/H_2O (1:1, v/v) solution. Excitation was provided at 640 nm.

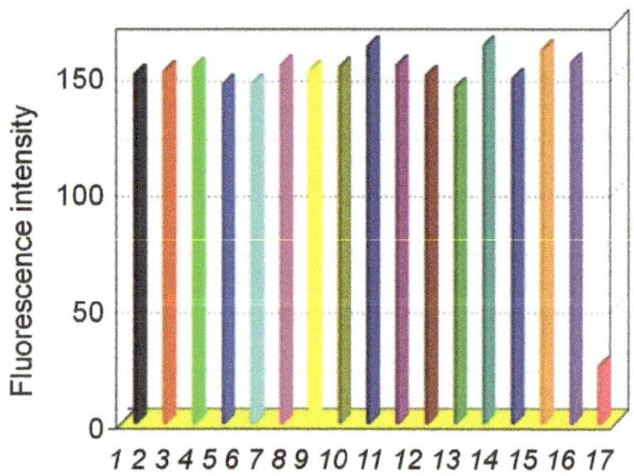

Figure 5. Emission change of 1 at 715 nm upon addition of each cation in CH_3CN/H_2O (1:1, v/v) solution at pH 7.40 {0.1 mol/L HEPES, I = 0.1 ($NaNO_3$)}: 1, none; 2, Zn^{2+}; 3, Cd^{2+}; 4, Mn^{2+}; 5, Ca^{2+}; 6, Mg^{2+}; 7, K^+; 8, Na^+; 9, Li^+; 10, Al^{3+}; 11, Co^{2+}; 12, Ni^{2+}; 13, Pb^{2+}; 14, Hg^{2+}; 15, Fe^{3+}; 16, Ag^+ 17, Cu^{2+}. With the exception of Cu^{2+}, Hg^{2+} and Al^{3+} where 10^{-4} mol/L of cation were added, each solution contained 10^{-3} mol/L of interest. The concentration of 1 was 10 μmol/L and excitation was provided at 640 nm.

Table 1. Recovery study of spiked determination of copper in Xiang River water with proposed Probe 1.

Sample	Cu^{2+} spiked (mol/L)	Cu^{2+} recovered (mol/L)	Recovery (%)
1	0	0	–
2	1.0×10^{-5}	$(1.02^a \pm 0.03^b) \times 10^{-5}$	102.0
3	5.0×10^{-6}	$(4.93^a \pm 0.04^b) \times 10^{-6}$	98.6

[a]Average were calculated with n = 3, [b]Standard deviations.

determined by the **1**-based probe showed satisfactory results. The present probe is useful for the determination of Cu^{2+} in real samples.

4. Conclusion

In summary, a new near-infrared fluorescent probe was designed and synthesized for the detection of Cu^{2+} based on quenching the fluorescence of tricarbocyanine chromophore with **1**-Cu^{2+} complexation. Compared to reported fluorescent probes, **1**-based probe showed high selectivity and large stokes shift over existing reagents and methods for the fluorescence determination of Cu^{2+} in neutral medium. And the proposed method can be used for the determination of Cu^{2+} in real samples.

5. Acknowledgements

This work was supported by the National Natural Science Foundation of China (Grant 20505008, 21105038), "973" National Key Basic Research Program of China (2007CB310500), Ministry of Education of China (NCET-07-0272), and Hunan Natural Science Foundation (06JJ4010, 07JJ3025).

REFERENCES

[1] A. P. de Silva, H. Q. Gunaratne, T. Gunnlaugsson, A. J. Huxley, C. P. McCoy, J. T. Rademacher and T. E. Rice, "Signaling Recognition Events with Fluorescent Sensors and Switches," *Chemical Reviews*, Vol. 97, No. 5, 1997, pp. 1515-1566.

[2] E. Vuori, A. Huunan-Seppälä and J. O. Kilpiö, "Biologically Active Metals in Human Tissues. I. The Effect of Age and Sex on the Concentration of Copper in Aorta, Heart, Kidney, Liver, Lung, Pancreas and Skeletal Muscle," *Scandinavian Journal of Work, Environment & Health*, Vol. 4, No. 2, 1978, pp. 167-175.

[3] D. Radisky and J. Kaplan, "Regulation of Transition Metal Transport across the Yeast Plasma Membrane," *The Journal of Biological Chemistry*, Vol. 274, 1999, pp. 4481-4484.

[4] B. M. Rode and Y. Suwannachot, "The Possible Role of Cu(II) for the Origin of Life," *Coordination Chemistry Reviews*, Vol. 190-192, 1999, pp. 1085-1099.

[5] C. Barranguet, F. P. van den Ende, M. Rutgers, A. M. Breure, M. Greijdanus, J. J. Sinke and W. Admiraal, "Copper-Induced Modifications of the Trophic Relations in Riverine Algal-Bacterial Biofilms," *Environmental Toxicology and Chemistry*, Vol. 22, No. 6, 2003, pp. 1340-1349.

[6] N. Pourreza and R. Hoveizavi, "Simultaneous Preconcentration of Cu, Fe and Pb as Methylthymol Blue Complexes on Naphthalene Adsorbent and Flame Atomic Absorption Determination," *Analytica Chimica Acta*, Vol. 549, No. 1-2, 2005, pp. 124-128.

[7] J. S. Becker, M. V. Zoriy, C. Pickhardt, N. Palomero-Gallagher and K. Zilles, "Imaging of Copper, Zinc, and Other Elements in Thin Section of Human Brain Samples (Hippocampus) by Laser Ablation Inductively Coupled Plasma Mass Spectrometry," *Analytical Chemistry*, Vol. 77, No. 10, 2005, pp. 3208-3216.

[8] Z. H. Li, L. Zhang, Z. P. Zang, X. J. Chang and X. J. Zou, "Attapulgite Modified with 2-Hydroxy-1-Naphthaldehyde as Selective Solid-Phase Extractant for Determination of Copper(II) in Environmental Samples by ICP-OES," *Microchim Acta*, Vol. 171, No. 1-2, 2010, pp. 161-168.

[9] J. J. Pinto, C. Moreno and M. García-Vargas, "A Very Sensitive Flow System for the Direct Determination of Copper in Natural Waters Based on Spectrophotometric Detection," *Talanta*, Vol. 64, No. 2, 2004, pp. 562-565.

[10] V. Beni, V. Ogurtsov, N. Bakunin, D. W. Arrigan and M. Hill, "Development of A Portable Electroanalytical System for the Stripping Voltammetry of Metals: Determination of Copper in Acetic Acid Soil Extracts," *Analytica Chimica Acta*, Vol. 552, No. 1-2, 2005, pp. 190-200.

[11] Y. Zheng, X. Cao, J. Orbulescu, V. Konka, F. M. Andre-

opoulos, A. M. Pham and R. M. Leblanc, "Peptidyl Fluorescent Chemosensors for the Detection of Divalent Copper," *Analytical Chemistry*, Vol. 75, No.7, 2003, pp. 1706-1712.

[12] L. A. Saari and W. R. Seitz, "Immobilized Calcein for Metal Ion Preconcentration," *Analytical Chemistry*, Vol. 56, No. 4, 1984, pp. 810-813.

[13] X. Q. Chen, T. Pradhan, F. Wang, J. S. Kim and J. Yoon, "Fluorescent Chemosensors Based on Spiroring-Opening of Xanthenes and Related Derivatives," *Chemical Reviews*, Vol. 112, No. 3, 2012, pp. 1910-1956.

[14] X. H. Qian, Y. Xiao, Y. F. Xu, X. F. Guo, J. H. Qian and W. P. Zhu, " 'Alive' Dyes as Fluorescent Sensors: Fluorophore, Mechanism, Receptor and Imagesin Living Cells," *Chemical Communications*, Vol. 46, 2010, pp. 6418-6436.

[15] X. H. Zhao, Q. J. Ma, X. B. Zhang, B. Huang, Q. Jiang, J. Zhang, G. L. Shen and R. Q. Yu, "A Highly Selective Fluorescent Sensor for Cu^{2+} Based on a Covalently Immobilized Naphthalimide Derivative," *Analytical Science*, Vol. 26, No. 5, 2010, pp. 585-590.

[16] R. Martínez, F. Zapata, A. Caballero, A. Espinosa , A. Tarraga and P. Molina, "2-Aza-1,3-butadiene Derivatives Featuring an Anthracene or Pyrene Unit: Highly Selective Colorimetric and Fluorescent Signaling of Cu^{2+} Cation," *Organic Letters*, Vol. 8, No. 15, 2006, pp. 3235-3238.

[17] H. J. Kim, J. Hong, A. Hong, S. Ham, J. H. Lee and J. S. Kim, "Cu^{2+}-Induced Intermolecular Static Excimer Formation of Pyrenealkylamine," *Organic. Letters*, Vol. 10, No. 10, 2008, pp. 1963-1966.

[18] C. L. He, F. L. Ren, X. B. Zhang, Y. Y. Dong and Y. Zhao, "A Fluorescent Chemosensor for Copper(II) Based on a Carboxylic Acid-functionalized Tris(2,2'-bipyridine)-ruthenium(II) Complex," *Analytical Science*, Vol. 22, No. 12, 2006, pp. 1547-1551.

[19] X. B. Zhang, J. Peng, C. L. He, G. L. Shen and R. Q. Yu, "A Highly Selective Fluorescent Sensor for Cu^{2+} Based on 2-(2'-Hydroxyphenyl)benzoxazole in a Poly(vinyl chloride) Matrix," *Analytica Chimica Acta*, Vol. 567, No. 2, 2006, pp. 189-195.

[20] H. Y. Luo, X. B. Zhang, J. H. Jiang, C. Y. Li, J. Peng, G. L. Shen and R. Q. Yu, "An Optode Sensor for Cu^{2+} with High Selectivity Based on Porphyrin Derivative Appended with Bipyridine," *Analytical Science*, Vol. 23, No. 5, 2007, pp. 551-555.

[21] Y. Q. Weng, F. Yue, Y. R. Zhong and B. H. Ye, "A Copper(II) Ion-Selective On-Off-Type Fluoroionophore Based on Zinc Porphyrin-Dipyridylamino," *Inorganic Chemistry*, Vol. 46, No. 19, 2007, pp. 7749-7755.

[22] N. Shao, Y. Zhang, S. M. Cheung, R. H. Yang, W. H. Chan, T. Mo, K. A. Li and F. Liu, "Copper Ion-Selective Fluorescent Sensor Based on the Inner Filter Effect Using a Spiropyran Derivative," *Analytical Chemistry*, Vol. 77, No. 22, pp. 2005, pp. 7294-7303.

[23] N. Shao, J. Y. Jin, H. Wang, Y. Zhang, R. H. Yang and W. H. Chan, "Tunable Photochromism of Spirobenzopyran via Selective Metal Ion Coordination: An Efficient Visual and Ratioing Fluorescent Probe for Divalent Copper Ion," *Analytical Chemistry*, Vol. 80, No. 9, 2008, pp. 3466-3475.

[24] X. Qi, E. J. Jun, L. Xu, S. J. Kim, J. S. Hong and Y. J. Yoon and J. Yoon, "New BODIPY Derivatives as OFF-ON Fluorescent Chemosensor and Fluorescent Chemodosimeter for Cu^{2+}: Cooperative Selectivity Enhancement toward Cu^{2+}," *The Journal of Organic Chemistry*, Vol. 71, No. 7, 2006, pp. 2881-2884.

[25] N. Narayanan and G. Patonay, "A New Method for the Synthesis of Heptamethine Cyanine Dyes: Synthesis of New Near-Infrared Fluorescent Labels," *The Journal of Organic Chemistry*, Vol. 60, No. 8, 1995, pp. 2391-2395.

[26] L. Yuan, W. Y. Lin, K. B. Zheng, L. W. He and W. M. He, "Far-red to Near Infrared Analyte-Responsive Fluorescent Probes Based on Organic Fluorophore Platforms for Fluorescence Imaging," *Chemical Society Reviews*, Vol. 42, 2013, pp. 622-661.

[27] P. Li, X. Duan, Z. Z. Chen, Y. Liu, T. Xie, L. B. Fang, X. R. Li, M. Yin and B. Tang, "A Near-Infrared Fluorescent Probe for Detecting Copper(II) with High Selectivity and Sensitivity and Its Biological Imaging Applications," *Chemical Communications*, Vol. 47, 2011, pp. 7755-7757.

[28] B. Tang, H. Huang, K. H. Xu, L. H. Tong, G. W. Yang, X. Liu and L. G. An, "Highly Sensitive and Selective Near-Infrared Fluorescent Probe for Zinc and Its Application to Macrophage Cells," *Chemical Communications*, Vol. 34, 2006, pp. 3609-3611.

[29] W. H. Liu, Y. Wang, J. H. Tang, G. L. Shen and R. Q. Yu, "An Optical Fiber Sensor for Berberine Based on Immobilized 1,4-Bis(Naphth[2,1-d]Oxazole-2-yl)Benzene in a New Copolymer," *Talanta*, Vol. 46, No. 4, 1998, pp. 679-688.

7

Preparation and Characterization of TiO_2 Photocatalytic Thin Film and Its Compounds by Micro-Arc Oxidation Technique

Qun Ma[1], Lili Ji[2], Yinchang Li[1], Tingting Jiang[1], Junxia Wang[1], Fei Li[1], Hongyun Jin[1], Yongqian Wang[1*]

[1]Engineering Research Center of Nano-Geo Materials of Education China University of Geosciences, Wuhan, China

[2]Jiangcheng College of China University of Geosciences, Wuhan, China

ABSTRACT

Mesoporous TiO_2 ceramic films have been prepared upon the Ti alloy substrate by the micro-arc oxidation (MAO) technology. To enhance the photo-catalytic property of the films, Eu_2O_3 particles were added into the electrolyte solution of Na_2CO_3/Na_2SiO_3. Scanning electron microscope (SEM), energy dispersive (EDS), X-ray photoelectron spectroscopy (XPS) and X-ray diffraction (XRD) are employed to characterize the modified films. Diffuse reflectance spectra (DRS) test, photo-generated current test and photo decomposition test are applied to evaluate the photo-catalytic property of the modified films. The results show that Eu_2O_3 transformed into one-dimensional (1-D) nano-wires embedded within the composite film, and the film has high photo-catalytic property.

Keywords: TiO_2; Eu_2O_3; Compound; Photo-Catalytic Property

1. Introduction

Rice Titanium dioxide (TiO_2) is one of the most important semi-conductor materials, which is widely used as an efficient photo-catalyst. In terms of environmental protection and new energy, TiO_2 has obtained widespread concern and research because of its chemical stability, biological compatibility and photo-catalytic ability since the report of photolysis of water on titanium dioxide (TiO_2) by Fujishima in 1972 [1]. However, the application of TiO_2 is limited. For the photo-catalytic applications, the band gap energy (Eg) of the semiconductor is critical. Because of the wideness (3.2 eV) of the Eg of TiO_2, and the ultra-violet (UV) light absorption and weak absorption of visible light [2], considerable efforts have been directed to extend the absorption edge of TiO_2 toward the visible part of the spectrum in the last three decades. So many researchers have been concentrating on finding an apt process and a modification technology to improve the photo-catalytic property of the TiO_2.

*Corresponding author.

TiO_2 powder and TiO_2 film are the most common existence forms of TiO_2. However, due to the hard recycle in polluted water and limited depth of photo-radiation, TiO_2 powder is not widely used, and thus many researches have been focusing on the preparation of TiO_2 film recent years [3-5]. It is well-known that there are many different methods for preparing TiO_2 film, such as sol-gel [6-9], liquid phase deposition (LPD) [10], hydrolysis precipitation [11] and chemical vapor deposition [12,13], etc. TiO_2 film can also be produced via micro-arc oxidation (MAO) technology [14,15]. Generally speaking, MAO is a useful approach for forming a ceramic coating on the surface of valuable metals, such as Al, Mg and Ti, and their alloys [16]. MAO is a process based on the anodic oxidation and occurs at potentials above the breakdown voltage of the oxide layer formed on the anode surface [17]. Now, micro-arc oxidation (MAO) is most frequently used to prepare TiO_2 film. The TiO_2 film that is prepared through MAO has much high quality, such as high wear resistance and corrosion resistance [18], favorable biological activity [19] and photo-catalytic activity. To enhance the efficiency of photo-catalytic property of TiO_2 film prepared by MAO, rare earth metals with incompletely occupied 4f and empty 5d orbit also often serve as catalysts or promote catalysis. Besides, it has been proved that metal Eu, as one of the rare earth metals, has the ability to improve the photo-catalytic activity of the TiO_2 films, because of the electrons trap effect supplied by the alterable chemical valence (Eu^{2+} and Eu^{3+}) sites [20].

In the present work, we have prepared TiO_2 film upon Ti alloy substrate via MAO technology. Meanwhile, in order to get high efficient photo-catalytic property, rare earth metal oxide Eu_2O_3 was compounded into the TiO_2 film, and methyl blue dye was used to test the photo-catalytic efficiency under UV-Vis irradiation. The grain could be combined on the surface of the TiO_2 layer directly by the MAO process of high temperature and fast solidification, which could improve its photo-catalytic property.

2. Experimental

All chemicals were of analytical grade and used without further purification. In our work, Eu_2O_3 particles were added into the electrolyte directly after ultrasonic dispersion in distilled water. We prepared uniform TiO_2/Eu_2O_3 film upon Ti alloy substrate by using the high temperature and fast solidification of MAO process. Meanwhile, we also prepared pour TiO_2 film, so as to make a contrast with films were produced without adding semiconductor compound into the electrolyte. Follow these steps:

At first, we made up 5 L electrolyte. The electrolyte was a mixed solution of sodium carbonate (20 g/L) and sodium silicate (8 g/L). The anode was TC4 substrate with reaction area of 20 × 20 × 5 mm. After polishing, the substrates were cleaned by ethanol, acetone and distilled water. A stainless steel container was used as electrolyte cell and cathode. MAO of titanium alloy substrate was carried out at a constant current using a 60 KW AC power supply. Different current densities were reached by adjusting the voltage between two working electrodes. The temperature was controlled below 40°C within a small circulating water channel. Oxidation time was ten minutes. And then, the pure TiO_2 film was prepared upon TC4 substrate when the voltage was 400 V and stable. Last, 10 g Eu_2O_3 particles were added into the same electrolyte (the concentration of Eu_2O_3 was 2 g/L). The TiO_2/Eu_2O_3 composite films were prepared at different voltage, 300 V, 350 V and 400 V. The detail experimental conditions were listed in **Table 1**.

The morphologies of the films and their chemical compositions were characterized by using scanning electron microscope (SEM), energy dispersive spectrometer (EDS) and X-ray photoelectron spectroscopy (XPS). The crystal structure was measured by using X-ray diffract meter (XRD) (D8 Advanced, Bruker AXS, Germany) with Cu Ka radiation (the scanning speed is 6° per minute). Photo-catalytic property was tested by diffuse reflectance spectra (DRS) test, photo-generated current test, and photo decomposition test.

Diffuse reflectance spectra (DRS) test UV-Vis diffuse reflectance spectra (DRS) of the TiO_2 layers were measured on the diffuse reflectance accessory of UV-Vis recording spectro-photometer (UV-2550, Shimadzu, Japan).

Photo-generated current test The TiO_2 layer, saturation calomel electrode and the platinum electrode, which were immersed in 1 mol/L NaOH aqueous solution, were chosen to the working electrode, reference electrode and counter electrode respectively. The working electrode was irradiated by using a high pressure mercury lamp (160 W) in 10 cm horizon distance away. The photocurrent was detected using an electrochemical workstation (CHI 660C, Chenhua, Shanghai).

Photo decomposition test The methyl blue solution of 12 mg/L was pour into two beakers. The TiO_2 layer without complex and the compounded layers were immersed in them, and the left was used as reference. All of them were illuminated by a high pressure mercury lamp (160 W) under 15cm vertical distance. The absorbance of methyl blue solution was measured by a UV-Vis spectrophotometer (UV-2550, Shimadzu, Japan).

Table 1. Experimental conditions of the TiO_2/Eu_2O_3 compounded film.

Samples	Electrical					Electrolyte		
	Voltage		Frequency	Duty cycle	Time	Na_2CO_3	$Na_2SiO_3 \cdot 9H_2O$	Eu_2O_3
	Forward	Negative						
TiO_2	400 V	10 V	1 KHz	20%	10 min	20 g/L	8 g/L	No
TiO_2/Eu_2O_3	300 V	10 V	1 KHz	20%	10 min	20 g/L	8 g/L	2 g/L
TiO_2/Eu_2O_3	350 V	10 V	1 KHz	20%	10 min	20 g/L	8 g/L	2 g/L
TiO_2/Eu_2O_3	400 V	10 V	1 KHz	20%	10 min	20 g/L	8 g/L	2 g/L

3. Results and Discussions

The SEM morphologies of the pure TiO_2 film at 400V are shown in **Figure 1**. From **Figure 1**, we can see clearly that the pure TiO_2 film is porous with diameter of pores (the diameter of pores is about 1 - 5 μm). **Figure 2** shows the SEM morphologies of the TiO_2/Eu_2O_3 composite films at 300 V, 350 V and 400 V. Meanwhile, **Figure 2** gives its two different magnifications photos at 400 V. It can be seen that the MAO TiO_2 surface exhibits a typical crater and porous microstructure with a diameter around 1 - 2 μm. At 300 V and 350 V, the surface of the TiO_2/Eu_2O_3 composite film present micro pore structure, as shown in (a) and (b), which is just ordinary film prepared by MAO technology. However, at the high voltage 400 V, there are many one-dimensional (1-D) structures homogeneously embedded within the film, as shown in (c-1) and (c-2). Due to the small size effect, these 1-D structures can enhance the photo-catalytic property of TiO_2 film. So, our researches were concentrated on the TiO_2/Eu_2O_3 composite film prepared at 400V. In order to get more detail information about the 1-D nano-wires materials on the surface of TiO_2/Eu_2O_3 film, we made size statistics about it. It can be noticed that these 1-D nano-wires have the average length of 1.26 μm and the average width of 87.5 nm from **Figure 3**.

In order to prove that the 1-D nano-wires were Eu_2O_3,

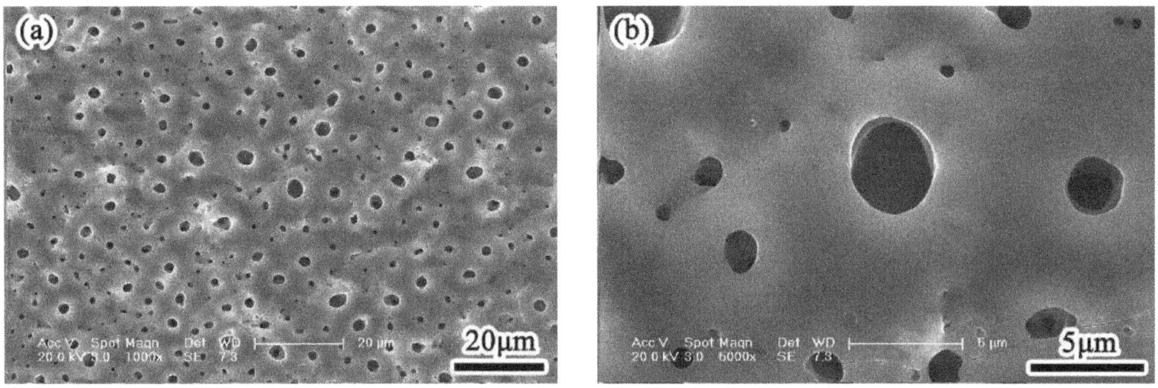

Figure 1. SEM morphologies of TiO_2 film without composites. (a) Low magnification; (b) High magnification.

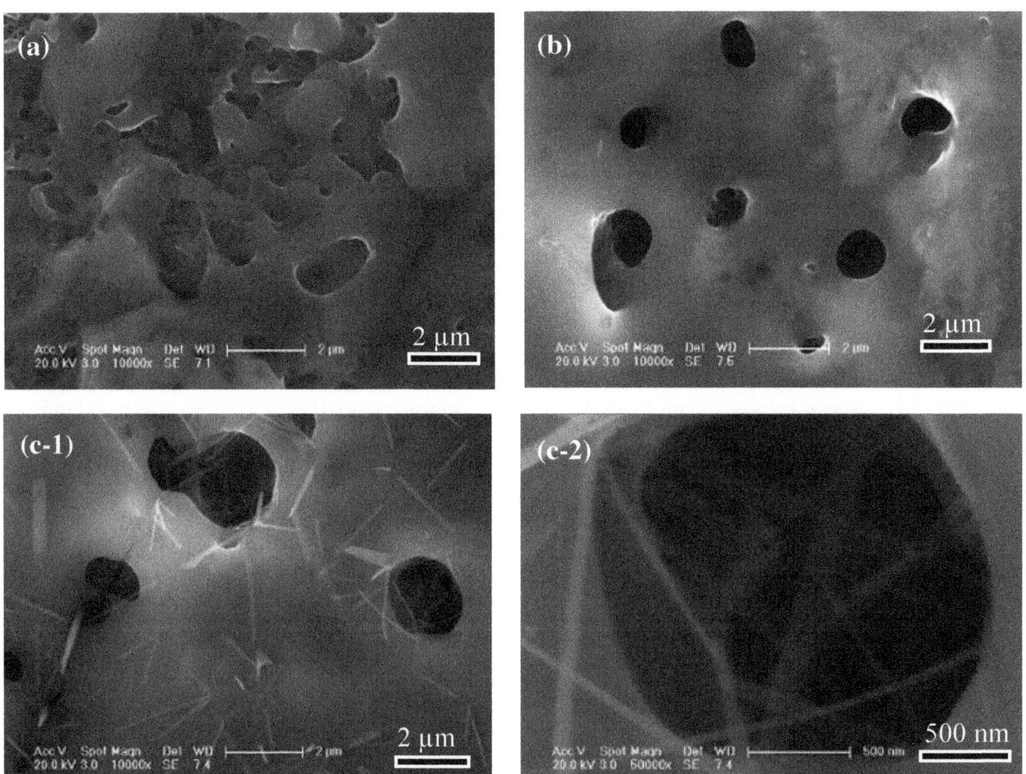

Figure 2. SEM morphologies of the TiO_2/Eu_2O_3 film by different voltage. (a) 300 V, (b) 350 V, (c-1) 400 V, (c-2) 400 V.

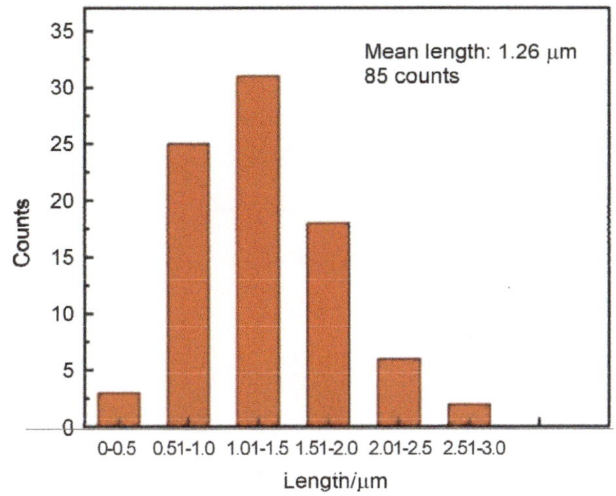

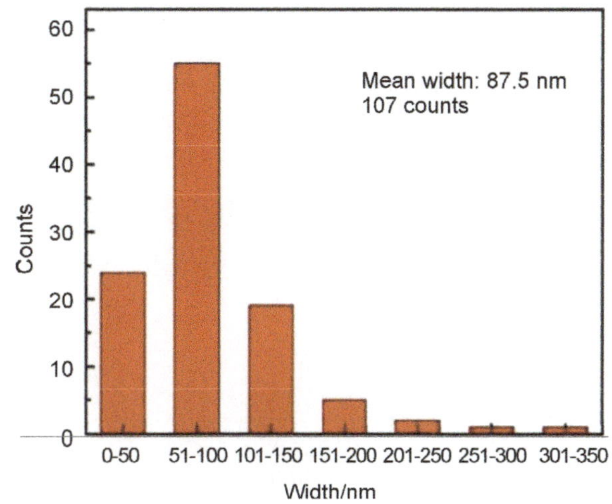

Figure 3. Statistical size distribution of the nano-wires on TiO_2/Eu_2O_3 composite film: (a) Length; (b) Width.

which were added into the electrolyte and entered into the micro-porous of the TiO_2 films, we made EDS test. EDS measurement revealed that these 1-D nano-wires were composed of Eu and O elements, as shown in **Figure 4**, which implies that original Eu_2O_3 particles have transformed into 1-D nano-wires during MAO process. At the same time, we know that Ti and some O elements come from the TiO_2 film, and Au elements come from SEM observation of injection of the sample, and Na elements come from the electrolyte.

XPS technique was applied to confirm Eu_2O_3 existence in the TiO_2/Eu_2O_3 composite film and investigate the chemical states of the elements, whose result is presented in **Figure 5**. It can be seen that the peaks mainly contain the Ti 2p, O 1s, C 1s and Eu $3d^5$, which also demonstrates the existence of Eu_2O_3 inside the film.

Figure 6(a) illustrates XRD patterns of the pure TiO_2 and the TiO_2/Eu_2O_3 composite film (400V). It can be noticed clearly that both of them mainly contain anatase phase with few rutile phase. At the same time, we know that 1-D nano-wires have little affection to the crystal structure of the films. **Figure 6(b)** displays the enlarged XRD peaks of anatase TiO_2 plans A (101) in the 2θ region of 24.0° - 27.0°. We measured the full width half maximum (FWHM) of the main peak of anatase TiO_2 plans (101) in the 25.3°. The result is that the FWHM of the pure TiO_2 film is 0.24 and the TiO_2/Eu_2O_3 composite film is 0.28. Using the Scherrer Formula, we calculated that in pure TiO_2 film, the grain size of anatase TiO_2 is about 33 - 34 nm but in the TiO_2/Eu_2O_3 composite film, the size is around 28 - 29 nm. From calculation results, we found that the grain size of anatase in the TiO_2/Eu_2O_3 composite film decreased obviously, compared with pure TiO_2 film. The reason is that rare earth metals has incompletely occupied 4f and empty 5d orbit and the growth of TiO_2 grains was inhibited after Eu_2O_3 was compounded in TiO_2 films.

For the sake of studying the photo-catalytic properties of the TiO_2/Eu_2O_3 composite film (400 V), diffuse reflectance spectra (DRS) test was employed and its result is presented in **Figure 7**. Obviously, the composite film exhibits a higher absorption of UV and visible light with an Einstein shift about 10 nm. Before Eu_2O_3 was compounded, the pure TiO_2 film only has little absorption of visible light. The results reveal that improved absorption in UV and visible light ranges is contributed from the addictive phase Eu_2O_3. Eu^{3+} possesses abundant energy levels, so Eu_2O_3 can increase the absorption of UV and visible light and improve the photo-catalytic properties of the composite film.

To study the photo-catalytic efficiency of the TiO_2/Eu_2O_3 composite film (400 V), we also measured the intensity of photo-generated current of it, as shown in **Figure 8**. The reason why the TiO_2/Eu_2O_3 composite film has a higher intensity of photon generated current is that in Eu_2O_3, iron Eu^{3+} has an incomplete 4f orbital track and an empty 5d orbital track, which tends to produce multi-electron configuration and therefore can effectively inhibit the recombination of photo-electrons and holes.

Figure 9 shows the result of the degradation experiments of methyl blue. It revealed that the photo-catalytic performance of the TiO_2/Eu_2O_3 composite film (400 V) exhibited two times higher than that of the pure TiO_2 film. We can see that after four hours, the degradation of methyl blue reaches to 90% when using the TiO_2/Eu_2O_3 composite film (400 V), while only 45% when using the film without Eu_2O_3.

4. Conclusion

In conclusion, the TiO_2 films were prepared upon Ti alloy substrate by micro-arc oxidation (MAO) process. The films had a porous structure with a rough surface which is suitable for modification. When rare earth metal oxide Eu_2O_3 was added into the electrolyte, a composite film was produced during MAO and Eu_2O_3 transformed into one-dimensional (1-D) nano-wires embedded within the composite film. Comparing with the pure films, the composite films exhibited high efficient photo-catalytic properties, such as high absorption in UV and visible light ranges, high photo-generated current intensity and high photo degradation rate.

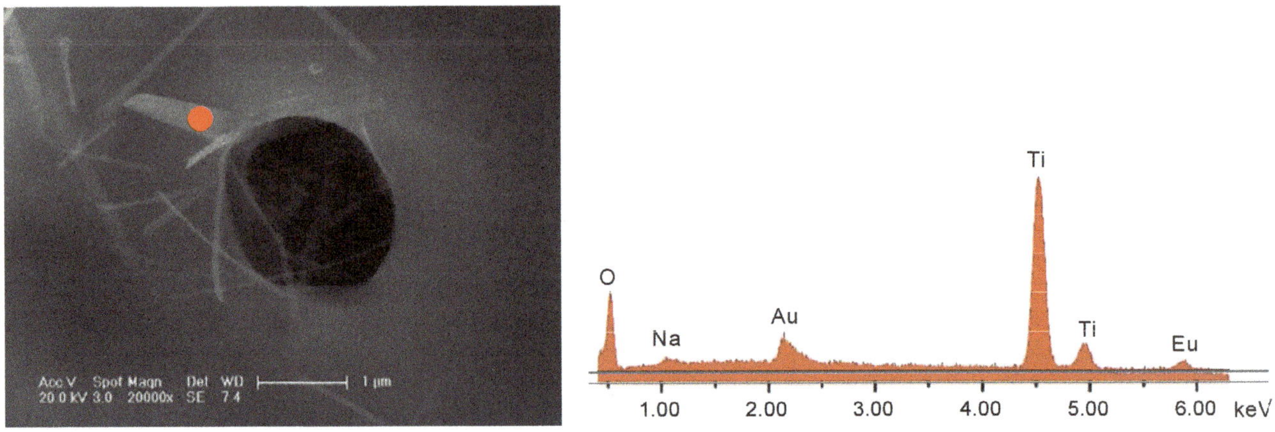

(a) (b)

Figure 4. EDS analysis of the TiO_2/Eu_2O_3 composite film.

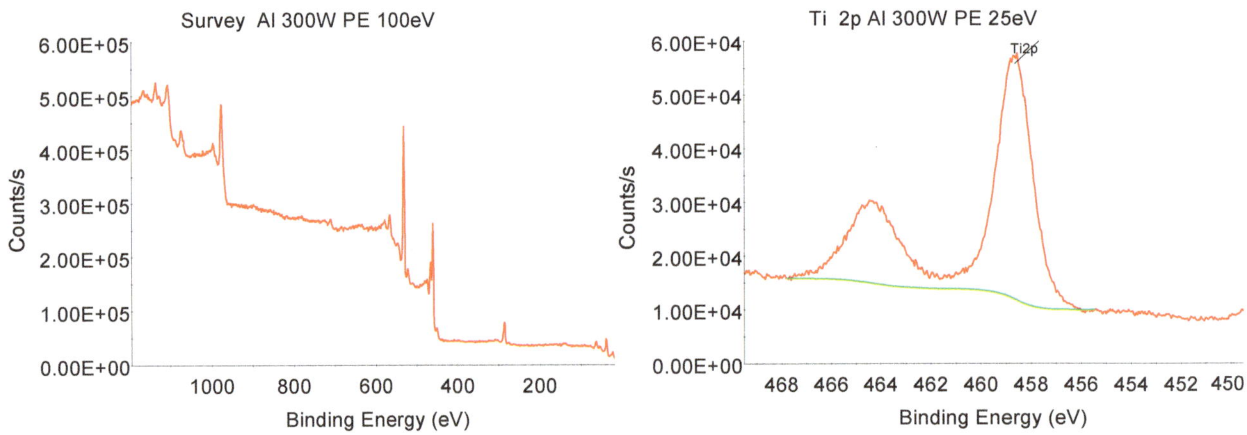

(a) (b)

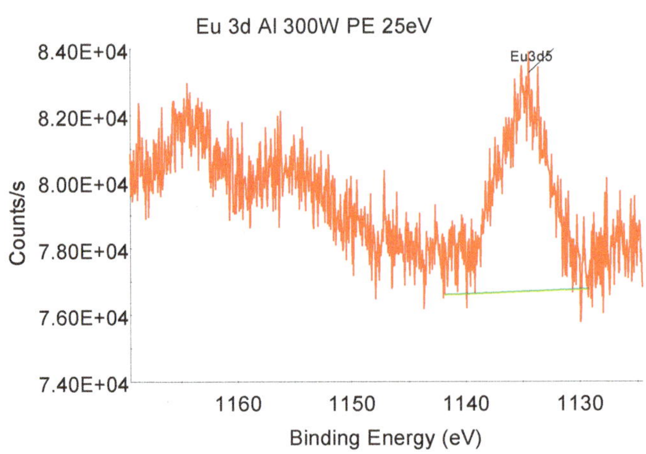

(c)

Figure 5. XPS spectra of the TiO_2/Eu_2O_3 composite film.

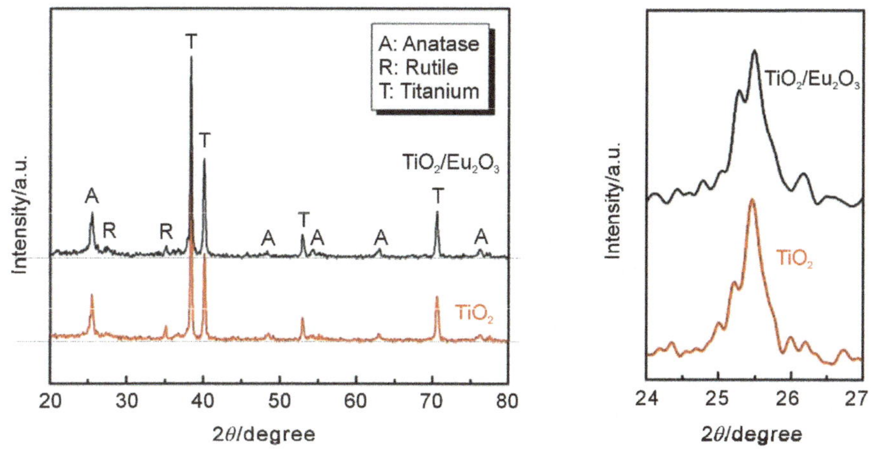

Figure 6. XRD patterns of the TiO_2 and the TiO_2/Eu_2O_3 composite films prepared by MAO (a)survey, (b) anatase (101) peak.

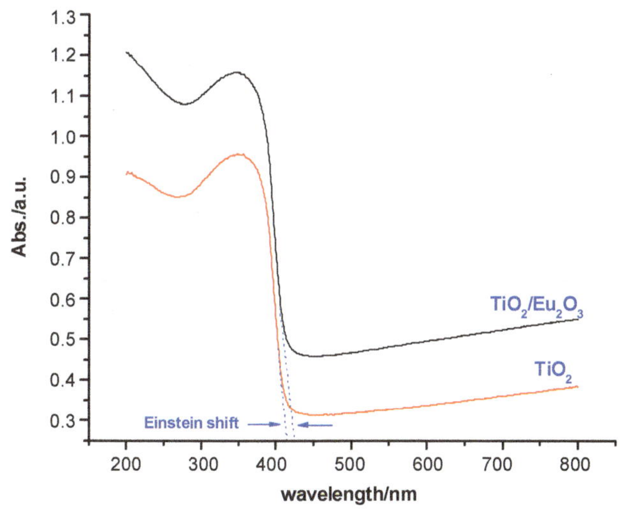

Figure 7. UV-Vis spectra of the TiO_2 film and the TiO_2/ Eu_2O_3 composite film.

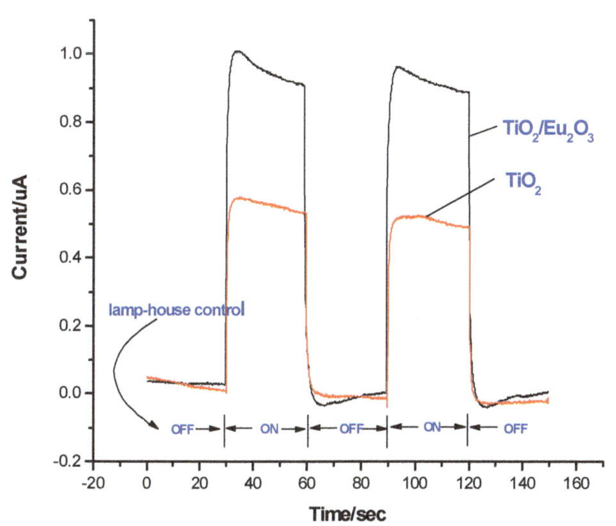

Figure 8. Photo-generated current of pure TiO_2 and TiO_2/ Eu_2O_3 composite film.

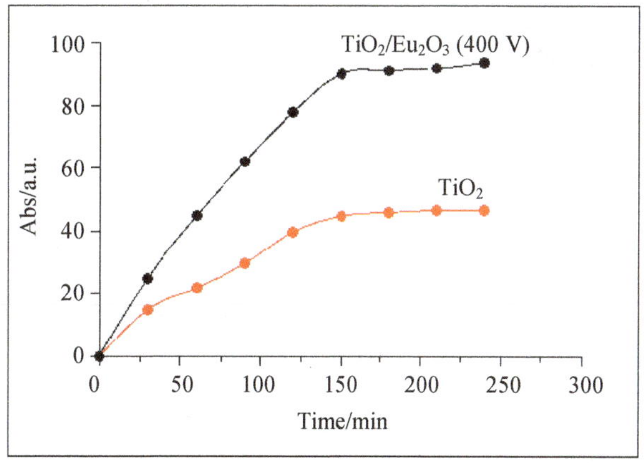

Figure 9. The concentration of methylene blue photo-degraded by pure TiO_2 and TiO_2/Eu_2O_3 film.

5. Acknowledgement

The work was supported by the National Natural Science Foundation of China (21203170, 21201156), the Hall of Hubei Province Science and Technology Project (2011CDB347) and the Fundamental Research Funds for the Central Universities, China University of Geosciences (Wuhan, CUG130401), National Training Programs of Innovation and Entrepreneurship for Undergraduates (201310491015). The financial support was gratefully appreciated.

REFERENCES

[1] A. Fujishima and K. Honda, "Electrochemical Photocatalysis of Water at a Semiconductor Electrode," *Nature*, Vol. 238, 1972, pp. 37-38.

[2] S. Grayer and M. Halmann, "Electrochemical and Photoelectron Chemical Reduction of Molecular Nitrogen to Ammonia," *Journal of electroanalytical chemistry and interfacial electrochemistry*, Vol. 170, 1984, pp. 363-368.

[3] Y. Han, D. H. Chen and L. Zhang, "Nano-Crystallized SrHA/SrHA-$SrTiO_3$/$SrTiO_3$-TiO_2 Multilayer Coatings Formed by Micro-Arc Oxidation for Photo-Catalytic Application," *Nanotechnology*, Vol. 33, No. 19, 2008, Artivle ID: 335705.

[4] F. D. Fonzo, C. S. Casari, V. Russo, M. F. Brumella and A. L. Bassi, *Nanotechnology*, Vol. 20, 2009, pp. 15604-15610.

[5] L. Wan, J. F. Li, J. Y. Feng, W. Sun and Z. Q. Mao, "Photo-Catalysts of Cr Doped TiO_2 Film Prepared by Micro Arc Oxidation," *Chinese Journal of Chemical Physics*, Vol. 5, No. 21, 2008, pp. 487-492.

[6] D. Chatterjee and S. Dasgupta, "Visible Light Induced Photocatalytic Degradation of Organic Pollutants," *Photochemistry and Photobiology*, Vol. 2-3, No. 6, 2005, pp. 186-205.

[7] V. Ramaswamy, N. B. Jagtap, S. Vijayanand, D. S. Bhange and P. S. Awati, "Photocatalytic Decomposition of Methylene Blue on Nano-Crystalline Titania Prepared by Different Methods," *Materials Research Bulletin*, Vol. 5, No. 43, 2008, pp. 1145-1152.

[8] A. Z. Moshfegh, "Nano-Particle Catalysts," *Journal of Physics D-Applied Physics*, Vol. 23, No. 42, 2009, Article ID: 233001.

[9] M. Janus, E. Kusiak and A. W. Morawski, "Carbon Modified TiO2 Photo-Catalyst with Enhanced Adsorptivity for Dyes from Water," *Catalysis Letters*, Vol. 3-4, No. 131, 2009, pp. 506-511.

[10] H. T. Feng, F. Wang and X. G. Tong, "LPD Preparation of Iron-Doped TiO2 Films and Its Performance Analysis," *Ceramics international*, Vol. 2, 2006, pp. 16-17.

[11] L.Y. Shi, H. C. Gu and C. Z. Li, "Preparation and Properties of SnO2-TiO2 Composite Photo-Catalysts," *Chinese Journal Catalysis*, Vol. 20, 1999, pp. 338-342.

[12] P. Evans, T. English, D. Hammond, M. E. Pemble and D. W. Sheel, "The Role of SiO_2 Barrier Layers in Determining the Structure and Photo-Catalytic Activity of TiO_2 Films Deposited on Stainless Steelv," *Applied Catalysis*, Vol. 2, No. 321, 2007, pp. 140-146.

[13] J. Mungkalasiri, L. Bedel, F. Emieux, J. Doré, F. N. R. Renaud and F. Maury, "DLI-CVD of TiO_2-Cu Antibacterial Thin Films: Growth and Characterization," *Surface and Coatings Technology*, Vol. 6-7, No. 204, 2009, pp. 887-892.

[14] E. Matykina, A. Berkani, P. Skeldon and G. E. Thompson, "Real-Time Imaging of Coating Growth during Plasma Electrolytic Oxidation of Titanium," *Electrochimica Acta*, Vol. 4, No. 53, 2007, pp. 1987-1994.

[15] F. Chen, H. Zhou, C. Chen and Y. J. Xia, "Study on the tri-Biological Performance of Ceramic Coatings on Titanium Alloy Surfaces Obtained through Micro-Arc Oxidation," *Progress in Organic Coatings*, Vol. 2-3, No. 64, 2009, pp. 264-267.

[16] W. Xue, Z. Deng, Y. Lai and R. Chen, "Analysis of Phase Distribution for Ceramic Coatings Formed by Micro-Arc Oxidation on Aluminum Alloy," *Journal of the American Ceramic Society*, Vol. 5, No. 81, 1998, pp. 1365-1368.

[17] S. Ikonopisov, "Theory of Electrical Breakdown during Formation of Barrier Anodic Film," *Electrochemical Acta*, Vol. 22, 1977, pp. 1077-1082.

[18] N. Schiffa, B. Grosgogeata, M. Lissaca and F. Dalardb, "Influence of Fluoride Content and pH on the Corrosion Resistance of Titanium and Its Alloys," *Biomaterials*, Vol. 9, No. 23, 2002, pp. 1995-2002.

[19] Y. K. Lee, "Effects of Electrical Parameters on Titania Film Grown by Micro Arc Oxidation," *Modern Physics Letters B*, Vol. 16, No. 23, 2009, pp. 2035-2040.

[20] P. Yang, C. Lu and N. Hua, "Titanium Dioxide Nano-Particles Co-Doped with Fe^{3+} and Eu^{3+} Ions for Photo-Catalysis," *Materials Letters*, Vol. 57, 2002, pp. 794-801.

$Y_3Fe_5O_{12}$ Prepared by Mechanosynthesis from Different Iron Sources

Claudia A. Cortés-Escobedo[1], Ana M. Bolarín-Miró[2], Félix Sánchez-De Jesús[2], Raúl Valenzuela[3], Elizabeth P. Juárez-Camacho[1], Iván L. Samperio-Gómez[2], Souad Ammar[4]
[1]CIITEC, Instituto Politécnico Nacional, Mexico City, Mexico
[2]Área Académica de Ciencias de la Tierra y Materiales, UAEH, Hidalgo, Mexico
[3]Departamento de Materiales Metálicos y Cerámicos, IIM, UNAM, Mexico City, Mexico
[4]ITODYS, UMR, CNRS 7086, Université de Paris-Diderot, Paris, France

ABSTRACT

Yttrium iron garnet, $Y_3Fe_5O_{12}$ (YIG) powders were synthesized by mechanochemical processing (MCP) from different iron sources (FeO, Fe_2O_3 and Fe_3O_4) mixed with Y_2O_3, followed by a heat treatment. The aim of this work is to demonstrate that MCP followed by annealing at very low temperatures (as compared with the classic solid state reaction) can induce the formation of nanostructured YIG. The effect of iron source on final structure was also studied. X-ray diffraction (XRD) and scanning electron microscopy (SEM) were used to characterize the synthesized powders. The precursors mixed in a stoichiometric ratio to obtain YIG were milled at room temperature in a shaker mixer mill with a ball:powder weight ratio of 10:1. A partial synthesis of YIG was achieved after 9 h of milling time by using the three sources of iron; however, a significant fraction of the product was the perovskite $YFeO_3$. The largest yield of YIG was obtained by using FeO. In all cases a single garnet phase could only be completely obtained after an annealing process at 900°C, around 400°C lower than the typical temperatures to prepare the material by solid state reaction. An analysis of the microstrain and lattice parameters associated with peak displacements is discussed.

Keywords: Mechanochemistry; YIG; Annealing; Iron Garnet

1. Introduction

Ferrimagnetic garnets have attracted attention as microwave device materials and magneto-optical recording medium for their unique magneto-optical properties [1]. In order to modify these properties, doping with different cations such as bismuth [2,3], lanthanum [4,5], cerium [6], titanium [7], gadolinium [6] and other rare earth cations has been reported.

Garnets are characterized by a compact oxygen array [4] and are assigned to space group Ia3d $8(O_h^{10})$ where the cations are located at the center oxygen polyhedron [2,8]. Due to the possibility to exchange the positions of the cations in the cell, these materials are the basis for many high-technology devices like telecommunications and microwave. The fundamental magnetic properties of YIG are originated from the magnetic ions and their relationship to the surrounding oxygen ions.

Yttrium iron garnet can be synthesized by several methods. The conventional and oldest one is the sintering of the corresponding oxides in a furnace at 1400°C (solid state reaction) [3,5]. This process produces large particles and consumes a great amount of energy [9]. This method can produce an intermediate phase, $YFeO_3$, which has an antiferromagnetic structure. Iron garnet powders can be produced by wet chemical methods such as sol-gel [6-10], coprecipitation [8], chemical vapor deposition [11], thermal plasma spraying [12], etc. A particular method is mechanochemical synthesis which promotes the formation of a new oxide by mechanical activation of precursor's oxides or salts [13]. In spinel ferrites, it leads to a change in the distribution of cations in interstitial sites, thereby modifying the magnetic properties [13,14]. Mechanosynthesis of yttrium iron garnet with subsequent heat treatment have been reported in the literature, using annealing temperatures higher than 900°C. Paesano *et al.* [15] reported the successful mechanosynthesis of YIG from a mixture of Y_2O_3 and Fe_2O_3 followed by an annealing at 1000°C. Widatallah *et al.* [16] presented a detailed study of the influence of the milling process on the formation of single phase YIG by mechanochemical synthesis, using similar experimental conditions to those reported by Paesano [15]. Also, M. Niyaifar *et al.* [2] used high energy milling to mechanically activate the powder mixture of BiO, Y_2O_3, and Fe_2O_3 and achieve doped Bi-YIG complete synthesis by subsequent annealing at high temperature. All these authors characterized magnetically the obtained powders.

In mechanochemical synthesis experiments in air involving metallic oxides, the reduction of iron oxide phases has been observed to occur in a closed stainless steel container after prolonged milling, and the reaction occurs in a steady-state manner during milling [14,17-20].

2. Experimental Procedure

Fe_2O_3 (Sigma Aldrich, 99% purity), FeO (Sigma Aldrich, 99% purity), Fe_3O_4 (Sigma Aldrich, 99% purity), La_2O_3 (Sigma Aldrich, 99.9% purity), Y_2O_3 (Sigma Aldrich, 99.9% purity) and Gd_2O_3 (Sigma Aldrich, 99.9% purity) powders were used as precursor materials. These powders were mixed in a stoichiometric ratio according to the following equations:

$$3/2Y_2O_3 + 5FeO + 0.0547O_2 \rightarrow Y_3Fe_5O_{12+\delta} \quad (1)$$

$$3/2Y_2O_3 + 5/2\,Fe_2O_3 + 0.0547O_2 \rightarrow Y_3Fe_5O_{12+\delta} \quad (2)$$

$$3/2Y_2O_3 + 5/3\,Fe_3O_4 + 0.0547O_2 \rightarrow Y_3Fe_5O_{12-\delta} \quad (3)$$

The oxygen (0.0547 mol) in these reactions was calculated from the air contained in the milling vial that was used in the milling process. A total of 5 g of the starting mixtures were loaded with steel balls of 1.27 cm in diameter into a steel cylindrical vial (50 cm^3) (steel/steel, S/S) at room temperature in air and milled during 9 hours. The charge ratio (CR) was 10:1. To prevent excessive heating of the vials, the experiments were carried out by alternating 90 min of milling followed by 30 min in standby. Annealing treatments were performed from 700°C to 1000°C in air atmosphere.

Milled powders were characterized by X-ray diffraction (XRD) using a Siemens D5000 diffractometer with $CoK\alpha_1$ (λ = 1.790300 Å) radiation. Patterns were collected in a 2θ interval of 20° - 120° with increments of 0.02 (2θ). Rietveld refinement was performed on the X-ray patterns. This method takes into account all of the information collected in a pattern, and it uses a least squares approach method [21] to refine the theoretical line profile until it matches the measured profile.

Morphologies of the milling powders were analyzed using a scanning electron microscope, JEOL JSM-6300, working at 15 kV.

3. Results and Discussion

3.1. Mechanosynthesis

Figure 1 shows the X-ray diffraction patterns of powder mixtures after 9 hours of milling time for the three iron sources: FeO, Fe_2O_3 and Fe_3O_4. This figure shows the partial formation of perovskite and garnet structures for all the samples.

On the other hand, a widening of the peaks is observed which is associated with a decrease in particle size. In particular, for the mixture FeO + Y_2O_3, a peak is observed after 9 hours of milling at 2θ = 52.3°. This reflection is related with the (110) planes of metallic Fe, and suggests the occurrence of the following redox semi-reactions:

$$2Fe^{2+} \rightarrow 2Fe^{3+} + 2e^- \quad (4)$$

$$Fe^{2+} + 2e^- \rightarrow Fe^0 \quad (5)$$

But the presence of metallic Fe due to milling media erosion is also possible, because of the hardness of the FeO, higher than those for the other precursors.

The starting stoichiometry corresponds to the YIG formation; therefore, there is an iron oxide excess in the mixture as far as orthoferrite is present.

In addition, results of Rietveld refinements for powder mixtures milled for 9 hours, presented in **Table 1**, show the presence of metallic Fe only for the mixture FeO + Y_2O_3. Nevertheless, it is believed that only with a lack of oxygen the last redox reactions are favored, *i.e.* with only 0.0547 mol of oxygen in the vial, or by heat treatment in inert atmosphere, as will be discussed below.

Besides, **Table 1** shows that YIG formation is favored by starting from Fe^{2+}, instead of Fe^{3+} or ($Fe^{2+}Fe^{3+}$); in contrast, the mixture with magnetite ($Fe^{2+}Fe^{3+}$) privileges the formation of orthoferrite and reactants are undetectable after 9 hours of milling. This perovskite phase is

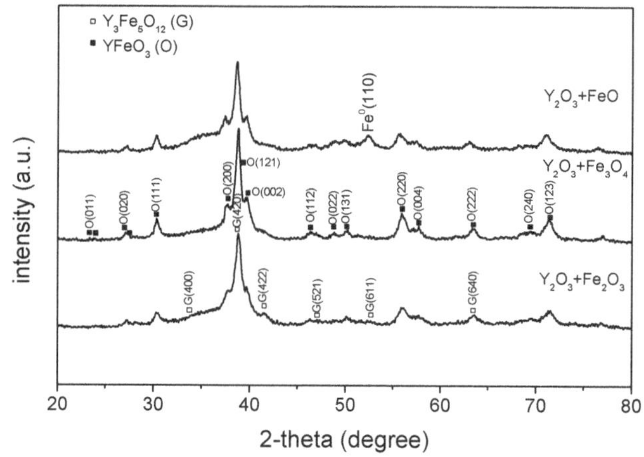

Figure 1. X-ray powder diffraction patterns of different mixtures milled at 9 h.

more stable than YIG according to Bolarín-Miró, *et al.* [13].

Cell parameters are reported in **Table 2** for milled mixtures obtained from different sources of iron. The values for the orthoferrite phase appear closer to reported values [20,22-27]; these results, together with calculated microstrains, as reported in **Table 1**, suggest that the stability of orthoferrite structure is higher than that of the garnet phase; this could mean that YIG structure absorbs all the mechanical strain produced by milling, while orthoferrite remains with the same lattice parameters.

Calculations of the Gibbs free energy [28], *i.e.*, the thermodynamic part of the process states that the most spontaneous reaction (presenting the lowest free Gibbs energy) occurs for the $FeO + Y_2O_3$ mixture to produce YIG in the presence of oxygen. Less spontaneous reaction, in contrast, corresponds to orthoferrite from Fe_2O_3 and Fe_3O_4.

There is however, the kinetic side which has to be considered. Some results [29,30] indicate that, even when starting from Fe^{3+} sources, Fe^{2+} ions are formed and have a larger diffusion coefficient than ferric ions. In most cases, the formation of the ferrite phase occurs by simultaneous diffusion of both iron cations. Under air atmosphere, Fe^{2+} ions are re-oxidized to the ferric state. It seems that in the case of FeO as a source, where ferrous ions are already present, it would be easier to enhance diffusion and formation of the resulting phases.

3.2. Annealing

The obtained results (**Figure 1**) show that MCP is unable to induce the full formation of YIG as a single phase, independently of the iron source (precursor used). It is therefore necessary to complete the synthesis by means of an annealing process. XRD pattern for mixtures $FeO + Y_2O_3$ milled and heat treated at temperatures from 700˚C to 1000˚C are shown in **Figure 2**. The garnet phase is fully formed at 900˚C; below this temperature the peaks of the orthoferrite phase are clearly observed while the peaks of the garnet phase just begin to appear at 700˚C.

The (110) Fe^0 peak reflection disappears when temperature is increased up to 800˚C. It is also observed an increase in relative intensities and narrowing for all the peaks, both indicators of grain and crystal growth.

In **Figure 3(a)** it can be observed for the starting mixture ($Fe_2O_3 + Y_2O_3$), that main peaks of both garnet and perovskite phases reach comparable intensities at 800˚C, while at 900˚C the peak of the perovskite almost completely disappears.

For the $Fe_3O_4 + Y_2O_3$ mixture shown in **Figure 3(b)**, it is only for $T > 900$˚C that peaks of orthoferrite phase decrease in intensity and disappear; at 800˚C, the perovskite phase still remains and shows a significant proportion in the milled material. Again, the width of the peaks decrease and its height increases for all the peaks, indicating grain and crystal growth.

Table 1. Data from Rietveld refinement of XRD patterns for the mixtures ball-milled 9 h.

Products	($Fe_2O_3 + Y_2O_3$) Fe_2O_3 ICDS # 43465		($FeO + Y_2O_3$) FeO ICDS # 76639		($Fe_3O_4 + Y_2O_3$) Fe_3O_4 ICDS # 20596	
	Weight %	μstrain	Weight %	μstrain	Weight %	μstrain
Fe_xO_x	5.6	0.0022	1.7	0.0008	0	0.0008
Y_2O_3	0.7	0.0008	0.6	0.0008	0.2	0
$Y_3Fe_5O_{12}$	36.5	0.0581	44.6	0.0465	28.2	0.0548
$YFeO_3$	57.2	0.0026	49.9	0.0006	71.6	0.0015
Fe	0	0	3.2	0.0008	0	0

Y_2O_3 ICDS # 43465; $Y_3Fe_5O_{12}$ ICDS # 2012; $YFeO_3$ ICDS# 43260; Fe ICDS # 64998.

Table 2. Lattice parameters for mixtures after mechanosynthesis.

Reactants	Lattice parameters (Å)				Rietveld parameters		
	YIG[a]	YFeO$_3$			Goodness of fit		
	a	a	b	c	Rw	Rexp	χ^2
$Fe_2O_3 + Y_2O_3$	12.174	5.268	5.559	7.614	19.0	18.2	1.10
$FeO + Y_2O_3$	12.574	5.295	5.599	7.640	24.9	19.1	1.69
$Fe_3O_4 + Y_2O_3$	12.290	5.270	5.551	7.612	18.7	18.2	1.06
Reference[a]	12.376	5.281	5.595	7.605	-	-	-

[a] $Y_3Fe_5O_{12}$ ICSD #2012 [23]; $YFeO_3$ ICSD #43260 [20].

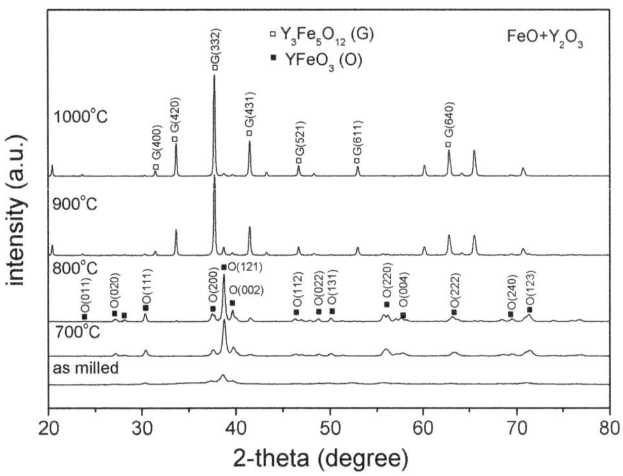

Figure 2. X-ray powder diffraction patterns of milled powders (FeO + Y_2O_3) annealed at different temperatures during 3 hours.

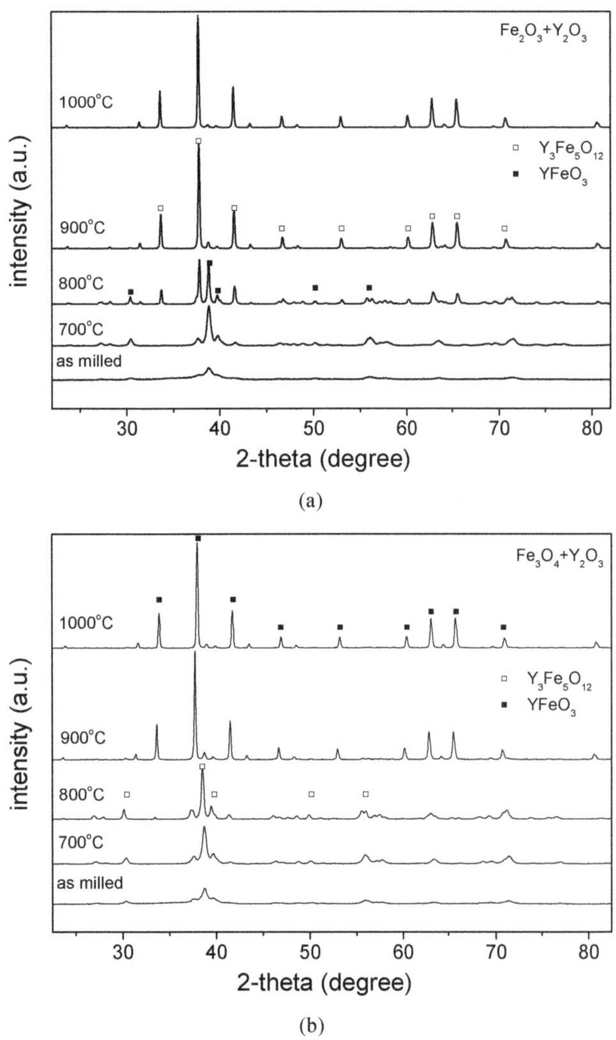

Figure 3. X-ray diffraction patterns of milled powders: (a) (Fe_2O_3 + Y_2O_3) mixture; (b) (Fe_3O_4 + Y_2O_3) mixture, annealed at different temperatures during 3 h.

3.3. Magnetic Characterization

Figure 4 shows the hysteresis loops of powders rich in orthoferrite from different oxide mixtures milled for 9 h.

There are differences in magnetic behaviors for each precursor. The highest saturation magnetization is obtained for FeO + Y_2O_3 milled powder mixture. This particular sample showed the presence of Fe^0 in X-ray diffraction analysis. So, if saturation magnetization is additive in weight proportion to each component (according to the mixture theory), an estimation of the total magnetic saturation for each obtained sample can then be carried out. Results of calculations are presented in **Table 3**, as well as the experimental results extracted from the hysteresis loops in **Figure 4**, where is presented the magnetic characterization of different oxides mixtures from different iron oxides after milling for 9 h.

If the experimental and calculated results (**Table 3**) are compared, we can conclude that the calculated values are always larger than the measured ones. This can be due to two possible reasons: first, the measured value is not strictly the saturation value (at the highest applied field there remains a noticeable slope), and second, the YIG content (which is the main component with high saturation magnetization) could be slightly overestimated. The experimental results for coercive field are in the 120 - 300 Oe range, these are characteristic values of small grains with a large concentration of defects.

Figure 5 is a plot of magnetic characterization of YIG obtained from different iron oxides after heat treatment at

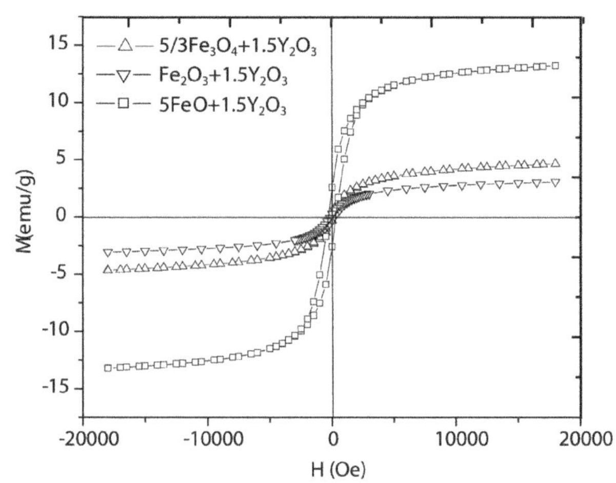

Figure 4. Hysteresis loops at H_{max} = 19 kOe measured at room temperature, for oxide powders milled from (FeO + Y_2O_3), (Fe_2O_3 + Y_2O_3) and (Fe_3O_4 + Y_2O_3) mixtures.

Table 3. Magnetization results (emu/g at 19 kOe) [29-32].

Sample	Fe_2O_3 + Y_2O_3	FeO + Y_2O_3	Fe_3O_4 + Y_2O_3
Experimental	3.05	13.23	4.63
Calculated	11.64	20.59	10.21

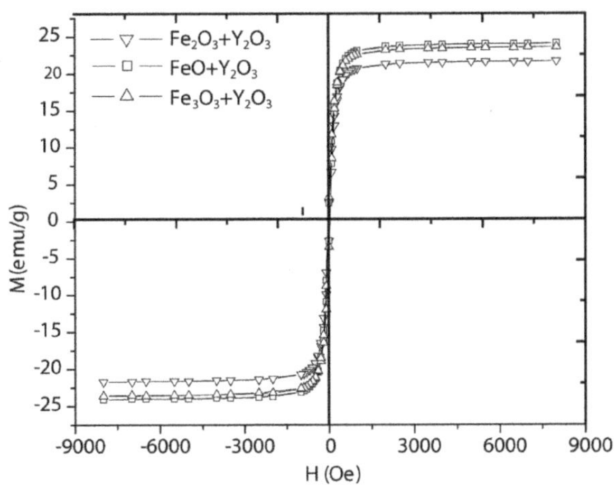

Figure 5. Hysteresis loops measured at room temperature for powders heat-treated at 900˚C.

900˚C. The highest value obtained is for FeO precursor with saturation magnetization of 24.07 emu/g and coercive force (H_c) of 23.93 Oe, for the others starting oxides the values are (Fe_2O_3) 21.68 emu/g and H_c is 26.8 Oe, and (Fe_3O_4) showed a magnetization of 23.58 emu/g and H_c of 27.37 Oe.

Diminution in values of coercive field with respect to powder samples indicates a relief of defects due to milling, and an increase of crystallite size, in agreement with XRD results.

4. Conclusion

Mechanosynthesis from mixtures of Y_2O_3 with Fe_2O_3, Fe_3O_4 and Fe_2O_3 leads to the formation of a mixture of orthoferrite and garnet phases; a single YIG phase is successfully obtained after annealing at 900˚C in all cases. A thermodynamic calculation showed that the ($FeO + Y_2O_3$) starting mixture produced a slightly larger fraction of YIG, as compared with the other two precursors. The magnetic properties of YIG samples produced by different sources showed minor differences in magnetic behavior; the YIG obtained from ($FeO + Y_2O_3$) exhibited a Ms value only slightly higher and very close to the reported one.

5. Acknowledgements

This project was financially assisted by the National Science and Technology Council of Mexico, CONACyT under grants No. 129910 and 130413. Authors are grateful to Adriana Tejeda-Cruz from Institute for Materials Research, UNAM, for technical support in XR diffraction.

REFERENCES

[1] Z. Cheng, H. Yang, L. Yu and W. Xu, "Saturation Magnetic Properties of $Y_{3-x}Re_xFe_5O_{12}$(Re: Gd, Dy, Nd, Sm and La) Nanoparticles Grown by a Sol-Gel Method," *Journal of Materials Science: Materials in Electronics*, Vol. 19, No. 5, 2008, pp. 442-447.

[2] M. N. Ramani, M. C. Radhakrishna, A. Hassnpour, M. Mozaffari and J. Amighian, "The Correlation of Lattice Constant with Superexchange Interaction in Bi-YIG Fabricated by Mechanochemical Processing," *Hyperfine Interactions*, Vol. 184, No. 1-3, 2008, pp. 161-166.

[3] M. N. Ramani, M. C. Radhakrishna, A. Hassnpour, M. Mozaffari and J. Amighian, "Magnetic Studies of $Bi_xY_{3-x}Fe_5O_{12}$ Fabricated Using Conventional Method," *Hyperfine Interactions*, Vol. 187, No. 1-3, 2008, pp. 137-141.

[4] Z. Cheng, Y. Cui, H. Yang and Y. Chang, "Effect Lanthanum Ions on Magnetic Properties of $Y_3Fe_5O_{12}$," *Journal of Nanoparticle Research*, Vol. 11, No. 5, 2009, pp. 1185-1192.

[5] S. R. Nimbore, D. R. Shengule, S. J. Shukla, G. K. Bichile and K. M. Jadhav, "Magnetic and Electrical Properties of Lanthanum Substitututed Yttium Iron Garnets," *Journal of Materials Science: Materials in Electronics*, Vol. 41, No. 19, 2006, pp. 6460-6464.

[6] H. Xu and H. Yang, "Magnetic Properties of YIG Doped with Cerium and Gadolinium Ions," *Journal of Materials Science: Materials in Electronics*, Vol. 19, No. 7, 2008, pp. 589-593.

[7] A. G. Arias, C. Torres, C. de Francisco, J. M. Muñoz, P. H. Gómez, O. Alejos, O. Montero and J. I. Iñiguez, "Defect Concentration in Ti-Subtituted YIG from Tg Curves," *Journal of Thermal Analysis and Calorimetry*, Vol. 86, No. 1, 2006, pp. 195-198.

[8] S.-M. Sim, K. A. Keller and T.-I. Mah, "Phase Formation in Yttrium Aluminum Garnet Powders Synthesized by Chemical Methods," *Journal of Materials Science*, Vol. 35, No. 3, 2000, pp. 713-717.

[9] M. Rozman, M. Drofenik and J. Stefan, "Sintering of Nanosized MnZn Ferrite Powders," *Journal of the American Ceramic Society*, Vol. 81, No. 7, 1998, pp. 1757-1764.

[10] S. H. Vajargah, H. R. M. Hosseini and Z. A. Nemati, "Preparation and Characterization of Yttrium Iron Garnet (YIG) Nanocrystalline Powders by Auto-Combustion of Nitrate-Citrate Gel," *Journal of Alloys and Compounds*, Vol. 430, No. 1-2, 2007, pp. 339-343.

[11] M. Mikami and K. Matsumi, "Epitaxial Growth of Yttrium Iron Garnet by Chemical Vapor Deposition," *Journal of Crystal Growth*, Vol. 37, No. 1, 1977, pp. 1-8.

[12] X. Z. Guo, B. G. Ravi, Q. Y. Yan, R. J. Gambino, S. Sampath, J. Margolies and J. B. Parise, "Phase and Microstructure Evolution in Precursor Plasma-Sprayed YIG Coatings," *Ceramics International*, Vol. 32, No. 1, 2006, pp. 61-66.

[13] A. M. Bolarín-Miró, P. Vera-Serna, F. Sánchez-De Jesús, C. A. Cortés-Escobedo and A. Martínez-Luévanos, "Mechanosynthesis and Magnetic Characterization of Nanocrystalline Manganese Ferrites," *Journal of Materials Science: Materials in Electronics*, Vol. 22, No. 8, 2011, pp. 1046-1052.

[14] T. Verdier, V. Nachbaur and M. Jean, "Mechanosynthesis of Zinc Ferrite in Hardened Steel Vials: Influence of ZnO on the Appearance of Fe(II)," *Journal of Solid State Chemistry*, Vol. 178, No. 11, 2005, pp. 3243-3250.

[15] A. Paesano Jr., S. C. Zanatta, S. N. de Medeiros, L. F. Cótica and J. B. da Cunha, "Mechanosynthesis of YIG and GdIG: A Structural and Mössbauer Study," *Hyperfine Interactions*, Vol. 161, No. 1-4, 2005, pp. 211-220.

[16] H. M. Widatallah, C. Johnson, S. H. Al-harthi, A. M. Gismelssed, A. D. Al-Rawas, S. J. Stewart, M. E. Elzain, I. A. Al-Omari and A. A. Yousif, "A Structural and Mössbauer Study of $Y_3Fe_5O_{12}$ Nanoparticles Prepared with High Energy Ball Milling and Subsequent Sintering," *Hyperfine Interactions*, Vol. 183, No. 1-3, 2008, pp. 87-92.

[17] P. M. Botta, P. G. Bercoft, E. F. Aglietti, H. R. Bertorello and J. M. P. López, "Two Alternative Synthesis Routes for MnZn Ferrites Using Mechanochemical Treatments," *Ceramics International*, Vol. 32, No. 8, 2006, pp. 857-863.

[18] M. Jalaly, M. H. Enayati, F. Karimzadeh and P. Kameli, "Mechanosynthesis of Nanostructured Magnetic Ni-Zn Ferrite," *Powder Technology*, Vol. 193, No. 2, 2009, pp. 150-153.

[19] V. Sepelak, M. Menzel, K. D. Becker and F. Krumeich, "Mechanochemical Reduction of Magnesium Ferrite," *The Journal of Physical Chemistry B*, Vol. 106, No. 26, 2002, pp. 20-25.

[20] P. Coppens and M. Eibschuetz, "Determination of the Crystal Structure of Yttrium Orthoferrite and Refinement of Gadolinium Orthoferrite," *Acta Crystallographica*, Vol. 19, 1965, pp. 524-531.

[21] D. Balzar and N. C. Popa, "Analyzing Microstructure by Rietveld Refinement," *Rigaku Journal*, Vol. 22, 2005, pp. 16-25.

[22] E. A. Owen and G. I. Williams, "A Low-Temperature X-Ray Camera," *Journal of Scientific Instruments*, Vol. 31, No. 2, 1954, pp. 49-54.

[23] M. Bonnet, A. Delapalme, H. Fuess and M. Thomas, "Refinement of the Structure of Yttrium Iron Garnet (YIG). A Case of Severe Extinction and Absorption," *Acta Crystallographica Section B: Structural Crystallography and Crystal Chemistry*, Vol. 31, 1975, pp. 2233-2240.

[24] R. L. Blake, R. E. Hessevick, T. Zoltai and L. W. Finger, "Refinement of the Hematite Structure," *American Mineralogist*, Vol. 51, 1966, pp. 123-129.

[25] M. G. Paton and E. N. Maslen, "A Refinement of the Crystal Structure of Yttria," *Acta Crystallographica*, Vol. 19, 1965, pp. 307-310.

[26] J. Benard, "Recherches Sur Les Variations de Composition du Protoxyde de Fer," *Bulletin de la Societe Chimique de France*, 1949, pp. 109-116.

[27] G. G. Dvoryankina and Z. G. Pinsker, "The Electron Diffraction Pattern Investigation of Fe_3O_4," *Doklady Akademii Nauk SSSR*, Vol. 132, 1960, pp. 110-113.

[28] Outokompu, "HSC Chemistry© for Windows," Pori, 2002.

[29] R. Valenzuela, "Magnetic Ceramics," Cambridge University Press, Cambridge, 2005, pp. 68-69.

[30] C. Kooy, "Material Transport in Solid State Reactions," *Fifth International Symposium on the Reactivity of Solids*, Munich, 1965, pp. 21-28.

[31] H. Danan, A. Herr and A. J. P. Meyer, "New Determinations of the Saturation Magnetization of Nickel and Iron," *Journal of Applied Physics*, Vol. 39, No. 2, 1968, pp. 669-670.

[32] M. Mozaffari, M. Gheisari, M. Niyaifar and J. Amighian, "Magnetic Properties of Mechanochemically Prepared Iron-Wüstite (Fe-FeyO) Nanocomposites," *Journal of Magnetism and Magnetic Materials*, Vol. 321, No. 19, 2009, 2981-2984.

The Effect of Co Incorporation into ZnO Nanoparticles

Nadia Febiana Djaja, Dionisius Agung Montja, Rosari Saleh*
Departemen Fisika, FMIPA-Universitas Indonesia, Depok, Indonesia

ABSTRACT

Co-doped ZnO nanoparticles are synthesized by co-precipitation method. The incorporation of Co ions into ZnO wurzite structure with increasing Co concentrations was followed by X-ray diffraction. Sample structures were further studied by infrared spectra, from which a broad and strong absorption band in the range of 400 - 700 cm^{-1} and -OH stretching vibrational mode approximately 3400 cm^{-1} were observed. Ultraviolet-visible measurements showed a red shift of the absorption edge and four peaks at 468, 515, 577 and 629 nm in its reflectance spectra with increasing Co content. Magnetic measurements showed a ferromagnetic behavior for all samples. From the ESR spectra the presence of Co-metal clusters and Co-oxide precipitates are not observed and can be excluded as the origin of ferromagnetic in our samples. The ESR results are supported by UV-Vis measurements data.

Keywords: Transition-Metal-Doped ZnO Nanoparticles; Room-Temperature-Ferromagnetic; Co-Precipitation

1. Introduction

Transition metal (TM)-doped ZnO nanoparticles are promising candidates for a variety of practical applications due to their potential applications due to the charge and the spin of electrons that lead to a new magnetic, optical and transport properties of these materials. Numerous studies of these materials grown by several methods have exhibited room temperature ferromagnetic (RTFM) [1-4]. A number of studies have been reported to exhibit room temperature ferromagnetism in V-, Mn-, Fe-, Co-, Ni-, and Cu-doped ZnO [5-10]. However, the origin of the room temperature ferromagnetism is still a matter of controversy [11-14] and raised questions about the role of the dopants. According to theoretical prediction transition metal doping in metal oxide such as ZnO would lead to ferromagnetism behavior. However, in the experiments this ferromagnetism behavior is also inconsistent. It seems that ferromagnetism behavior is very sensitive to the preparation method. There is some evidence that the observed ferromagnetism in TM-doped ZnO may arises from precipitation of secondary phase not from the replacement of Zn ion by TM ion in ZnO lattice [15,16]. On the other hand, there are results that exhibited the absence of secondary phase or magnetic clusters, supporting the intrinsic ferromagnetic origin. In addition, several groups [17-19] observed RTFM in undoped ZnO, TiO_2 and HfO_2 which did not contain transition metal ions and proposed that the defects at Zn sites and oxygen vacancies are responsible for the observed ferromagnetic behavior. Indeed, some authors [20-25] report that the presence of defects like oxygen vacancies play an important role in the indirect exchange mechanism. Recently, theoretical study suggested that oxygen vacancies can make a significant contribution to the ferromagnetism by forming bound magnetic polaron (BMP). From theoretical and experimental point of view the presence of ferromagnetism in transition metal doped-ZnO still continues to be an open topic.

In this work we report an investigation of structural, optical and magnetic properties of ZnO particles doped with Co. The co-precipitation method was used due to the method being cost effective, which requires low temperatures for processing and a high degree of solubility can be achieved [26]. The magnetic, structural, and electronic properties were investigated using a vibrating sample magnetometer (VSM), X-ray diffraction (XRD), energy dispersive X-ray (EDX) spectroscopy, Fourier-transform infrared spectroscopy, UV-Vis spectro-photometry and electron spin resonance (ESR).

2. Experimental

The starting materials, zinc sulfate heptahydrate ($ZnSO_4 \cdot 7H_2O$, 99%, Merck) and cobalt chloride hexahydrate ($CoCl_2 \cdot 6H_2O$, 99%, Aldrich) were used without further purification. Specific amounts of $ZnSO_4 \cdot 7H_2O$ and $CoCl_2 \cdot 6H_2O$ were dissolved in distilled water depending on the desired degree of Co-doping. For convenience, these solutions are designated as solution A.

*Corresponding author.

Solution A was then placed in an ultrasonic cleaner operating at a 57 kHz frequency for 2 h. Simultaneously, 44 mmol NaOH was added to 440 mL of de-ionized water to generate solution B. After sonication solution A was subsequently stirred with a magnetic stirrer at room temperature and then solution B was added until the final pH of the solution reached 12. The solution was further stirred for 0.5 h under constant magnetic stirring and then allowed to remain at room temperature for 18 h. Subsequently, the solution was centrifuged and washed several times with ethanol and distilled water to remove residual and unwanted impurities. The final product was dried in a vacuum oven at 200°C for 1 h yielding Co-doped ZnO powders.

The existence and composition of Co dopants were determined by energy-dispersive X-ray spectroscopy (EDX) LEO 420. To evaluate the phase purity of the samples, X-ray diffraction (XRD) measurements were performed using a Philips PW 1710 and monochromatic Cu-Kα (λ = 1.54060 Å) radiation operated at 40 kV and 20 mA in the range of 10° to 80°. The instrumental broadening including the instrumental symmetry was calibrated using a Si powder standard sample. The X-ray diffraction patterns were analyzed by means of the MAUD program using the Rietveld whole profile fitting method to determine the crystal structure and lattice parameters.

Structural properties were further examined using Fourier transform infra red (FTIR) spectroscopy. FTIR spectra of the powder samples were recorded using a Shimadzu Fourier transform spectrometer. All spectra wereobtained using pressed pellets of the prepared samples in potassium bromide (KBr) in the range of 400 - 4000 cm^{-1} with a resolution of 4 cm^{-1}.

To study the electronic interaction near the optical band gap resulting from the addition of dopant atoms diffuse reflectance UV-Vis measurements were performed using a Shimadzu UV-Vis spectrophotometer with an integrating sphere and a spectral reflectance standard in the wavelength range of 200 - 800 nm.

Magnetic measurements were carried out at room temperature using an Oxford Type 1.2 T vibrating sample magnetometer (VSM). Powder samples were tightly packed in a clear plastic drinking straw. Magnetization data was recorded as a function of applied magnetic field from 0 to ±1 Tesla. The backgound signal from sample holder has been corrected for the data reported in this study.

To obtain information on oxidation state and site occupancy of the Co and Mn ions in the ZnO matrix electron spin resonance (ESR) was carried out using X-band JEOL JES-RE1X at room temperature and an X-band spectrometer equipped with 9.1 GHz field modulation unit. The resonance was optimized for modulation amplitude, receiver gain, time constant and scan time. The amount of sample used in all measurements was kept constant. DPPH was used as the standard. The shape and area of the ESR spectra were analyzed using standard numerical methods.

3. Results and Discussion

EDX analysis confirmed the presence of doping elements in the samples. **Table 1** provides representative EDX spectra for the Co-doped ZnO nanoparticles. Elemental analysis revealed that the samples contain target elements within the detection limit of EDX. Quantitative characterization of the Co/Zn ratio was performed by calculating the area of the corresponding spectral K lines of EDX spectra. The amounts of Co in the ZnO particles were found to vary between 3 - 18 at%. These results are an average of four values from different regions. **Table 1** illustrates how the Co incorporation in ZnO nanoparticles varies as a function of the initial cations ratio used in the synthesis. It appears that the amounts of Co incorporated in the samples are slightly lower than the amounts of Co introduced in the synthesis.

XRD measurements were employed to investigate the structural properties and average crystallite size of the Co-doped ZnO samples, in addition to evaluate the phase purity. Representative XRD patterns shown in **Figure 1** reveal the presence of the wurtzite ZnO phase and no other impurity phases even for samples doped with the highest Co concentrations of approximately 18 at%. This indicates that the wurzite structure is not influenced by incorporation of Co. No secondary phases such as CoO, Co_2O_3 or $ZnCo_2O_4$ were observed in the samples, as shown in **Figure 1**, suggesting that Co occupy the zinc sites in the Co-doped ZnO samples. In all XRD patterns, the (101) peak is dominant, and its intensity is higher than that of the other peaks indicated that the preferential growth of [101] orientation along c-axis for all samples. Compared to diffraction pattern of undoped ZnO, the positions of diffraction peaks of Co-doped ZnO gradually change with doping, revealing interesting variations in d values, lattice parameters a, and c and lattice volume V.

The lattice constants were calculated from the Rietveld refinement of the diffraction pattern using the MAUD program and are summarized in **Table 1**. The results are comparable with those of undoped ZnO [27] and CoO [29].

A gradual decrease at the lattice parameter has been observed, dependent on the doping Co atoms in ZnO lattice. Although the change is very small, the variation of lattice parameters with the concentration of dopant gives an indication for the Co ions substituting in the

Table 1. The Co/Zn ratio introduced in the synthesis and from EDX spectra, lattice parameters, lattice volume, ratio of lattice parameters, bond length *l*, average crystallite size <*D*>, and band gap of Co-doped ZnO nanoparticle.

Sample	[Co/Zn]	at%	a = b (Å)	c (Å)	volume (Å^3)	c/a	*l* (Å)	<*D*> nm	Eg (eV)
ZnO	-	-	-	-	-	-	-	-	3.52 [27]
Co doped ZnO	0.03	3	3.2548	5.2178	47.8704	1.6031	3.7425	27	3.34
	0.06	6	3.2534	5.2161	47.8136	1.6033	3.7413	22	3.31
	0.1	12	3.2490	5.2078	47.6085	1.6029	3.7353	20	3.28
	0.2	18	3.2467	5.2057	47.5219	1.6034	3.7339	18	3.26
CoO	-	-	-	-	-	-	-	-	2.3 [28]

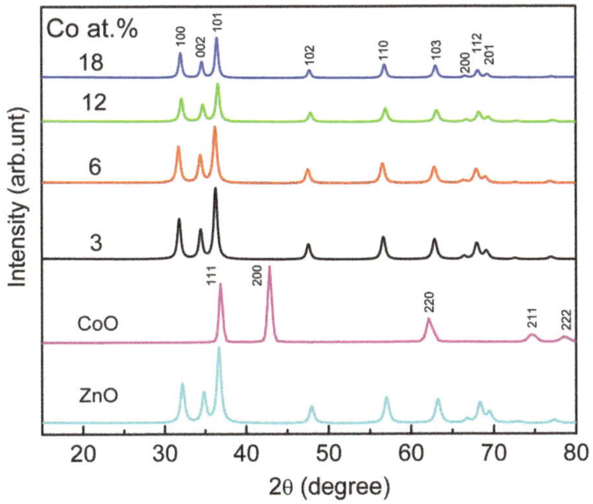

Figure 1. XRD patterns of undoped and Co-doped ZnO nanoparticles synthesized with different concentrations of Co.

ZnO lattice structure since the ionic radii of Co^{2+} (0.58 Å) are less than that of Zn^{2+}. The ratio of lattice parameters (c/a) as a function of dopant concentration is also shownon **Table 1**. The values have a good correlation with the standard value (1.60) and indicated that Co ions are able to substitute the Zn ions in the ZnO lattice without influence the overall crystal structure. It is also expected that with increasing dopant concentrations the Zn-O bond length *l* could be changed due to the change in lattice parameters. The bond length as a function of dopant concentrations is tabulated in **Table 1**. It is observed that the substitutional doping does not change significantly the Zn-O bond length *l* and the mean values are ≈3.74 Å. It is known that the breadth of the XRD peak can be linked to the average crystallite size, microstrain and defects or dislocations.

The average crystallite size using XRD measurements is not generally the same as the particle size due to powder aggregates. The average crystallite size as related to the line broadening can be calculated using Scherrer's equation, as given in Equation (2):

$$\langle D \rangle = \frac{K\lambda}{(\beta \cos\theta)} \quad (2)$$

where <*D*> = volume weighted crystallite size, *K* = shape factor (close to unity in our work, it was set to 0.9), λ = wavelength of Cu-Kα, $\beta_{hkl} = \left[\beta^2_{hkl\ measured} - \beta^2_{instrumental}\right]^{1/2}$ = instrumental corrected integral breadth of the reflection located at 2θ, and θ = angle of reflection. The average crystallite size for the Co-doped ZnO calculated using Scherrer's equation is presented in **Table 1**.

The formation of ZnO wurtzite structures in the Co-doped ZnO samples was further supported by FTIR measurements. Similar spectra were obtained for the undoped ZnO as well as the Co-doped ZnO samples (**Figure 2**). Analysis of the FTIR spectra was done based on the results of nanoparticles published in the literature [30-32]. For all doped and undoped ZnO samples the absorption peaks in the range of 400 - 700 cm^{-1} could be attributed to the ZnO stretching modes [33]. In our FTIR spectra the main peaks observed were: 1) absorption peaks in the range of 1100 - 1600 cm^{-1} corresponding to the Zn-OH bending mode [34], and this band could be normally reduced by calcinations process at higher temperature [35,36] and to C-OH plane bending and C-OH out-of-plane bending [37]; 2) a broad band in the 2900 - 3700 cm^{-1} region which can be explained as overlapping O-H stretching modes and C-H stretching modes. C-H local vibrational modes between 2800 and 3100 cm^{-1} have been observed in a number of semi-conductors [38-40]. In these materials, the local vibrational modes are assigned to the symmetric and antisymmetric C-H stretching modes.

To study the electronic interactions near the optical band gap region due to the presence of dopants diffuse-reflectance measurements were performed on the samples in the UV-Vis region at room temperature. All spectra were obtained in the range of 200 - 800 nm. **Figure 3** shows the diffuse-reflectance spectra, R, as a function of wavelength for samples shown in **Figure 1**. The spectra of Co-doped ZnO particles show peaks at

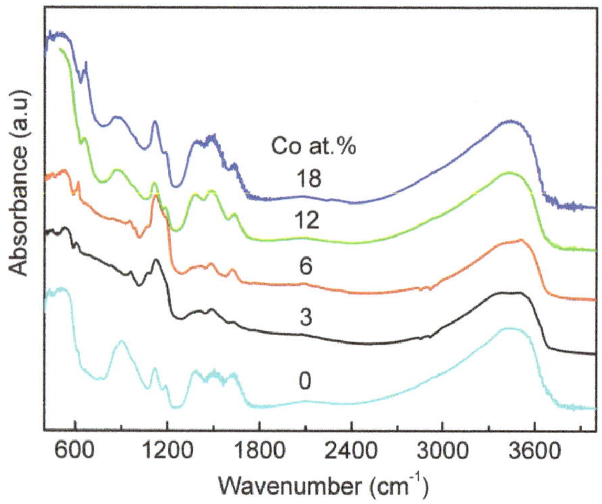

Figure 2. FTIR spectra of the Co-doped ZnO nanoparticles at various doping concentrations. The spectra are shifted vertically for clarity.

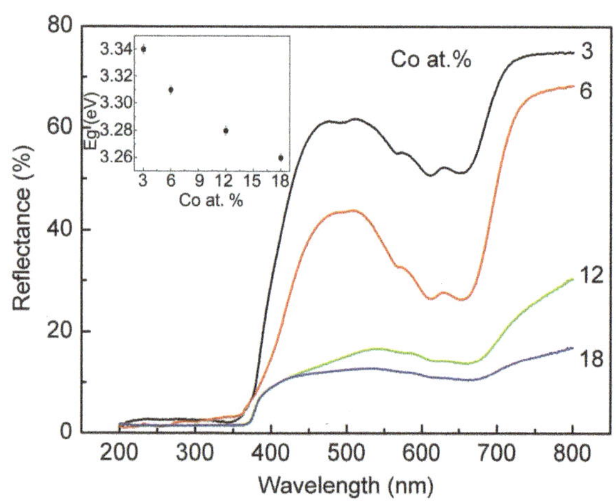

Figure 3. Diffuse-reflectance spectra of the Co-doped ZnO nanoparticles synthesized at various doping concentrations. The inset shows the correlated optical band gap of the Co-doped ZnO nanoparticles as a function of the doping concentrations.

468, 515, 577 and 629 nm. It is reported, that these peaks are only possible in a Co^{2+} ion with $3d^7$ high-spin configuration in a tetrahedral crystal field [3,41-49]. They were ascribed to $^4A_2$ (F) → $^2A_1$ (G), $^4A_2$ (F) → $^4T_1$ (P), and $^4A_2$ (F) → $^2T_1$ (G) transitions in high spin states (S = 3/2). The appearance of these transitions suggested that Co^{2+} ions substituted the Zn^{2+} in ZnO lattice. It may be noted that the peaks smeared out and the intensity of the reflection spectra decreased at Co concentration higher than 6 at%, indicated that additional Co ion incorporated into ZnO occupied sites with a different local crystal field [3].

The bandgap energy of the doped Co-doped ZnO samples was calculated from the diffuse-reflectance spectra by plotting the square of the Kubelka-Munk function $F(R)^2$ vs. the energy in electron volts. The linear part of the curve was extrapolated to $F(R)^2 = 0$ to calculate the direct band gap energy. The inset of **Figure 3** shows the bandgap as a function of the doping concentrations.

The absorption edge shifts to lower energies/longer wavelength. A similar shift in the absorption edge and band gap energy upon transition-metal (TM) doping was reported in Co-, Ni-, and Cr-doped ZnO nanoparticles [50-52] as well as other TM-doped semiconductor nanoparticles systems [53-55]. In addition to the reduction in the band gap energy, a decrease in the diffuse reflectance was also observed with increasing dopant concentration. The redshift of the band gap energy with incorporation of dopant into ZnO lattice was interpreted as mainly due to the *sp-d* exchange interactions between the band electrons and the localized d electrons of the dopant ions substituting Zn ions [32]. It is believed that the *s-d* and *p-d* exchange interactions lead to a negative and a positive correction to the conduction band as well as the valence band edges resulting in a band gap narrowing [32,45].

Information about the magnetic properties of Co-doped ZnO, was obtained by room temperature VSM measurements. The results are displayed in **Figure 4**. Plotted is the magnetization as a function of magnetic field for different contents of dopant. It is interesting to note that all the samples of the present investigation show ferromagnetic behavior. As is seen in **Figure 4** the Co-doped sample has a well-defined hysteresis at room temperature with coercive field (H_C) and the remnant magnetization (M_R) increased with increasing doping concentrations. The coercive field (H_C) and the remnant magnetization (M_R) are found to decrease. The values of M_R and H_C are summarized in **Table 2**. Accordingly, M_R

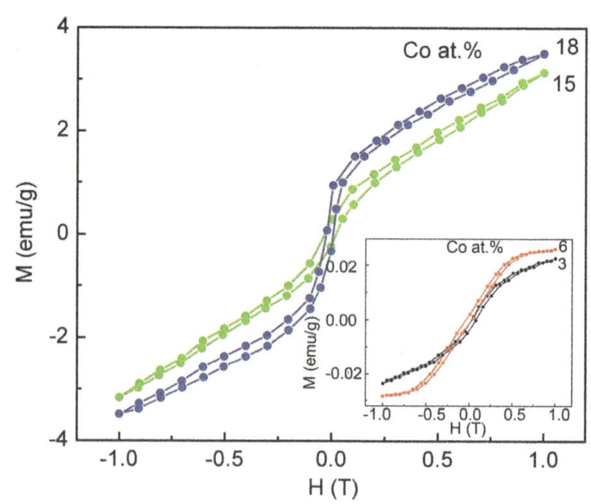

Figure 4. Room temperature M-H curves for Co-doped ZnO nanoparticles synthesized with various doping concentrations.

Table 2. The g value, line width (ΔH_{pp}), peak area, coercive field (H_C) and the remnant magnetization (M_R) of Co-doped ZnO nanoparticle.

Sample	at%	g value	ΔH_{PP}	Area ($\times 10^6$)	M_R (emu/g)	H_C (Oe)
Co doped ZnO	3	2.2005	328	319	0.002	530
	6	2.2558	352	675	0.001	350
	12	2.2902	392	1160	0.284	275
	18	2.3298	488	2670	0.664	235

value for all concentrations lies in the range of 0.002 - 0.664 emu/g. The results obtained from VSM measurements were further confirmed by room temperature ferromagnetic resonance signals observed in the ESR measurements.

In **Figure 5** the ESR spectra of Co-doped ZnO samples with different Co concentration in the range of 3 - 18 at% are shown. It is known that the Co^{2+} ion has $3d^7$ and the electronic ground state is $S_{7/2}$. The only natural isotope is ^{59}Co and the ESR spectrum is expected to show eight fine transitions. However, the ESR spectra of Co-doped ZnO samples investigated here showed only a broad signal without splitting into additional peaks.

The broad signal was attributed to ferromagnetic resonance, which arises from long range exchange interaction and from transitions within the ground sate of the ferromagnetic domain [3,55-57]. For all Co-doped samples the shape of ESR spectrum can be fitted with Gaussian in the entire range of magnetic field. The peak position and the linewidth (ΔH_{pp}) vary strongly with dopant concentrations. The peak position shifted to the lower magnetic field while the ΔH_{pp} shows monotonous increase with increasing dopant concentration as expected for a ferromagnetic material [58-60]. It is believed that the observed broadening of the signal is due to the presence of a non-homogeneous local magnetic field, which modifies the resonance field as well as the line shape of the signal [4]. The corresponding g-values increased from 2.2005 to 2.3298 as Co concentration increased from 3 to 18 at% (**Table 2**).

The mechanism responsible for the observed ferromagnetism at room temperature in transition metal-doped ZnO is also not clear and has been debated over the years. Nevertheless a few researchers have claimed to observe ferromagnetic behavior arising only from a secondary phase and not from the intrinsic property of the material itself. The results of the XRD and EDX measurements in our Co-doped samples demonstrate that the dopant ion was incorporated into the wurtzite lattice at Zn sites forming a solid solution instead of precipitates. However, a secondary phase might exist in the sample even though it was not detected in our XRD spectra. Thus, it is useful to consider all possible ferromagnetic impurity phases

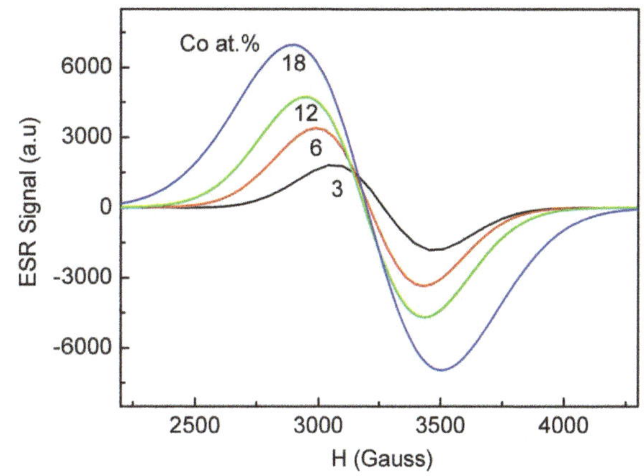

Figure 5. Experimental ESR spectra of the Co-doped ZnO nanoparticles prepared at various Co concentrations.

that might be present in Co-doped ZnO. In Co-doped ZnO samples the possible secondary phase is due to Co-metal clusters or Co-oxide precipitates. The possible Co-based oxides, such as CoO and Co_2O_3 are antiferromagnetic with T_N values of 293 and 30 K, respectively [61-63].

Moreover, from the ESR spectra the presence of Co-metal clusters and Co-oxide precipitates are not observed and can be excluded as the origin of RTFM in our Co-doped samples. The ESR results are supported by UV-Vis measurements data which show the characteristic transition of Co^{2+} ions. Therefore, according to the characterization methods applied in the present study, the ferromagnetism behavior observed in our Co-doped samples studied here does not seem to be related with the presence of any secondary phases. Therefore ferromagnetism is expected to arise from the intrinsic exchange interaction of ferromagnetic in our transition metal doped ZnO samples.

There are several mechanisms proposed in the literature, such as ferromagnetism induced by structural changes [64-66], Ruderman-Kittel-Kasuya-Yosida (RKKY) interaction, super-exchange, double exchange between the *d* states of transition metal ions, free-carried-mediated exchange and *sp-d* exchange mechanism etc. [67,68]. However, all of these proposed theories cannot well accord with this present experimental results. In a recent paper, Coey *et al.* [20] proposed a ferromagnetic exchange mechanism involving oxygen vacancies (V_O), which form F-centers with trapped electron, for the observed ferromagnetism in transition metal-doped oxide semiconductors. Overlap of the F-center electron orbitals with the *d* orbitals of the neighboring transition metal spins (TMs) to form TMs-V_O-TMs groups is crucial for the proposed ferromagnetism coupling [20]. From the discussion above, the ferromagnetic exchange mechanism involving oxygen vacancies model is an appropriate explanation for the observed RTFM behavior in our Co-doped samples. It would indicate that the oxygen vacancies are present in our transition metal doped ZnO.

4. Summary

In summary, Co-doped ZnO particles have been synthesized using the co-precipitation method and we have characterized their structural, optical and magnetic properties as a function of doping concentrations. No secondary phase was observed within the sensitifity of X-ray diffraction method. The optical band gap was found to decrease with increasing doping concentrations, indicating a clear red shift. All samples were found to exhibit room-temperature ferromagnetism. The specific magnetization-values were found to increase with increasing Co concentrations. From the ESR spectra in our dopant range, we obtained a broad line with g-value range from 2.23 to 2.17 attributed to ferromagnetic resonance arising from Co^{2+} ions. This result is consistent with the result obtained from VSM and UV-Vis measurements.

REFERENCES

[1] P. Sharma, A. Gupta, K. V. Rao, F. J. Owens, R. Sharma, R. Ahuja, J. M. Osorio and B. Johansson, "Ferromagnetism above Room Temperature in Bulk and Transparent Thin Films of Mn-Doped ZnO," *Nature Matter*, Vol. 2, No. 23, 2003, pp. 673-677.

[2] H.-W. Zhang, Z.-R. Wei, Z.-Q. Li and G.-Y. Dong, "Room-Temperature Ferromagnetism in Fe-Doped, Fe- and Cu-Codoped ZnO Diluted Magnetic Semiconductor," *Materials Letters*, Vol. 61, No. 17, 2007, pp. 3605-3607.

[3] J. Hays, K. M. Reddy, N. Y. Graces, M. H. Engelhard, V. Shutthanandan, M. Luo, C. Xu, N. C. Giles, C. Wang, S. Thevuthasan and A. Punnoose, "Effect of Co Doping on the Structural, Optical and Magnetic Properties of ZnO Nanoparticles," *Journal of Physics: Condensed Matter*, Vol. 19, No. 26, 2007, Article ID: 266203.

[4] K. Srinivas, S. M. Rao and P. V. Reddy, "Preparation and Properties of $Zn_{0.9}Ni_{0.1}O$ Diluted Magnetic Semiconductor Nanoparticles," *Journal of Nanoparticle Research*, Vol. 13, No. 2, 2011, pp. 817-837.

[5] N. Tahir, S. T. Hussain, M. Usman, S. K. Hasanain and A. Mumtaz, "Effect of Vanadium Doping on Structural, Magnetic, and Optical Properties of ZnO Nanoparticles," *Applied Surface Science*, Vol. 255, No. 20, 2009, pp. 8506-8510.

[6] J. L. Costa-Krämer, F. Briones, J. F. Fernández, A. C. Caballero, M. Villegas, M. Díaz, M. A. García and A. Hernando, "Nanostructure and Magnetic Properties of the MnZnO System, a Room Temperature Magnetic Semiconductor?" *Nanotechnology*, Vol. 16, No. 2, 2005, p. 214.

[7] D. Karmakar, S. K. Mandal, R. M. Kadam, P. L. Paulose, A. K. Rajarajan, T. K. Nath, A. K. Das, I. Dasgupta and G. P. Das, "Ferromagnetism in Fe-Doped ZnO Nanocrystals: Experiment and Theory," *Physical Review B*, Vol. 75, No. 14, 2007, Article ID: 144404.

[8] Y. K. Lakshmi, K. Srinivas, B. Sreedhar, M. M. Raja, M. Vithal and P. V. Reddy, "Structural, Optical, and Magnetic Properties of Nanocrystalline $Zn_{0.9}Co_{0.1}O$-Based Diluted Magnetic Semiconductors," *Materials Chemistry and Physics*, Vol. 113, No. 2-3, 2009, pp. 749-755.

[9] J. J. Lu, T. C. Lin, S. Y. Tsai, T. S. Mo and K. J. Gan, "Structural, Magnetic and Transport Properties of Ni-Doped ZnO Films," *Journal of Magnetism and Magnetic Materials*, Vol. 323, No. 6, 2011, pp. 829-832.

[10] D. B. Buchholz, R. P. H. Chang, J.-Y. Song and J. B. Ketterson, "Room-Temperature Ferromagnetism in Cu-Doped ZnO Thin Films," *Applied Physics Letters*, Vol. 87, No. 8, 2005, Article ID: 082504.

[11] J. H. Li, D. Z. Shen, J. Y. Zhang, D. X. Zhao, B. S. Li, Y. M. Lu, Y. C. Liu and X. W. Fan, "Magnetism Origin of Mn-Doped ZnO Nanoclusters," *Journal of Magnetism and Magnetic Materials*, Vol. 302, No. 1, 2006, pp. 118-121.

[12] S. Banerjee, K. Rajendran, N. Gayathri, M. Sardar. S. Senthilkumar and V. Sengodan, "Change the Room Temperature Magnetic Property of ZnO upon Mn Doping," *Journal of Applied Physics*, Vol. 104, No. 4, 2008, Article ID: 043913.

[13] K. Sato and H. Katayama-Yoshida, "Ferromagnetism in a Transition Metal Atom Doped ZnO," *Physica E*, Vol. 10, No. 1-3, 2001, pp. 251-255.

[14] T. Dietl, H. Ohno, F. Matsukura, J. Cibert and D. Ferrand, "Zener Model Description of Ferromagnetism in Zinc-Blende Magnetic Semiconductors," *Science*, Vol. 287, No. 5455, 2000, pp. 1019-1022.

[15] J. H. Park, M. G. Kim, H. M. Jang, S. Ryu and Y. M. Kim, "Co-Metal Clustering as the Origin of Ferromagnetism in Co-Doped ZnO Thin Films," *Applied Physics Letters*, Vol. 84, No. 8, 2004, pp. 1338-1340.

[16] S. A. Chambers, T. Droubay, C. M. Wang, A. S. Lea, R. F. C. Farrow, L. Folks, V. Deline and S. Anders, "Clusters and Magnetism in Epitaxial Co-Doped TiO_2 Anatase," *Applied Physics Letters*, Vol. 82, No. 8, 2003, p.

1257.

[17] M. Venkatesan, C. B. Fitzgerald and J. M. D. Coey, "Thin Films: Unexpected Magnetism in a Dielectric Oxide," *Nature*, Vol. 430, No. 7000, 2004, p. 630.

[18] A. Sundaresan, R. Bhargavi, N. Rangarajan, U. Siddesh and C. N. R. Rao, "Ferromagnetism as a Universal Feature of the Otherwise Nonmagnetic Oxides," *Physical Review B*, Vol. 74, No. 16, 2006, Article ID: 161306.

[19] N. H. Hong, J. Sakai and V. Brize, "Observation of Ferromagnetism at Room Temperature in ZnO Thin Films," *Journal of Physics: Condensed Matter*, Vol. 19, No. 3, 2007, Article ID: 036219.

[20] J. M. Coey, M. Venkatesan and C. B. Fitzgerald, "Donor Impurity Band Exchange in Dilute Ferromagnetic Oxides," *Nature Matter*, Vol. 4, No. 4, 2005, pp. 173-179.

[21] L. B. Duan, G. H. Rao, J. Yu, Y. C. Wang, W. G. Chu and L. N. Zhang, "Structural and Magnetic Properties of $Zn_{1-X}Mn_XO$ ($0 \leq X \leq 0.40$) Nanoparticles," *Journal of Applied Physics*, Vol. 102, No. 10, 2007, Article ID: 103907.

[22] M. El-Hilo and A. A. Dakhel, "Structural and Magnetic Properties of Mn-Doped ZnO Powders," *Journal of Magnetism and Magnetic Materials*, Vol. 323, No. 16, 2011, pp. 2202-2205.

[23] S. J. Han, J. W. Song, C. H. Yang, S. H. Park, Y. H. Jeong and K. W. Rhie, "A Key to Room-Temperature Ferromagnetism in Fe Doped ZnO:Cu," *Applied Physics Letters*, Vol. 81, No. 22, 2002, p. 4212.

[24] X. X. Liu, F. T. Lin, L. L. Sun, W. J. Cheng, X. M. Ma and W. Z. Shi, "Doping Concentration Dependence of Room-Temperature Ferromagnetism for Ni-Doped ZnO Thin Films Prepared by Pulsed-Laser Deposition," *Applied Physics Letters*, Vol. 88, No. 6, 2006, Article ID: 062508.

[25] M. El-Hilo, A. A. Dakhel and A. Y. Ali-Mohamed, "Room Temperature Ferromagnetism in Nanocrystalline Ni-Doped ZnO Synthesized by Co-Precipitation," *Journal of Magnetism and Magnetic Materials*, Vol. 321, No. 24, 2009, pp. 2279-2283.

[26] O. D. Jayakumar, H. G. Salunke, R. M. Kadam, M. Mohapatra, G. Yaswant and S. K. Kulshreshtha, "Magnetism in Mn-Doped ZnO Nanoparticles Prepared by A Co-Precipitation Method," *Nanotechnology*, Vol. 17, No. 5, 2006, pp. 1278-1285.

[27] S. P. Prakoso and R. Saleh, "Hydrogen Incorporation in Undoped ZnO Nanoparticles," *World Journal of Condensed Matter Physics*, Vol. 1, No. 4, 2011, pp. 130-136.

[28] M. A. White, S. T. Ochsenbein and D. R. Gamelin, "Colloidal Nanocrystals of Wurtzite $Zn_{1-X}Co_XO$ ($0 \leq x \leq 1$): Models of Spinodal Decomposition in an Oxide Diluted Magnetic Semiconductor," *Chemistry of Materials*, Vol. 20, No. 22, 2008, pp. 7107-7116.

[29] Y. Yu, G. B. Ji, J. M. Cao, J. S. Liu and M. B. Zheng, "Facile Synthesis, Characterization and Electrochemical Properties of Cuspate Deltoid CoO Crystallites," *Journal of Alloys and Compounds*, Vol. 471, No. 1-2, 2009, pp. 268-271.

[30] C. J. Cong, L. Liao, Q. Y. Liu, J. C. Li and K. L. Zhang, "Effects of Temperature on the Ferromagnetism of Mn-Doped ZnO Nanoparticles and Mn-Related Raman Vibration," *Nanotechnology*, Vol. 17, No. 5, 2006, p. 1520.

[31] S. Senthilkumaar, K. Rajendra, S. Banerjee, T. K. Chini and V. Sengodan, "Influence of Mn Doping on the Microstructure and Optical Property of ZnO," *Material Scince Semiconductor Processing*, Vol. 11, No. 1, 2008, pp. 6-12.

[32] S. Maensiri, P. Laokul and S. Phokha, "A Simple Synthesis and Magnetic Behavior of Nanocrystalline $Zn_{0.9}Co_{0.1}O$ Powders by Using Zn and Co Acetates and Polyvinyl Pyrrolidone as Precursors," *Journal of Magnetism and Magnetic Materials*, Vol. 305, No. 2, 2006, pp. 381-387.

[33] A. Hernández, L. Maya, E. Sánchez-Mora and E. M. Sánchez, "Sol-Gel Synthesis, Characterization and Photocatalytic Activity of Mixed Oxide ZnO-Fe_2O_3," *Journal of Sol-Gel Science and Technology*, Vol. 42, No. 1, 2007, pp. 71-78.

[34] J. Das, I. R. Evans and D. Khushalani, "Zinc Glycolate: A Precursor to ZnO," *Inorganic Chemistry*, Vol. 48, No. 8, 2009, pp. 3508-3510.

[35] S. Maensiri, P. Laokul and J. Klinkaewnarong, "A Simple Synthesis and Room-Temperature Magnetic Behavior of Co-Doped Anatase TiO_2 Nanoparticles," *Journal of Magnetism and Magnetic Materials*, Vol. 302, No. 2, 2006, pp. 448-453.

[36] Y. F. Chen, C. Y. Lee, M. Y. Yeng and H. T. Chiu, "The Effect of Calcination Temperature on the Crystallinity of TiO_2 Nanopowders," *Journal of Crystal Growth*, Vol. 247, No. 3-4, 2003, pp. 363-370.

[37] S. P. Prakoso and R. Saleh, "Synthesis and Spectroscopic Characterization of Undoped Nanocrytalline ZnO Particles Prepared by Co-Precipitation," *Materials Sciences and Applications*, Vol. 3, No. 8, 2012, pp. 530-537.

[38] R. Saleh, L. Munisa and W. Beyer, "Infrared Absorption in a-SiC:H Alloy Prepared by d.c. Sputtering," *Thin Solid Films*, Vol. 426, No. 1-2, 2003, pp. 117-123.

[39] M. O. Manasreh, J. M. Baranowski, K. Pakula, X. X. Jiang and J. Lin, "Localized Vibrational Modes of Carbon-Hydrogen Complexes in GaN," *Applied Physics Letter*, Vol. 75, No. 5, 1999, pp. 659-661.

[40] D. M. Joseph, R. Balagopal, R. F. Hicks, L. P. Sadwick and K. L. Wang, "Observation of Carbon Incorporation during Gallium Arsenide Growth by Molecular Beam Epitaxy," *Applied Physics Letter*, Vol. 53, No. 22, 1988,

p. 2203.

[41] M. Tortosa, M. Mollar, B. Mari and F. Lloret, "Optical and Magnetic Properties of ZnCoO Thin Films Synthesized by Electrodeposition," *Journal of Applied Physics*, Vol. 104, No. 3, 2008, Article ID: 033901.

[42] A. Dinia, J. P. Ayoub, G. Schmerber, E. Beaurepaire, D. Muller and J. J. Grob, "Effect of Ion Irradiation on the Structural and the Magnetic Properties of $Zn_{0.75}Co_{0.25}O$ Magnetic Semiconductors," *Physics Letters A*, Vol. 333, No. 1-2, 2004, pp. 152-156.

[43] G. L. Liu, Q. Cao, J. X. Deng, P. F. Xing, Y. F. Tian, Y. X. Chen, S. S. Yan and L. M. Mei, "High TC Ferromagnetism of $Zn_{1-x}Co_xO$ Diluted Magnetic Semiconductors Grown by Oxygen Plasma-Assisted Molecular Beam Epitaxy," *Applied Physics Letters*, Vol. 90, No. 6, 2007, Article ID: 052504.

[44] S. Ramachandran, A. Tiwari and J. Narayan, "$Zn_{0.9}Co_{0.1}O$-Based Diluted Magnetic Semiconducting Thin Films," *Applied Physics Letters*, Vol. 84, 2004, p. 5255.

[45] P. Koidl, "Optical Absorption of Co^{2+} in ZnO," *Physical Review B*, Vol. 15, No. 5, 1977, pp. 2493-2499.

[46] S. D. Kshirsagar, D. Inamdar, I. K. Gopalakrishnan, S. K. Kulshreshtha and S. Mahamuni, "Formation of Room-Temperature Ferromagnetic $Zn_{1-X}Co_XO$ Nanocrystals," *Solid State Communications*, Vol. 143, No. 10, 2007, pp. 457-460.

[47] Y. Z. Yoo, T. Fukumura, Z. Jin, K. Hasegawa, M. Kawasaki, P. Ahmet, T. Chikyow and H. Koinuma, "ZnO-CoO Solid Solution Thin Films," *Journal of Applied Physics*, Vol. 90, No. 8, 2001, p. 4246.

[48] C. J. Cong, J. H. Hong and K. L. Zhang, "Effect of Atmosphere on the Magnetic Properties of the Co-Doped ZnO Magnetic Semiconductors," *Materials Chemistry and Physics*, Vol. 113, No. 1, 2009, pp. 435-440.

[49] S. Yang, B. Y. Man, M. Liu, C. S. Chen, X. G. Gao, C. C. Wang and B. Hu, "Structural, Optical and Magnetic Properties of $Zn_{1-X}Co_XO$ Dilute Magnetic Semiconductors Thin Films by Pulsed Laser Deposition," *Physica B*, Vol. 405, No. 18, 2010, pp. 4027-4031.

[50] H. Yang and S. Nie, "Preparation and Characterization of Co-Doped ZnO Nanomaterials," *Materials Chemistry and Physics*, Vol. 114, No. 1, 2009, pp. 279-282.

[51] R. Elilaressi and G. Chandrasekaran, "Structural, Optical, and Magnetic Properties of Nanoparticles of ZnO:Ni—DMS Prepared by Sol-Gel Method," *Materials Chemistry Physics*, Vol. 123, No. 2-3, 2010, pp. 450-455.

[52] R. Bhargava, P. K. Sharma, A. K. Chawla, S. Kumar, R. Chandra, A. C. Pandey and N. Kumar, "Variation in Structural, Optical and Magnetic Properties of $Zn_{1-X}Cr_XO$ (x = 0.0, 0.10, 0.15, and 0.20) Nanoparticles: Role of Dopant Concentration on Non-Saturation of Magnetization," *Materials Chemistry and Physics*, Vol. 125, No. 3, 2011, pp. 664-671.

[53] M. Bouloudenine, N. Viart, S. Colis and A. Dinia, "Bulk $Zn_{1-X}Co_XO$ Magnetic Semiconductors Prepared by Hydrothermal Technique," *Chemical Physics Letters*, Vol. 397, No. 1-3, 2004, pp. 73-76.

[54] K. J. Kim and Y. R. Park, "Spectroscopic Ellipsometry Study of Optical Transitions in $Zn_{1-X}Co_XO$ Alloys," *Applied Physics Letters*, Vol. 81, No. 8, 2002, pp. 1420-1422.

[55] A. J. Hays, A. Punnoose, R. Baldner, M. H. Engelhard, J. Peloquin and K. M. Reddy, "Relationship between the Structural and Magnetic Properties of Co-Doped SnO_2 Nanoparticles," *Physical Review B*, Vol. 72, No. 7, 2005, pp. 075203-1-075203-7.

[56] J. C. Pivin, G. Socol, I. Mihailescu, P. Berthet, F. Singh, M. K. Patel and L. Vincent, "Structure and Magnetic Properties of ZnO Films Doped with Co, Ni or Mn Synthesized by Pulsed Laser Deposition under Low and High Oxygen Partial Pressures," *Thin Solid Films*, Vol. 517, No. 2, 2008, pp. 916-922.

[57] P. Sati, R. Hayn, R. Kuzian, S. Regnier, S. Schafer, A. Stepanov, C. Morhain, C. Deparis, M. Laugt, M. Goiran and Z. Golacki, "Magnetic Anisotropy of Co^{2+} as Signature of Intrinsic Ferromagnetism in ZnO:Co," *Physical Review Letters*, Vol. 96, No. 1, 2006, Article ID: 017203.

[58] O. D. Jayakumar, H. G. Salunke, R. M. Kadam, M. Mohapatra, G. Yaswant and S. K. Kulshreshtha, "Magnetism in Mn-Doped ZnO Nanoparticles Prepared by a Co-Precipitation Method," *Nanotechnology*, Vol. 17, No. 5, 2006, pp. 1278-1285.

[59] P. Sharma, A. Gupta, K. V. Rao, F. J. Owens, R. Sharma, R. Ahuja, J. M. O. Guillen, B. Johansson and G. A. Gehring, "Ferromagnetism above Room Temperature in Bulk and Transparent Thin Films of Mn-Doped ZnO," *Natuew Materials*, Vol. 2, 2003, pp. 673-677.

[60] N. S. Norberg, K. R. Kittilstved, J. E. Amonette, R. K. Kukkadapu, D. A. Schwartz and D. R. Gamelin. *Journal of the American Chemical Society*, Vol. 12, 2004, Article ID: 69387.

[61] K. Ueda, H. Tabata and T. Kawai, "Magnetic and Electric Properties of Transition-Metal-Doped ZnO Films," *Applied Physics Letters*, Vol. 79, No. 7, 2001, p. 988.

[62] D. R. Lide, "CRC Handbook of Chemistry and Physics," 84th Edition, CRC Press, Boca Raton, 2003, pp. 12-125.

[63] K. Ariga, M. Li, G. J. Richards and J. P. Hill, "Nanoarchitectonics: A Conceptual Paradigm for Design and Synthesis of Dimension Controlled Functional Nanomaterials," *Journal of Physical Chemistry*, Vol. 11, No. 1,

2011, pp. 1-13.

[64] N. H. Duc and T. Goto, "Handbook on the Physics and Chemistry of Rare Earthsvol," K. A. Gschneidner and L. Eyring, Eds., Elsevier, Amsterdam, 1999, p. 178.

[65] T. Goto, K. Fukamichi and H. Yamada. *Physica B*, Vol. 3, 2001, Article ID: 00167.

[66] E. P. Wohlfarth and P. Rhodes, "Collective Electron Metamagnetism," *Philosophical Magazine*, Vol. 7, No. 83, 1962, pp. 1817-1824.

[67] B. Pal and P. K. Giri, "Defect Mediated Magnetic Interaction and HighTc Ferromagnetism in Co Doped ZnO Nanoparticles," *Journal of Nanoscience and Nanotechnology*, Vol. 11, 2011, pp. 1-8.

[68] V. A. Fonoberov, K. A. Alim, A. A. Balandin, F. Xiu and J. Liu, "Photoluminescence Investigation of the Carrier Recombination Processes in ZnO Quantum Dots and Nanocrystals," *Physical Review B*, Vol. 73, No. 16, 2006, Article ID: 165317.

10

Preparation of Titanium Phosphates with Sodium Triphosphate in Hydrothermal Process and Their Powder Properties for Cosmetics

Hiroaki Onoda*, Taisuke Yamaguchi
Department of Informatics and Environmental Sciences, Kyoto Prefectural University, Kyoto, Japan

ABSTRACT

As a white pigment, titanium oxide is used for cosmetic application. This oxide is well known to have the photo catalytic activity. Therefore, a certain degree of sebum on the skin is decomposed by the ultraviolet radiation in sunlight. In this work, titanium phosphates were prepared with titanium chloride and sodium triphosphate in hydrothermal process as a novel white pigment. Their chemical composition, powder properties, photo catalytic activity, color phase, and smoothness were studied. The obtained phosphates had little photo catalytic activity. The particle size of phosphate pigments consisted of a normal distribution at 2 μm by the hydrothermal treatment for 4 and 8 hours. The slipping resistance of samples became small by the hydrothermal treatment.

Keywords: Titanium Phosphates; Condensed Phosphate; Hydrothermal Process; Photo Catalytic Activity; Smoothness

1. Introduction

As a white pigment, titanium dioxide is used for cosmetic application [1]. This oxide is well known to have the photo catalytic activity. Therefore, a certain degree of sebum on the skin is decomposed by the ultraviolet radiation in sunlight. To repress this effect, some kinds of the technical process are investigated and used. For example, as one of these techniques, the composite particles with silicon oxide are used [2]. However, these particles are hard materials for human face. The mild materials for human face are required as a white pigment. In addition, there was a report that microfine titanium dioxide would be adsorbed through the skin [3]. The novel white pigment that is not adsorbed is required.

Phosphates have been used for ceramic materials, catalysts, adsorbent, fluorescent materials, dielectric substances, biomaterials, for metal surface treatment, as fertilizer, detergents, food additives, in fuel cells, pigments, and other items [4-8]. It is well known that phosphate materials have high affinity for a living body. Therefore, as a novel white pigment, phosphates have a possibility for the cosmetics.

As a pigment for cosmetics, the particle shape and size distribution are important. Sphere homogenized particle are expected to spread well on the skin. However, too much small particles are not suitable because pigments got into the pores in the skin. The hydrothermal process is generally used to control the powder properties of inorganic phosphate materials [9-11]. The time of the hydrothermal treatment had important factor such as temperature, reactant concentration, and pH values. In previous work [12], we prepared the titanium phosphate pigment that had no catalytic activity. However, the obtained particles had the wide distribution from 1 μm to 300 μm in size. Therefore, the purpose in this work is to obtain the homogenized titanium phosphate particles.

Phosphates are transformed to other forms of phosphates by hydrolysis and dehydration reactions at elevated temperatures. These condensed phosphates have the possibility to have novel functional properties. Orthophosphate has been mainly investigated for many applications, on the other hand, condensed phosphates have been little studied.

In this work, titanium phosphates were prepared from titanium chloride and sodium triphosphate solution in hydrothermal process for several hours. Their chemical composition, powder properties, photo catalytic activity, color phases, and smoothness of the obtained precipitates and their thermal products were studied for the cosmetic application.

2. Experimental

Sodium triphosphate, $Na_5P_3O_{10}$, was synthesized by heating the mixture of disodium hydrogen phosphate,

*Corresponding author.

Na_2HPO_4, and sodium dihydrogen phosphate, NaH_2PO_4, at 400˚C for 5 hours. 0.1 mol/l of titanium chloride solution was mixed with 0.1 mol/l of the sodium triphosphate solution in a molar ratio of Ti/P = 5/12. This ratio of Ti/P was determined from the theoretical chemical composition of $Ti_5(P_3O_{10})_4$. The mixed solution was set in the autoclave for 0, 2, 4, 6, 8, and 12 hours at 120˚C and 100 kPa (hydrothermal treatment). Then, the precipitates were filtered off, washed with water, and dried.

The chemical composition of these materials was analyzed by XRD. XRD patterns were recorded on a Rigaku Denki RINT 2000 X-Ray diffractometer using monochromated CuKα radiation. TG and DTA curves of precipitates were measured with a Shimadzu DTG-60H at a heating rate of 10 K minute^{-1} in air. Samples were heated at 100˚C, 200˚C, 400˚C, 600˚C in air conditions. These thermal products were also analyzed by XRD patterns.

The powder properties of the precipitates and their thermal products were estimated from SEM images and particle size distributions. Scanning electron microscopy (SEM) images of titanium phosphates were observed using JGM-5510LV, JEOL Ltd. The particle size distributions of these materials were measured with laser diffraction/scattering particle size distribution HORIBA LA-910.

The cosmetic properties were estimated by the photo catalytic activity, the color phase, and the smoothness. The photo catalytic activity of samples was estimated with the decomposition of methylene blue by 365 nm radiation [13,14]. The color of phosphate pigments was estimated by ultraviolet-visible (UV-Vis) reflectance spectra with a Shimadzu UV2550. The smoothness of particles was measured on an artificial leather with KES-SE objective evaluation of surface friction property, KATO Tech Co. Ltd.

3. Results and Discussion

3.1. Chemical Composition of Precipitates

Figure 1 shows XRD patterns of samples prepared in hydrothermal process for several hours. All samples had XRD peaks of TiP_3O_9 in spite of the hydrothermal hours [15]. This chemical composition had a different ratio of Ti/P from those prepared under experimental conditions. The chemical composition $P_3O_9^{3-}$ represents a *Cyclo*-triphosphate anion described in a recent report [16]. *Cyclo*-triphosphate was probably not formed in this condition. As inferred from the regularity in phosphate chemistry, titanium *Cyclo*-triphosphate was not synthesized by heating [17]. Another fault was that tetravalent titanium cation reduced to trivalent cation. We infer that this compound was a mixture of $TiO(PO_3)_2$ and $Ti(PO_3)_4$. The previous work was unable to distinguish between TiP_3O_9 and the mixture of $TiO(PO_3)_2$ and $Ti(PO_3)_4$ (for example $Ti_2O(PO_3)_6$. The peak intensity became weak by the hydrothermal treatment. All samples had a certain degree of amorphous phase. By heating at 600˚C, the peaks of TiO_2 appeared in XRD patterns.

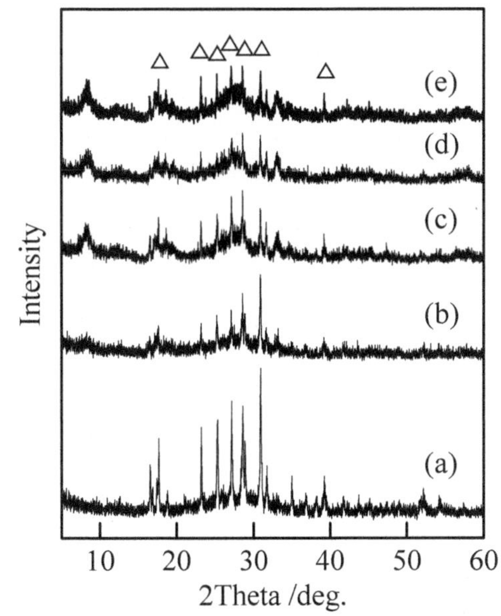

Figure 1. XRD patterns of samples prepared in hydrothermal treatment for various hours, (a) 0, (b) 2, (c) 4, (d) 8, and (e) 12 hours, △; TiP_3O_9.

Figure 2 shows TG curves of samples prepared in hydrothermal process for several hours. The weight losses at 600˚C were 14.4%, 16.7%, 17.9%, 20.1%, and 20.2% in TG curves of samples prepared for 0, 2, 4, 8, and 12 hours, respectively. With the increase of hydrothermal hours, these weight losses below 600˚C became larger. Especially, samples prepared for 8 and 12 hours had the step at 230˚C in TG curve. By the hydrothermal treatment, titanium phosphates became easy to contain water, hydroxide anion, and so on.

Figure 3 shows DTA curves of samples prepared in hydrothermal process for several hours. The endothermic peaks at 70˚C and 110˚C were observed in DTA curves of all samples. These peaks were due to the volatilization of adsorbed and crystalline water. Sample prepared without the hydrothermal treatment had an exothermic peak at 550˚C and an endothermic peak at 590˚C in DTA curve. These peaks were considered from the transition of chemical phases. By the hydrothermal treatment, the transition took place in the exothermic reactions at 590˚C and 680˚C. A small endothermic peak at 270˚C appeared in DTA curves in samples prepared with the hydrothermal treatment. This peak was due to the volatilization of water from the condensation of hydroxide anions.

3.2. Powder Properties of Titanium Phosphates

From the viewpoint of particle shape, spherical particles

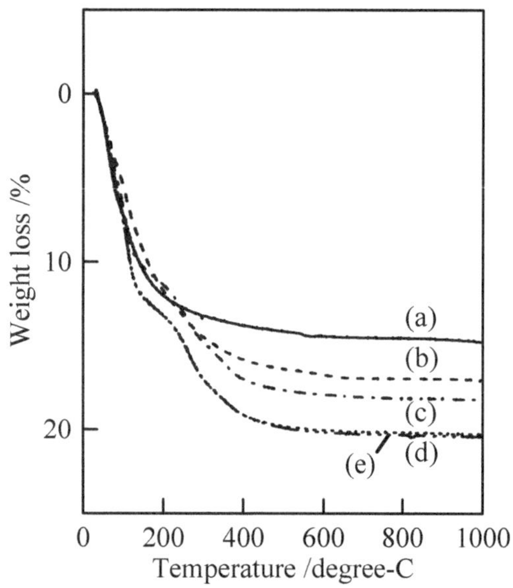

Figure 2. TG curves of samples prepared in hydrothermal treatment for various hours, (a) 0, (b) 2, (c) 4, (d) 8, and (e) 12 hours.

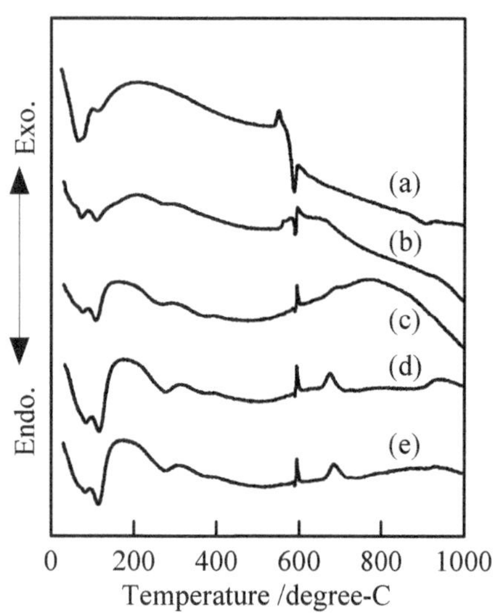

Figure 3. DTA curves of samples prepared in hydrothermal treatment for various hours, (a) 0, (b) 2, (c) 4, (d) 8, and (e) 12 hours.

are suitable for the cosmetic applications. **Figure 4** shows SEM images of samples prepared in hydrothermal process for several hours. All samples had no specified shape in spite of the hydrothermal time. The particle size in SEM images became smaller with the hydrothermal hours.

Figure 5 shows the particle size distribution of samples prepared in hydrothermal process for several hours. The main part of particles was from 0.5 to 400 μm. The ratios at 3 and 30 μm were high in the distribution of sample prepared without the hydrothermal treatment (**Figure 5(a)**). Sample prepared with the hydrothermal treatment

Figure 4. SEM images of samples prepared in hydrothermal treatment for various hours, (a) 0, (b) 2, (c) 4, (d) 8, and (e) 12 hours.

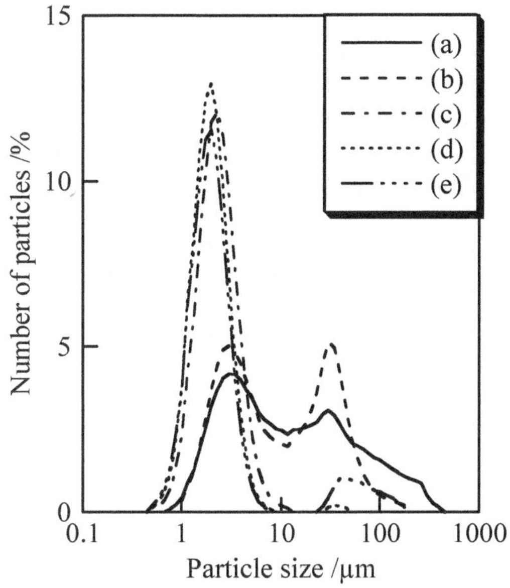

Figure 5. Particle size distribution of samples prepared in hydrothermal treatment for various hours, (a) 0, (b) 2, (c) 4, (d) 8, and (e) 12 hours.

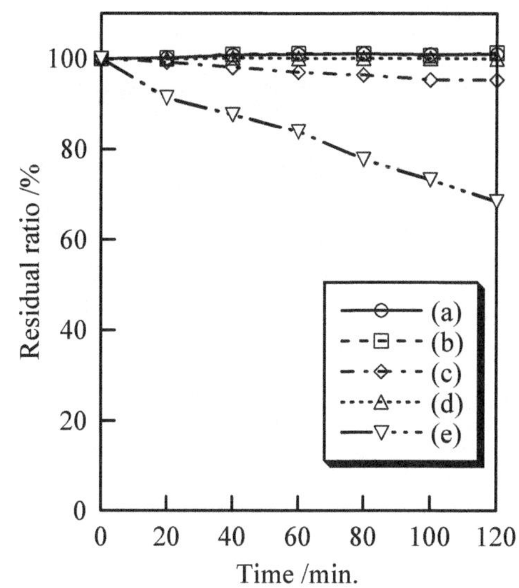

Figure 6. Photo catalytic activity of samples prepared in hydrothermal treatment for various hours and then heated, (a) blank, (b) 0 hour, R.T., (c) 8 hours, R.T., (d) 8 hours, 400°C, and (e) TiO_2.

for 2 hours consisted of two normal distributions at 3 and 30 μm (**Figure 5(b)**). On the other hand, samples treated for 4 and 8 hours had a normal distribution at 2 μm (**Figures 5(c)** and **(d)**). The large particles at 40 μm appeared in particle size distribution of sample treated for 12 hours (**Figure 5(e)**). By heating, the particle size became larger than 10 μm (not shown).

For cosmetic application, small and homogenized particles are suitable. However, too much smaller particles had trouble to get into the pores in the skin [3]. The standard size of the white pigment for cosmetics is difficult to determine, because the size of pore in the skin is determined from age, gender, climate, and so on. Further, too much larger particles are also inappropriate because of the cracking of coating on the skin. It is important to control the particle size of pigments. The hydrothermal treatment has the possibility to control the particle size of titanium phosphate pigments for the cosmetics.

3.3. Cosmetic Properties of Titanium Phosphates

Figure 6 shows the photo catalytic activity of samples prepared in various conditions. When samples had the photo catalytic activity, the residual ratio of methylene blue became lower. This residual ratio was calculated from the adsorption at 660 nm. Because titanium dioxide is used as a white pigment in cosmetics, this compound was evaluated for the comparison with uncoated titanium phosphate [1]. The methylene blue was decomposed with titanium dioxide by UV radiation (**Figure 6(e)**). On the other hand, titanium phosphate had little photo catalytic activity in spite of the hydrothermal time and heating temperature (**Figures 6(b)-(d)**). Titanium phosphate was mild material to protect the sebum on the skin.

Figure 7 shows UV-Vis reflectance spectra of samples prepared in hydrothermal treatment for various hours. All samples and titanium oxide indicated high reflectance at the range of visible light. Titanium oxide adsorbed the ultraviolet light that was shorter than 400 nm in wavelength. On the other hand, samples prepared in the hydrothermal treatment for 4, 8, and 12 hours absorbed the light shorter than 360 nm in wavelength.

As mentioned above, the pigment with high smoothness spreads well on the skin. The smoothness of the powders is also important in cosmetics [18]. **Table 1** shows the smoothness of samples prepared in hydrothermal process for several hours. The values of MIU and MMD represented the slipping resistance and roughness of powders, respectively. Sample powders spread on the article leather, and then a sensor ran on these powders. The values of MIU and MMD were calculated from the power to move a sensor and the pitching of a sensor, respectively. The value of MIU became small with the hydrothermal treatment. Samples had less change in the value of MMD by the hydrothermal treatment. **Table 2** shows the smoothness of samples heated at several temperatures. By heating, the value of MIU became larger, because the heated samples had large and dispersed particles in size. The value of MMD was less affected from the heating temperatures. As a pigment for cosmetics, the heated phosphates were less suitable than sample without heating.

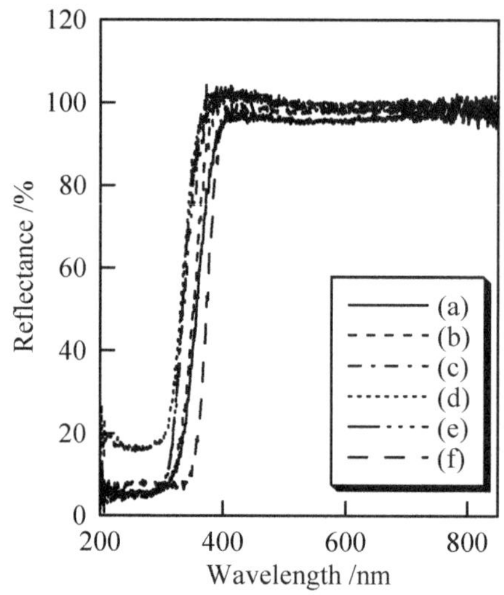

Figure 7. UV-Vis reflectance spectra of samples prepared in hydrothermal treatment for various hours, (a) 0, (b) 2, (c) 4, (d) 8, (e) 12 hours, and (f) TiO_2.

Table 1. Smoothness of samples prepared for various hydrothermal hours.

Hydrothermal time/h	MIU	MMD
0	0.250	0.008
2	0.204	0.006
4	0.201	0.006
8	0.193	0.006
12	0.231	0.006

Table 2. Smoothness of samples (hydrothermal time; 8 hours) heated at several temperatures.

Temperature/°C	MIU	MMD
R.T.	0.193	0.006
100	0.224	0.007
200	0.263	0.007
400	0.258	0.008
600	0.231	0.006

4. Conclusion

Titanium phosphates were obtained from titanium chloride and sodium triphosphate solution. The obtained materials had a certain degree of amorphous phase. These titanium phosphates had less photo catalytic activity to protect the sebum on the skin. The particle size of phosphate pigments consisted of a normal distribution at 2 μm by the hydrothermal treatment for 4 and 8 hours. The slipping resistance of sample powders became small by the hydrothermal treatment.

5. Acknowledgements

The authors gratefully thank Dr. Takeshi Toyama, Nihon University, Japan, for the measurements of the smoothness.

REFERENCES

[1] U. Diebold, "The Surface Science of Titanium Dioxide," *Surface Science Report*, Vol. 48, No. 5-8, 2003, pp. 53-229.

[2] M. Senzuki, T. Tamura, K. Miura, Y. Ikarashi, Y. Watanabe and M. Fujii, "Study on Penetration of Titanium Dioxide (TiO_2) Nanoparticles into Intact and Damaged Skin *in Vitro*," *The Journal of Toxicological Sciences*, Vol. 35, No. 1, 2010, pp. 107-113.

[3] A. O. Gamer, E. Leibold and B. Van Ravenzwaay, "The *in Vitro* Absorption of Microfine Zinc Oxide and Titanium Dioxide through Porcine Skin," *Toxicology in Vitro*, Vol. 20, No. 3, 2006, pp. 301-307.

[4] H. Onoda, H. Nariai, A. Moriwaki, H. Maki and I. Motooka, "Formation and Catalytic Characterization of Various Rare Earth Phosphates," *Journal of Materials Chemistry*, Vol. 12, No. 6, 2002, pp. 1754-1760.

[5] H. Onoda, T. Ohta, J. Tamaki and K. Kojima, "Decomposition of Trifluoromethane over Nickel Pyrophosphate Catalysts Containing Metal Cation," *Applied Catalysis A; General*, Vol. 288, No. 1-2, 2005, pp. 98-103.

[6] D. J. Jones, G. Aptel, M. Brandhorst, M. Jacquin, J. Jimenez-Jimenez, A. Jimenez-Lopez, P. Maireles-Torres, I. Piwonski, E. Rodrigues-Castellon, J. Zajac and J. Roziere, "High Surface Area Mesoporous Titanium Phosphate: Synthesis and Surface Acidity Determination," *Journal of Materials Chemistry*, Vol. 10, No. 8, 2000, pp. 1957-1963.

[7] H. Onoda, H. Matsui and I. Tanaka, "Improvement of Acid and Base Resistance of Nickel Phosphate Pigment by the Addition of Lanthanum Cation," *Materials Science and Engineering B*, Vol. 141, No. 1-2, 2007, pp. 28-33.

[8] A. Bhamik and S. Inagaki, "Mesoporous Titanium Phosphate Molecular Sieves with Ion-Exchange Capacity," *Journal of American Chemical Society*, Vol. 123, No. 4, 2001, pp. 691-696.

[9] C. Xu, J. Lee and A. S. Teja, "Continuous Hydrothermal Synthesis of Lithium Iron Phosphate Particles in Subcritical and Supercritical Water," *Journal of Supercritical Fluids*, Vol. 44, No. 1, 2008, pp. 92-97.

[10] R. Vani, E. K. Girija, K. Elayaraja, S. P. Parthiban, R. Kesavamoorthy and S. N. Kalkura, "Hydrotehrmal Synthesis of Porous Triphasic Hydroxyapatite/(Alpha and Beta) Tricalcium Phosphate," *Journal of Materials Science: Materials in Medicine*, Vol. 20, Suppl. 1, 2009, pp. 43-48.

[11] M. Tomozawa and S. Hiromoto, "Microstructure of Hy-

droxyapatite- and Octacalcium Phosphate-Coatings Formed on Magnesium by a Hydrothermal Treatment at Various pH Values," *Acta Materialia*, Vol. 59, No. 1, 2011, pp. 355-363.

[12] H. Onoda and T. Yamaguchi, "Preparation of Titanium Phosphate with Additives and Powder Properties for Cosmetics," *Materials Science and Applications*, Vol. 3, No. 1, 2012, pp. 18-23.

[13] V. Ramaswamy, N. B. Jagtap, S. Vijayanand, D. S. Bhange and P. S. Awati, "Photocatalytic Decomposition of Methylene Blue on Nanocrystalline Titania Prepared by Different Methods," *Materials Research Bulletin*, Vol. 43, No. 5, 2008, pp. 1145-1152.

[14] P. Du, A. Bueno-Lopez, M. Verbaas, A. R. Almeida, M. Makkee, J. A. Moulijn and G. Mui, "The Effect of Surface OH-Population on the Photocatalytic Activity of Rare Earth-Doped P25-TiO_2 in Methylene Blue Degradation," *Journal of Catalysis*, Vol. 260, No. 1, 2008, pp. 75-80.

[15] W. T. A. Harrison, T. E. Gier and G. D. Stucky, "Titanium (III) Tris(metaphosphate)," *Acta Crystallographica, Section C*, Vol. 50, No. 11, 1994, pp. 1643-1646.

[16] H. Maeda, M. Matsushita, M. Tsuhako, H. Nariai and H. Nakayama, "Phosphorylation of Polyamidoamine Dendrimers with *Cyclo*-Triphosphate in Aqueous Solution," *Phosphorus Research Bulletin*, Vol. 23, 2009, pp. 52-56.

[17] M. T. Averbuch and A. Durif, "Topics in Phosphate Chemistry," World Scientific Publishing, Singapore, 1996.

[18] S. Y. Cheng, C. W. M. Yuen, C. W. Kan, K. K. L. Cheuk, J. C. O. Tang and S. Y. Li, "A Comprehensive Study of Silicone-Based Cosmetic Textile Agent," *Fibers and Polymers*, Vol. 10, No. 1, 2009, pp. 132-140.

11

Optimizing the Conditions for Residual Stress Measurement Using a Two-Dimensional XRD Method with Specimen Oscillation

Osamu Takakuwa, Hitoshi Soyama
Department of Nanomechanics, Graduate School of Engineering, Tohoku University, Sendai, Japan

ABSTRACT

In order to optimize the conditions for residual stress measurement using a two-dimensional X-ray diffraction (2D-XRD) in terms of both efficiency and accuracy. The measurements have been conducted on three stainless steel specimens in this study. The three specimens were processed by annealing, a cavitating jet in air and a disc grinder, with each method introducing different residual stresses at the surface. The specimens were oscillated in the ω-direction, representing a right-hand rotation of the specimen about the incident X-ray beam. The range of the oscillation, $\Delta\omega$, was varied and optimum $\Delta\omega$ was determined. Moreover, combinations of the tilt angle between the specimen surface normal and the diffraction vector, ψ, with the rotation angle about its surface normal, ϕ, have been studied with a view to find the most optimum condition. The results show that the use of ω oscillations is an effective method for improving analysis accuracy, especially for large grain metals. The standard error rapidly decreased with increasing range of the ω oscillation, especially for the annealed specimen which generated strong diffraction spots due to its large grain size.

Keywords: Two-Dimensional XRD; Diffraction Ring; Oscillation; Residual Stress; Stainless Steel

1. Introduction

X-ray diffraction (XRD) is an effective method for the measurement of residual stress in polycrystalline metals. Such measurements are important since residual stress affects the mechanical property of metallic materials, e.g., resistance to fatigue strength [1-3], stress corrosion cracking [4] and hydrogen embrittlement [5-7]. The residual stress measurements need to be conducted with both high accuracy and efficiency both in the laboratory and in situations of practical application. Several XRD methodologies have been developed so as to satisfy the needs mentioned above.

The $\sin^2\psi$ method is one of the most simple and common XRD methods for residual stress measurements and standardized approaches for this method have already been suggested [8,9]. On problem for the $\sin^2\psi$ method, however, is that it is not easy to conduct accurate residual stress measurements for large grain metals when employing only a one-dimensional position sensitive proportional counter (PSPC) or scintillation counter. With such a detector only a few grains are irradiated resulting in a sparse X-ray diffraction pattern and shortage of ψ angular information. In addition, the conventional $\sin^2\psi$ method can measure residual stresses only in one direction per measurement and is therefore only really suitable for metals which do not have a complex stress field, *i.e.*, not including shear stresses. In order to solve this problem, several studies have been conducted on the conventional $\sin^2\psi$ method and based on these a useful additional methodology has been proposed for evaluating the biaxial stress including shear stress [10,11].

Recently, an XRD method for residual stress measurements using a two-dimensional PSPC (2D-PSPC) has been developed. The method focuses on the direct relationship between the stress tensor and diffraction conic section distortion [12]. The method is known as the two-dimensional XRD method (2D-XRD or XRD^2). The 2D-XRD method can measure six components of the stress tensor, *i.e.*, biaxial stress, including shear stress, from the direct relationship between the stress tensor and the distortion of the diffraction conic section obtained directly by the 2D-PSPC. Moreover, this method can also measure the local residual stress in textured material, weakly diffracting material and in regions of small area down to mm^2, since the diffracted X-ray are efficiently detected by the 2D-PSPC.

The fundamentals and details of 2D-XRD have been described well in the literature [13,14] and in a book [15]. However a standard methodology for 2D-XRD has not been established yet, since this method has only been relatively recently applied to residual stress measurements. Therefore the precise conditions used for the measurement, e.g., the rotation angles and exposure times needed to detect the diffraction ring, need to be optimized. The overall optimization process needs to take into account both measurement efficiency and accuracy.

The diffracted X-ray from large grain materials, such as annealed specimens, produces strongly scattered spots in the X-ray diffraction profile and, consequently, a lack of accuracy in the residual stress measurement. This occurs even if the measurement time is prolonged. This can be a problem for the measurements using 2D-XRD. In order to solve this problem, an oscillatory method should be effective and indeed in the conventional $\sin^2\psi$ method, an oscillation of the ψ angle is generally used. However when the ψ angle is oscillated, the range of this oscillation needs to be as small as possible because the incident angle about the lattice plane directly changes during the ψ oscillation. This affects the relationship between the diffraction angle and the ψ angle, *i.e.*, the 2θ-$\sin^2\psi$ relationship. For this reason an oscillation which keeps the ψ angle constant is more preferable.

In this paper residual stress measurements using the 2D-XRD method have been carried out on three specimens made of stainless steel in order to optimize several of the conditions for residual stress measurements, e.g., sample rotation angles and exposure time, with respect to both measurement efficiency and accuracy. The specimens had three types of residual stress as follows: a low stress value with a large grain (by Annealing), an equibiaxial compressive residual stress (by Cavitating jet in air) and an anisotropic tensile residual stress (by Disc grinder). The conditions for the stress measurements were optimized, in terms of the detection time and the number of diffraction ring measurements taken from various angles of the specimen. The optimization process took into account the achievement of both high accuracy and good measurement efficiency.

2. Experimental Apparatus and Procedures

2.1. Introduction of Various Residual Stresses

The material under test was made of JIS SUS316L austenitic stainless steel. The geometry and dimensions of the specimen were 35 mm square and 3 mm thick. Three types of treatment were chosen which introduce different residual stresses into the specimens. The specimens were prepared by annealing, the use of a cavitating jet in air and by using a disc grinder. The annealing treatment releases residual stresses introduced by shape forming and increases the grain size. The use of a cavitating jet in air introduces high equibiaxial compressive residual stresses into the surface layer by impacts due to cavitation bubble collapse. The method is highly effective and has now been applied for practical usage. The disc grinder is one of a number of surface finishing methods that are frequently applied to metals. Its characteristics are that it introduces high anisotropic tensile residual stress at the near surface of the specimen. For the annealed specimen, the sample was heated at 1000 degree Celsius for 1 h, and then furnace cooled. For the specimen prepared using a cavitating jet in air, the conditions for the treatment were same as previously reported [16,17] and for this treatment the processing time per unit length was 1 s/mm. For the sample prepared using the disc grinder, the rotation speed was set to 11,000 rpm and the disc, manufactured by NIPPON RESIBON COOPRATION (A/W36P), had a diameter of 100 mm. The direction of residual stress caused by the disc grinder was defined by the rotating and the scanning direction of the disc as x and y, respectively. After the preparation of the specimens, residual stress measurements were carried out in accordance with following conditions.

2.2. Residual Stress Measurements

The X-ray diffraction measurements were carried out using Cr Ka X-rays from a tube operated at 35 kV and 40 mA through a 0.5 mm diameter total reflection collimetor and with an incident monochromator (D8 DISCOVER, Bruker AXS Inc.). The lattice plane, $(h\ k\ l)$, used was the γ-Fe (2 2 0) plane and the diffraction angle without strain was 128 degrees. The specimen was placed in a diffractometer, the geometry of which is shown in **Figure 1**. The diffraction ring from the specimen was detected at several angles by the 2D-PSPC, with the angles denoted as ϕ and ψ as shown on the figures. These angular values were used to calculate the stress tensor. The fundamentals of the calculation of biaxial stress using the 2D-XRD method are described in detail in ref [15].

2.2.1. Optimizing the Range of Oscillation and Exposure Time

An initial ω angle of 110 degrees was chosen and the specimen was oscillated over a range of $\Delta\omega$ during the measurement. The parameter $\Delta\omega$ was varied between 0 (without oscillation) and 2, 4, 6, 8 and 10 degrees in order to verify the effect of the ω oscillation on the diffraction ring and ultimately on the stress calculation. In addition, the exposure time per frame for the detection of the diffraction ring at a single position (ϕ, ψ), t_e, was also varied as t_e = 60, 90, 120, 150 and 180 sec for each value of

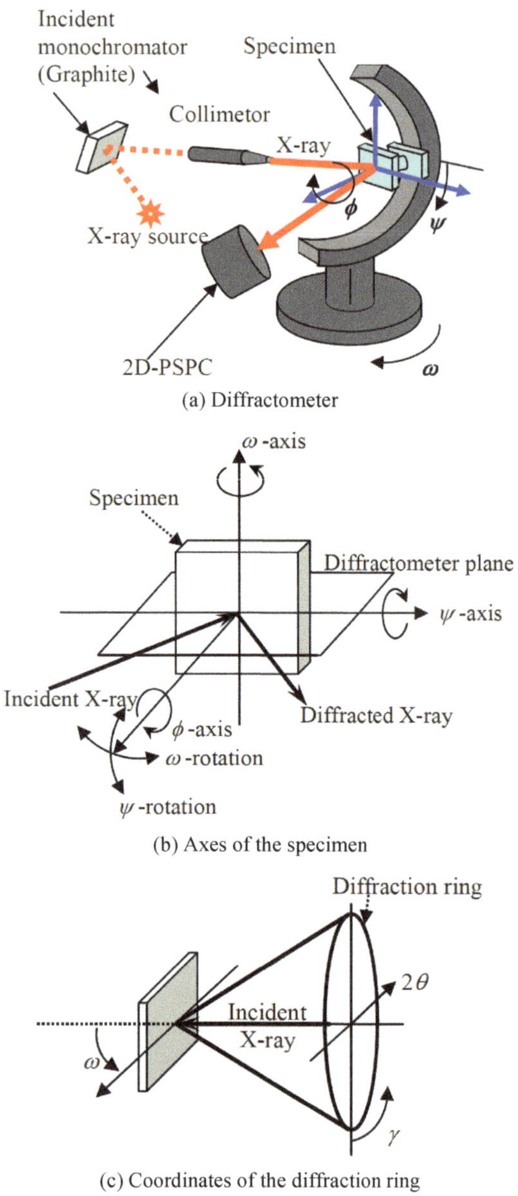

Figure 1. Geometry of the diffractometer, showing the axes of the specimen and the coordinates of the diffraction ring.

$\Delta\omega$ and for each of the three specimens. The ϕ and ψ angles were chosen as shown in **Table 1**. For the stress calculation, the part of the diffraction ring that was extracted ranged from 2θ = 120 - 135 degrees in the radial direction and from γ = 70 - 110 degrees in the circumferential direction of the diffraction ring, as shown in **Figure 2**. This range was divided into 15 sub regions, which were then used to calculate the diffraction ring distortion [13]. One data point on the distorted diffraction ring is generated from each subregion. The peak position 2θ value of the diffraction profile was obtained by fitting the profile with a Pearson type VII function in each subregion. A total of 495 (15 sub regions × 33 frames) data points were taken and the peak position 2θ values

Table 1. Conditions of the specimen rotation angles.

ψ	ϕ
0	0
15	0, 45, 90, 135, 180, 225, 270, 315
30	0, 45, 90, 135, 180, 225, 270, 315
45	0, 45, 90, 135, 180, 225, 270, 315
60	0, 45, 90, 135, 180, 225, 270, 315

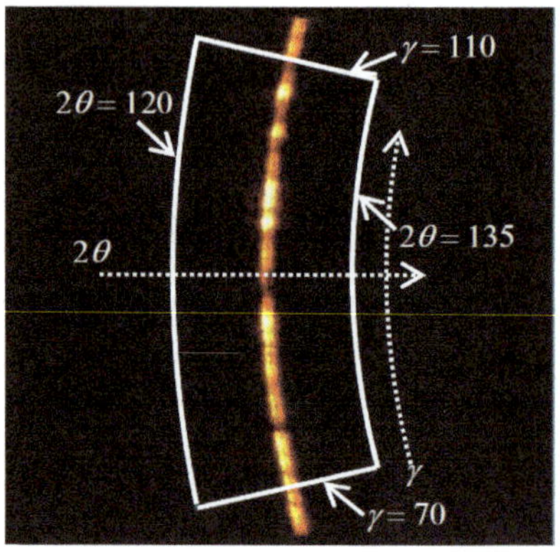

Figure 2. Area of the diffraction ring used in the stress calculation.

were obtained from 33 frames of this measurement. The biaxial stress was calculated from these data points. From these measurements, the optimum range of ω oscillation, $\Delta\omega_{opt}$, and exposure time, $t_{e\ opt}$, were determined.

2.2.2. Optimizing the Range of Oscillation and Exposure Time

For the biaxial residual stress measurement at least six different diffraction rings are needed, taken in appropriate directions [15]. Needless to say, an increase in the number of frames, *i.e.*, diffraction rings, obtained from multiple angles is a way to realize accurate measurements. The number of frames was 33 for the combination of the ϕ and the ψ angles, as shown in **Table 1** in Subsection 2.2.1. This is a substantial number of frames compared to typical measurements [13]. However, the increase in the number of frames directly increases the total measurement time, resulting in a decrease in the measurement efficiency. The number of frames and the ϕ and ψ angles need to be chosen with a consideration of the efficiency, whilst also taking account of accuracy required. Therefore, the residual stress was calculated with several combinations of the ϕ and ψ angles, as shown in **Table 2** and **Figure 3**, using the diffraction data obtained at $\Delta\omega = \Delta\omega_{opt}$ and $t_e = t_{e\ opt}$. The black points in **Figure 3** were chosen for the stress calculations and the white ones represent points that were removed from the calculation. The stress calculations were done using the method mentioned in Subsection 2.2.1. The combination of ϕ and ψ angles was optimized by comparing the residual stress value of the baseline calculated

Table. 2. Conditions of the specimen rotation angles.

(a) Condition 1 (25 frames)

ψ	ϕ
0	0
15	0, 45, 90, 135, 180, ~~225~~, 270, ~~315~~
30	0, 45, 90, 135, 180, ~~225~~, 270, ~~315~~
45	0, 45, 90, 135, 180, ~~225~~, 270, ~~315~~
60	0, 45, 90, 135, 180, ~~225~~, 270, ~~315~~

(b) Condition 2 (21 frames)

ψ	ϕ
0	0
15	0, 45, 90, 135, 180, ~~225~~, ~~270~~, ~~315~~
30	0, 45, 90, 135, 180, ~~225~~, ~~270~~, ~~315~~
45	0, 45, 90, 135, 180, ~~225~~, ~~270~~, ~~315~~
60	0, 45, 90, 135, 180, ~~225~~, ~~270~~, ~~315~~

(c) Condition 3 (17 frames)

ψ	ϕ
0	0
15	0, ~~45~~, 90, ~~135~~, 180, ~~225~~, 270, ~~315~~
30	0, ~~45~~, 90, ~~135~~, 180, ~~225~~, 270, ~~315~~
45	0, ~~45~~, 90, ~~135~~, 180, ~~225~~, 270, ~~315~~
60	0, ~~45~~, 90, ~~135~~, 180, ~~225~~, 270, ~~315~~

(d) Condition 4 (13 frames)

ψ	ϕ
0	0
15	0, ~~45~~, 90, ~~135~~, 180, ~~225~~, ~~270~~, ~~315~~
30	0, ~~45~~, 90, ~~135~~, 180, ~~225~~, ~~270~~, ~~315~~
45	0, ~~45~~, 90, ~~135~~, 180, ~~225~~, ~~270~~, ~~315~~
60	0, ~~45~~, 90, ~~135~~, 180, ~~225~~, ~~270~~, ~~315~~

(e) Condition 5 (17 frames)

ψ	ϕ
0	0
30	0, 45, 90, 135, 180, 225, 270, 315
60	0, 45, 90, 135, 180, 225, 270, 315

(f) Condition 6 (13 frames)

ψ	ϕ
0	0
30	0, 45, 90, 135, 180, ~~225~~, 270, ~~315~~
60	0, 45, 90, 135, 180, ~~225~~, 270, ~~315~~

(g) Condition 7 (11 frames)

ψ	ϕ
0	0
30	0, 45, 90, 135, 180, ~~225~~, ~~270~~, ~~315~~
60	0, 45, 90, 135, 180, ~~225~~, ~~270~~, ~~315~~

(h) Condition 8 (9 frames)

ψ	ϕ
0	0
30	0, ~~45~~, 90, ~~135~~, 180, ~~225~~, 270, ~~315~~
60	0, ~~45~~, 90, ~~135~~, 180, ~~225~~, 270, ~~315~~

(i) Condition 9 (7 frames)

ψ	ϕ
0	0
30	0, ~~45~~, 90, ~~135~~, 180, ~~225~~, ~~270~~, ~~315~~
60	0, ~~45~~, 90, ~~135~~, 180, ~~225~~, ~~270~~, ~~315~~

by the conditions shown in **Table 1**.

3. Results and Discussion

3.1. Optimum Range of Oscillation and Exposure Time

Figures 4-6 plot the biaxial residual stress, σ_{Rx}, σ_{Ry}, and the standard errors, $\Delta\sigma_{Rx}$, $\Delta\sigma_{Ry}$, varying with the exposure time, t_e, as a function of the range of the ω oscillation, $\Delta\omega$, for the annealed specimen (AN), the specimen treated by means of a cavitating jet in air (CJA) and the sample processed using a disc grinder (DG), respectively. The negative value on the plots represents a compressive stress. The standard error values $\Delta\sigma_{Rx}$ and $\Delta\sigma_{Ry}$ are almost the same for all the specimens, since the condition for these measurements can obtain the diffraction ring distortion from multi-symmetrical directions. In **Figure 4**, the residual stresses σ_{Rx} and σ_{Ry} are quite variable, for instance, σ_{Rx} varies from −150 to 30 MPa without oscillation ($\Delta\omega$ = 0 degree) and the standard error is over 200 MPa. This error does not decrease in spite of an increase in the exposure time, e.g., in the case of t_e = 180 s. In contrast, the values of σ_{Rx} and σ_{Ry} gradually converge with σ_{Rx} = −20 MPa and σ_{Ry} = −25 MPa with an increase in the range of the oscillation, $\Delta\omega$. Moreover, the standard error, $\Delta\sigma_{Rx}$, and, $\Delta\sigma_{Ry}$, rapidly decreases along with increasing $\Delta\omega$ and then saturates at $\Delta\omega$ = 8 degrees for each t_e. These results will be discussed later.

In **Figures 5** and **6** it can be seen that high equibiaxial compressive residual stresses, e.g., σ_{Rx} = −381 and σ_{Ry} = −354 MPa at $\Delta\omega$ = 8 degree for t_e = 120 s, were introduced by CJA processing and high anisotropic tensile residual stresses, e.g., σ_{Rx} = 657 and σ_{Ry} = 196 MPa at $\Delta\omega$ = 8 degree for t_e = 120 s, were introduced by DG processing into each specimen. The values of σ_{Rx} and σ_{Ry} for both the CJA and DG specimens does not fluctuate regardless of the values of $\Delta\omega$ and t_e, unlike those of the AN specimen in **Figure 4**. For the CJA specimen shown in **Figure 5**, $\Delta\sigma_{Rx}$ and $\Delta\sigma_{Ry}$ slightly decrease with increasing $\Delta\omega$ for each t_e. The standard error exhibits a small value, e.g., $\Delta\sigma_{Rx}$ = 6 MPa at $\Delta\omega$ = 8 degree for t_e = 120 s without oscillation, except for the case of $\Delta\sigma_{Rx}$ = 28 MPa at t_e = 60 s. For the DG specimen as shown in **Figure 6**, the $\Delta\sigma_{Rx}$ and $\Delta\sigma_{Ry}$ values stay constant regardless of an increase in $\Delta\omega$ for each t_e. The standard error also exhibits a low value, e.g., $\Delta\sigma_{Rx}$ = 6 MPa at $\Delta\omega$ = 8 degree for t_e = 120 s in the case without oscillation. The ω oscillation has a favorable effect on the residual stress measurements, considerably improving those for the AN specimen and then showing a decreasing improvement for the CJA and DG specimens, respectively. From these results, the optimum range of the ω oscillation, $\Delta\omega_{opt}$, and of the exposure time, $t_{e\ opt}$, can be determined as $\Delta\omega_{opt}$ = 8 degree and $t_{e\ opt}$ = 120 s, respectively in this

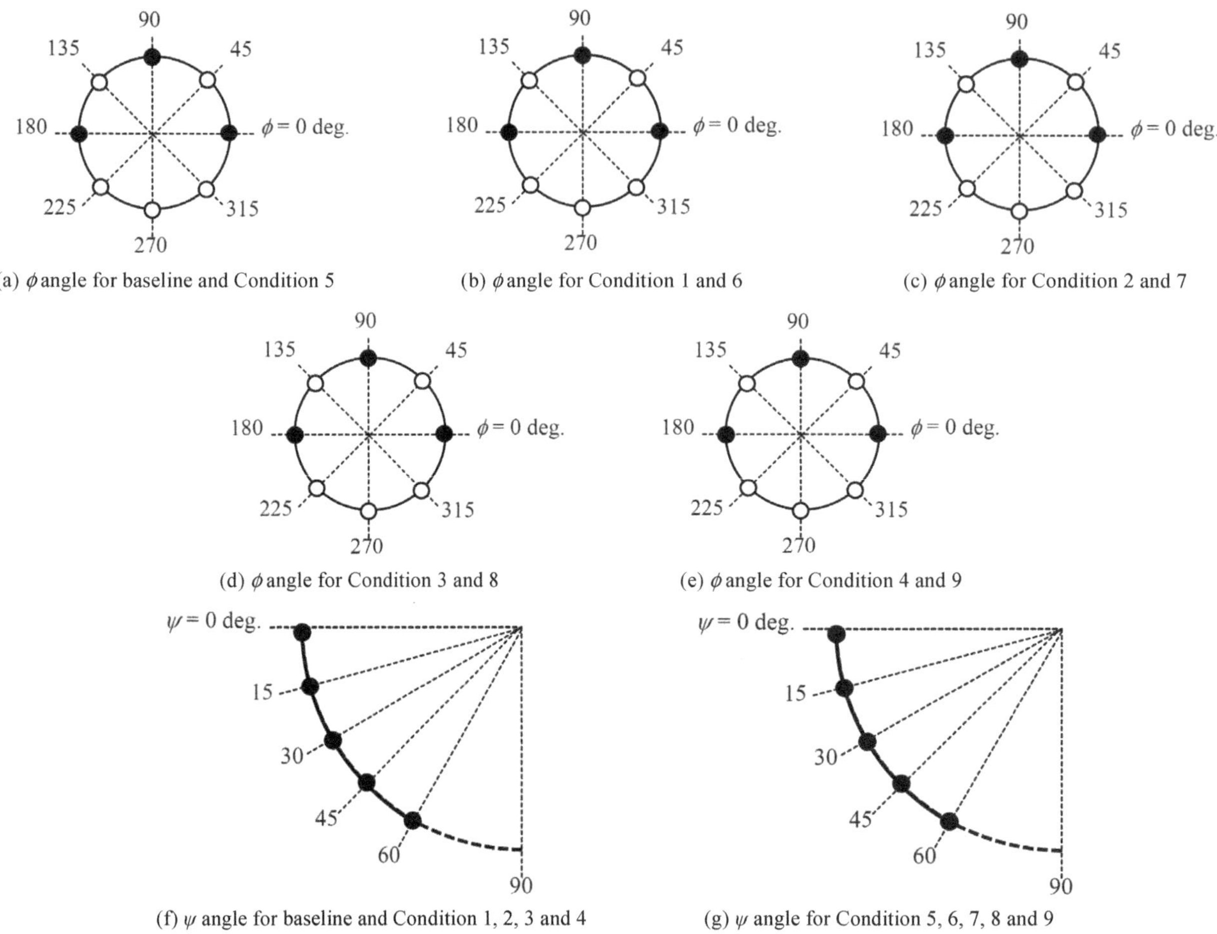

Figure 3. Schematic illustration of conditions 1 - 9.

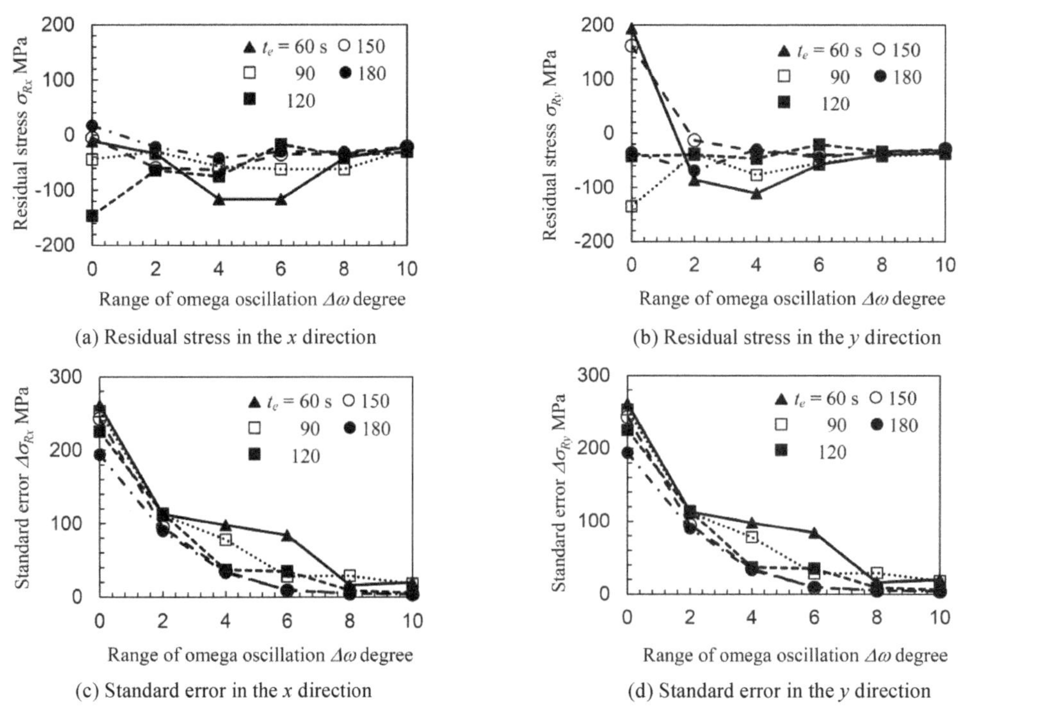

Figure 4. Variation of the biaxial residual stress with the range of the ω oscillation for the annealed specimen (AN).

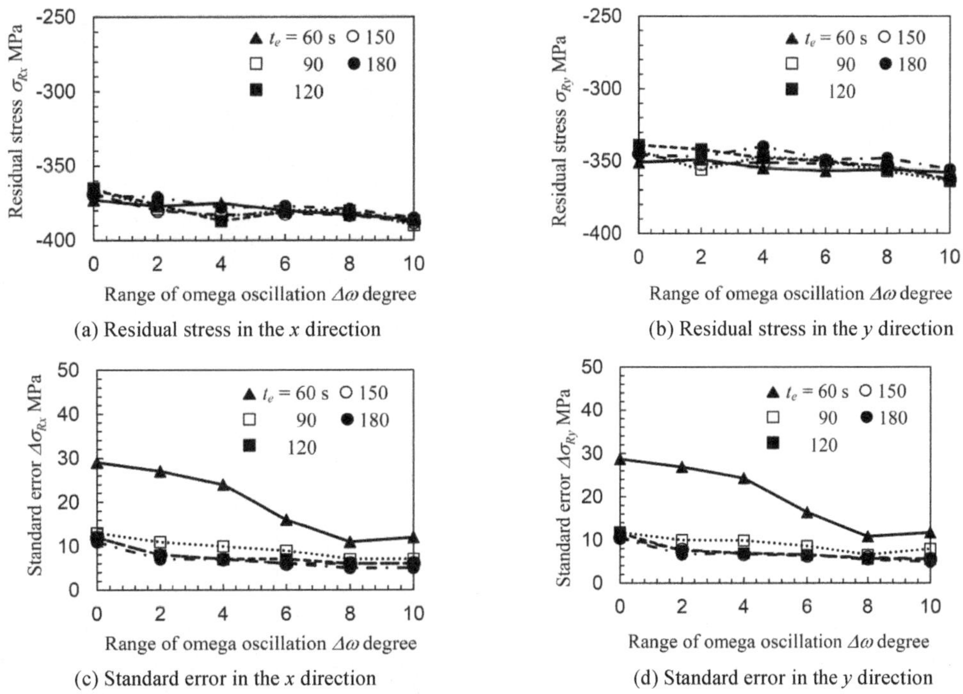

Figure 5. Variation of the biaxial residual stress with the range of the ω oscillation for the specimen treated by the cavitating jet in air (CJA).

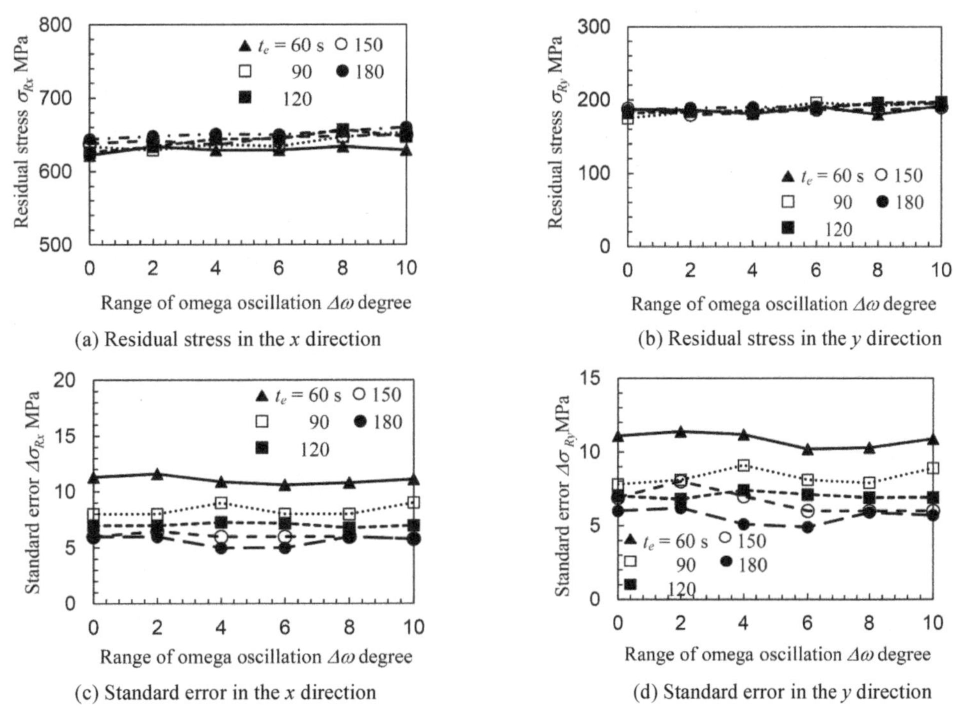

Figure 6. Variation of the biaxial residual stress with the range of the ω oscillation for the specimen treated by the disc grinder (DG).

study.

The optimum exposure time needs to be generalized in order to apply it to other metals. The standard error, $\Delta\sigma_{Rx}$, is plotted as a function of the number of X-ray counts, N, for the AN specimen in **Figure 7**. In the figure, $\Delta\sigma_{Rx}$ decreases with an increasing number of counts, but then saturates at around N = 200,000 for at t_e = 120 s. For these measurements it can be concluded that an exposure time which realizes $N \geq 200{,}000$ is enough to calculate accurately the distortion of the diffraction ring due to

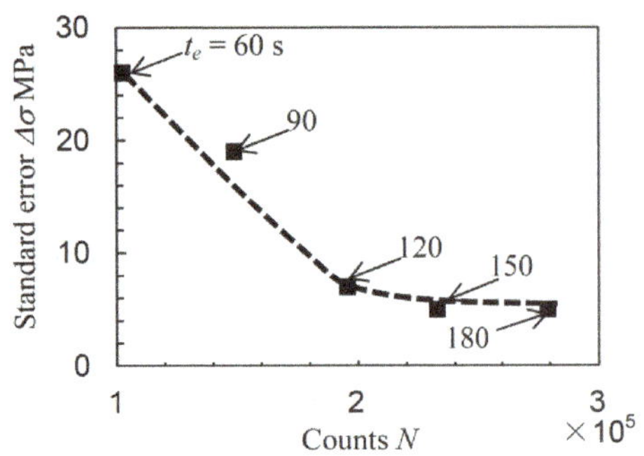

Figure 7. Decrease of the standard error with increasing number of diffracted X-ray counts.

residual stress. This rule can be applied just as much to other metals as it can be to SUS316L.

Figure 8 shows the diffraction rings obtained at $\Delta\omega = 0$ (without oscillation) and at $\Delta\omega = 10$ degrees for $t_e = 60$ and 180 s. For the AN specimen shown in **Figure 8(a)**, many strongly diffracting spots from the γ-Fe (2 2 0) plane with a random 2θ position and γ direction were detected in the case of $\Delta\omega = 0$. In contrast, the diffracted X-ray forms a diffraction ring at $\Delta\omega = 10$ degree for both the $t_e = 60$ and 180 s cases. This makes the distortion of diffraction ring due to residual stresses much easier to detect, leading to an improved accuracy for the residual stress measurements as shown **Figure 4**. Each spot detected at $\Delta\omega = 0$ shown in **Figure 8(a)** is attributed to diffraction from a large grain in the irradiated area which has grown during annealing. In principle, the diffraction angle, 2θ, of these spots should correspond to the stress tensor. The diffraction angles are however scattered, since the diffraction angles depend on the wavelengths of the X-ray which satisfy the Bragg angle for each large grain [18]. In other words this problem is attributed to the wavelength distribution for each large grain. In order to solve this large grain effect, the effect of the wavelength distribution has to be effectively suppressed by oscillating the specimen in the same direction of wavelength distribution [19], *i.e.*, the ω direction in this 2D-XRD measurement. For the CJA specimen, whilst there are some strong diffraction spots, these are not as strong as for the AN specimen. This is because the large grains within the sample have been refined due to impacts caused by cavitation bubble collapse. These impacts induce plastic deformation at the surface and, as a result, a high equibiaxial compressive residual stress is introduced. Whilst the effectiveness of the angular oscillation is smaller than that for large grain metals such as the AN specimen, it is still effective, especially for a short exposure time. On the other hand, for the DG processed specimen, the diffracting X-rays uniformly form a diffraction ring without strongly diffracting spots regardless of whether the measurement is with and without ω oscillation, as shown in **Figure 8(c)**. The DG process grinds

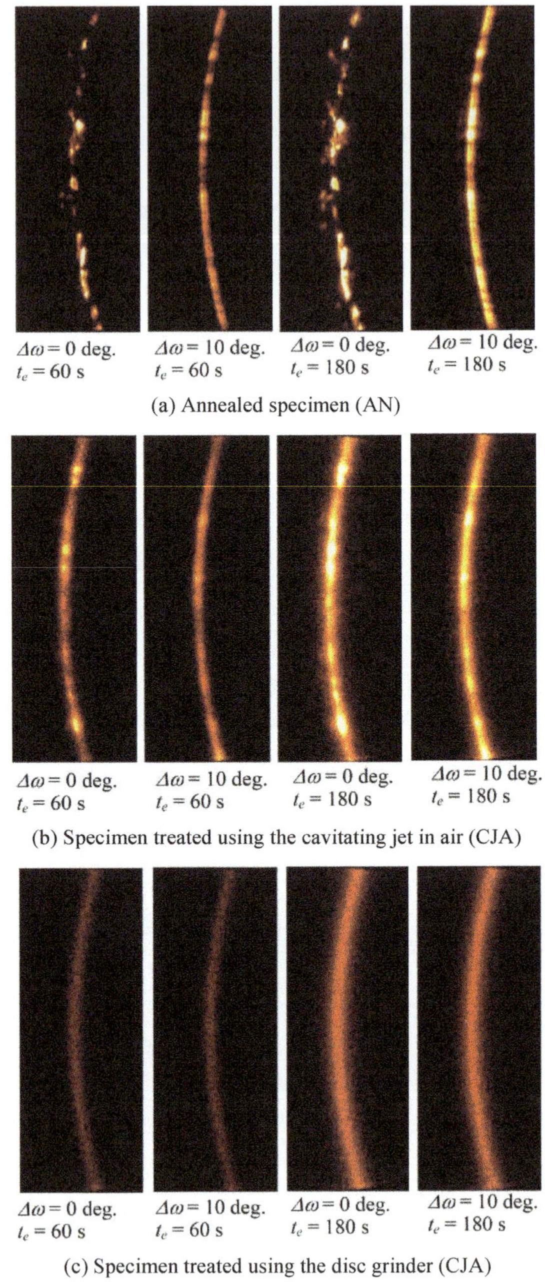

Figure 8. Comparison of the diffraction ring, with and without ω oscillation, for exposure times of 60 and 180 seconds.

the surface and has a tendency to minimize the grain size, which can be instantly visible through the ease of formation of the diffraction ring.

Summarizing the points above, the use of ω oscillation is quite an effective way of suppressing the large grain size effect, which occurs from annealed metals. The use of ω oscillation does not require an additional time and helps to detect X-rays diffracted from other grains, which have same crystal orientation. In general terms the use of ω oscillation is effective in residual stress measurements to improve the accuracy.

3.2. Optimum Angles of Specimen Rotation

It is important to verify the effect of choosing different angles for specimen rotation, *i.e.*, for the ϕ and ψ angles, on the residual stress value and on the standard error value. **Figures 9-11** plot the ratio of the residual stress value and the standard error to the baseline value, $\sigma_R/\sigma_{R\,B}$, and, $\Delta\sigma_R/\Delta\sigma_{R\,B}$, for conditions 1 to 9 for the AN, CJA and DG specimens, respectively. The baseline value was calculated from the 33 frames shown in **Table 1**, *i.e.*, from 495 data points. The black and white bar represents the σ_{Rx} and σ_{Ry} values, respectively. The baseline values are −38 ± 9 MPa in σ_{Rx} and −34 ± 9 MPa in σ_{Ry} for the AN specimen, −381 ± 6 MPa in σ_{Rx} and −354 ± 6 MPa in σ_{Ry} for the CJA specimen and 657 ± 6 MPa in σ_{Rx} and 196 ± 7 MPa in σ_{Ry} for the DG specimen, respectively. The value which nearly equals to 1 represents that the stress calculation has been precisely done as well as the baseline data.

The $\sigma_R/\sigma_{R\,B}$ and $\Delta\sigma_R/\Delta\sigma_{R\,B}$ values for each specimen are highly dependent upon the combination of ϕ and ψ angles chosen. Using conditions 2, 4, 7 and 9 shows inaccurate results for $\sigma_R/\sigma_{R\,B}$, especially for the AN and the DG specimens. For instance, the $\sigma_R/\sigma_{R\,B}$ and $\Delta\sigma_R/\Delta\sigma_{R\,B}$ values calculated using condition 7 are 3.4 and 5.8, respectively with regard to σ_{Rx} for the AN specimen. These values are quite large. The reason for this is that these

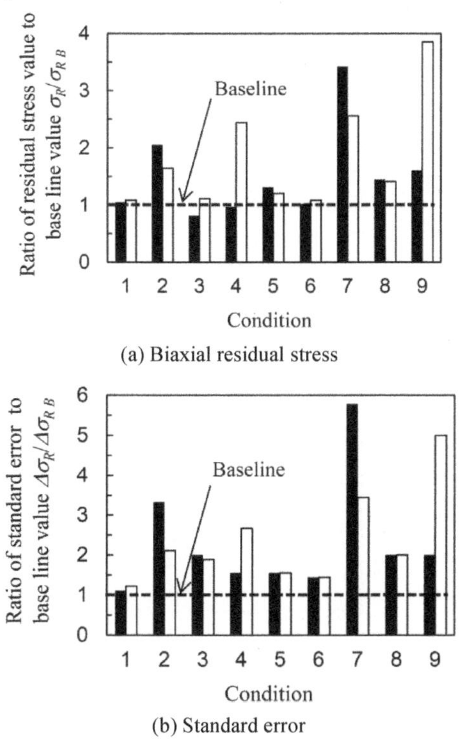

Figure 9. Variation of the ratio of the biaxial residual stress value and standard error to baseline data for various conditions representing a different choice of ϕ and ψ angles for the annealed specimen (AN).

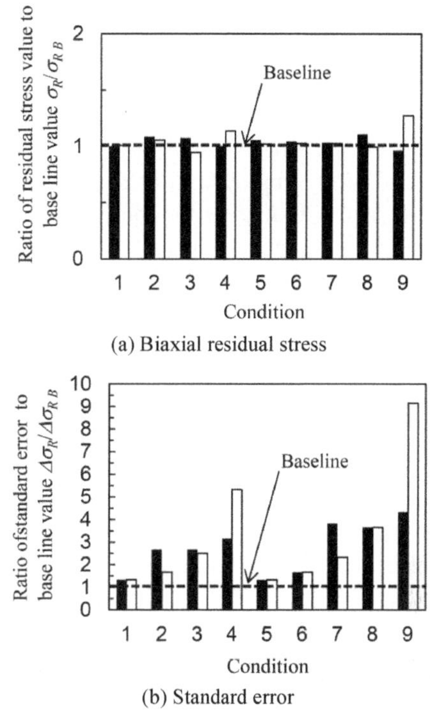

Figure 10. Variation of the ratio of the biaxial residual stress value and standard error to baseline data for various conditions representing a different choice of ϕ and ψ angles for the specimen treated using the cavitating jet in air (CJA).

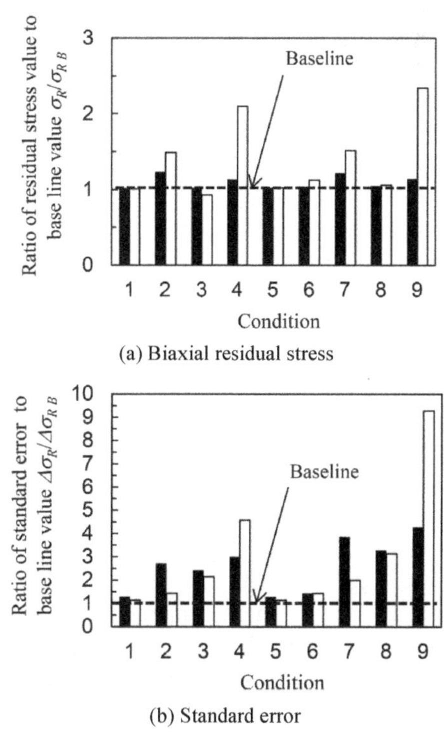

Figure 11. Variation of the ratio of the biaxial residual stress value and standard error to baseline data for various conditions representing a different choice of ϕ and ψ angles for the specimen treated using the disc grinder (DG).

conditions can obtain a diffraction ring from only one side, e.g., for ϕ = 0, 45, 90, 135 and 180 degrees in conditions 2 and 7. In contrast, accurate results for $\sigma_R/\sigma_{R\,B}$ for each specimen are obtained using conditions 1, 3, 5, 6 and 8. This occurs since several angles of ϕ, which are not one-sided, were employed in these conditions, e.g., ϕ = 0, 90, 180, and 270 degrees for conditions 3 and 9. The general rule is that a lack of diffraction information from a opposite side causes variability in the determination of the stress vector. This effect appears in any stress direction and depends precise combination of ω, ϕ and ψ angles used [15]. Moreover, this effect becomes smaller when the residual stress measurements are carried out to assess equibiaxial stress fields such as that found in the CJA specimen, since the diffraction ring is distorted symmetrically due to the equibiaxial stress field. In order to choose the optimum conditions, several comparisons will be made regarding conditions 1, 3, 5, 6 and 8 in the following paragraph.

Comparing conditions 3 and 5, both having same number of frames of 17, the $\Delta\sigma_R/\Delta\sigma_{R\,B}$ for σ_{Rx} values obtained were 2.0 and 1.6 for the AN specimen, 2.7 and 1.3 for the CJA specimen and 2.4 and 1.3 for the DG specimen, respectively. Overall therefore, the standard error obtained using condition 5 is smaller than that using condition 3. Focusing on the combination of ϕ and ψ angles, the intervals between these, $\Delta\phi$ and $\Delta\psi$, are 45 and 30 degrees, respectively, in condition 5 and are 90 and 15 degrees, respectively, in condition 3. A $\Delta\psi$ value of 15 degrees is more than enough, since a large area can be covered with regard to the ψ direction using the 2D-PSPC. The area in the ψ direction covered by the 2D-PSPC in the case of both $\Delta\psi$ = 15 and $\Delta\psi$ = 30 degrees is plotted in **Figure 12**. For $\Delta\psi$ = 15, a large overlap occurs compared to that for $\Delta\psi$ = 30. Needless to say, the accuracy increases with decreasing $\Delta\psi$ when $\Delta\phi$ is same, due to a direct increase in the number of frames. However, when the number of frames is the same, decreasing $\Delta\phi$ tends to make the measurements more accurate. In comparison between conditions 5 and 6, having same $\Delta\psi$ value of 30 degrees, the results indicate that using condition 6 produced results hardly different to that obtained using condition 5 for each of the specimens, this in spite of relatively small number of frames. This result indicates that one diffraction ring measured from an opposing side, e.g., from ϕ = 270 degrees, might be sufficient for high accuracy measurements.

In order to determine the optimum conditions, **Figure 13** plots the ratio of the standard error of the biaxial residual stress to the baseline value, $\Delta\sigma_R/\Delta\sigma_{R\,B}$, as a function of total measurement time, t_T, for the AN specimen. In **Figure 13**, a $\Delta\sigma_R/\Delta\sigma_{R\,B}$ value of close to 1 represents a measurement result that shows the same high accuracy as the baseline. The lower t_T time indicates that the meas-

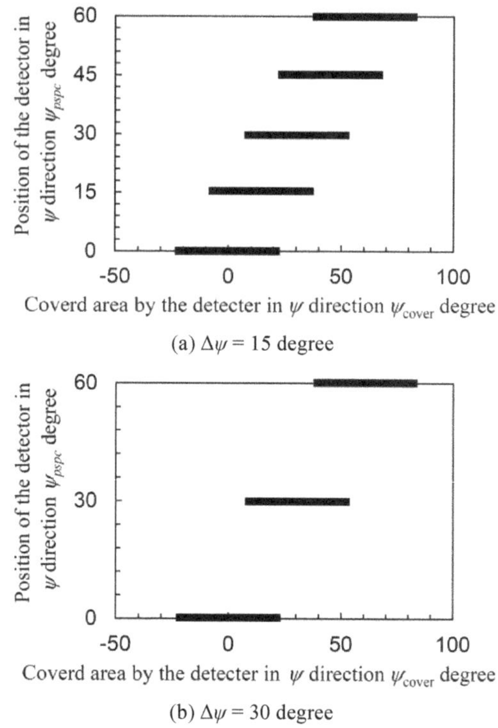

Figure 12. Comparison of the area covered by the 2D-PSPC with regard to an ψ angular range of between $\Delta\psi$ = 15 and 30 degrees.

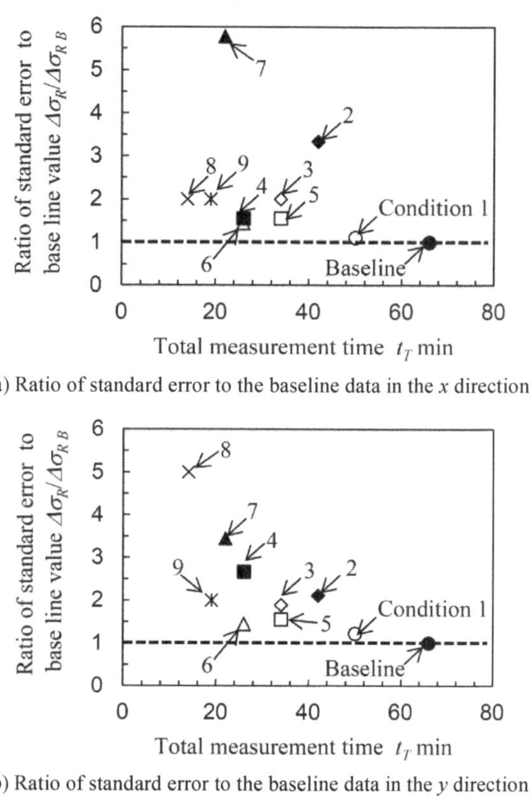

Figure 13. Determination of the optimum conditions taking account of the efficiency and accuracy based on a comparison with the baseline data.

urements become quicker. Therefore, the condition realized by the lower left point in **Figure 13** indicates a point at which the measurement becomes more accurate and effective. From the figure it can be seen that condition 6 represents the most optimum condition, taking account of both high efficiency and measurement accuracy. In this study, the measurement takes 66 and 50 minutes respectively for the baseline, which requires 33 frames and for condition 1, which requires 25 frames. In contrast, the measurement takes only 26 minutes for condition 6, which requires only 13 frames. In summary, when an accurate residual stress measurement is required regardless of measurement time, then baseline or condition 1 should be chosen. However when a quick measurement is required which also has a good accuracy then condition 6 may be used. This condition takes about half the time of the most accurate (condition 1) measurement.

4. Conclusions

In order to optimize the conditions for residual stress measurement using 2D-X-ray diffraction (2D-XRD) in terms of both efficiency and accuracy, residual stress measurements were conducted on three specimens made of austenitic stainless steel. The specimens were processed by annealing, a cavitating jet in air and a disc grinder introducing different residual stresses at the surface. The specimens were oscillated in the ω direction, representing a right-hand rotation of the specimen about incident X-ray with a varying range of the oscillation, $\Delta\omega$ and taken for several exposure times, te. Moreover, the combination of the tilt angle between the specimen surface normal and the diffraction vector, ψ, with the rotation angle about its surface normal, ϕ, has been optimized in terms of the measurement efficiency and accuracy. The conclusions obtained in the present study are summarized as follows.

1) The standard error rapidly decreases with an increase in the range of the ω oscillation, $\Delta\omega$. This occurs especially for the annealed specimen, which creates strongly diffracted spots due to the presence of large grain sizes within the sample. The use of ω oscillations is quite an effective way of suppressing the problems posed by large grain metals. For the specimens processed by means of the cavitating jet in air and the disc grinder, the effectiveness of the ω oscillation method is less than that for the annealed specimen due to the minimization of grain size caused by these processes. However the oscillation still helps by detecting additional X-rays diffracted from other grains with the same crystal orientation. Therefore, the use of ω oscillations is generally effective for improving the accuracy of residual stress measurements. The optimum range of the ω oscillation, $\Delta\omega$, is around 8 degrees.

2) The greater the number of diffraction rings, the greater the accuracy of the result. When the number of diffraction rings is same, the detection of the diffraction ring at several ϕ angles makes the measurements more accurate than detecting at several ψ angles. The combination of using ψ and ϕ values of $\phi = 0$ degree and $\psi = 0$ degree with $\phi = 0, 45, 90, 135, 180$ and 270 degrees and $\psi = 30$ and 60 degrees can lead to an optimized quick and accurate result.

5. Acknowledgements

This work was partly supported by the Japan Society for the Promotion of Science under the Grant-in-Aid for Scientific Research (B) 24360040, the Young Scientists (Start-up) 24860004.

REFERENCES

[1] Y. Sano, K. Akita, K. Masaki, Y. Ochi, I. Altenberger and B. Scholtes, "Laser Peening without Coating as a Surface Enhancement Technology," *Journal of Laser Micro Nanoengineering*, Vol. 1, No. 3, 2006, pp. 161-166.

[2] H. Soyama and Y. Sekine, "Sustainable Surface Modification Using Cavitation Impact for Enhancing Fatigue Strength Demonstrated by a Power Circulating-Type Gear Tester," *International Journal of Sustainable Engineering*, Vol. 3, No. 1, 2010, pp. 25-32.

[3] A. Naito, O. Takakuwa and H. Soyama, "Development of Peening Technique Using Recirculating Shot Accelerated by Water Jet," *Materials Science and Technology*, Vol. 28, No. 2, 2012, pp. 234-239.

[4] Y. Sano, M. Obata, T. Kubo, N. Mukai, M. Yoda, K. Masaki and Y. Ochi, "Retardation of Crack Initiation and Growth in Austenitic Stainless Steel by Laser Peening without Protective Coating," *Materials Science and Engineering A*, Vol. 417, No. 1-2, 2006, pp. 334-340.

[5] O. Takakuwa, M. Nishikawa and H. Soyama, "Numerical Simulation of the Effects of Residual Stress on the Concentration of Hydrogen around a Crack Tip," *Surface & Coatings Technology*, Vol. 206, No. 11-12, 2012, pp. 2892-2898.

[6] O. Takakuwa and H. Soyama, "Suppression of Hydrogen-Assisted Fatigue Crack Growth in Austenitic Stainless Steel by Cavitation Peening," *International Journal of Hydrogen Energy*, Vol. 37, No. 6, 2012, pp. 5268-5276.

[7] O. Takakuwa and H. Soyama, "Using an Indentation Test to Evaluate the Effect of Cavitation Peening on the Invasion of the Surface of Austenitic Stainless Steel by Hydrogen," *Surface & Coatings Technology*, Vol. 206, No. 18, 2012, pp. 3747-3750.

[8] P. S. Prevey, "X-Ray Diffraction Residual Stress Tech-

niques," *ASM Handbook*, Vol. 10, 1986, pp. 380-392.

[9] The Society of Materials Science, "Standard for X-Ray Stress Measurement (2002): Iron and Steel," Japan.

[10] H. Dölle, "The Influence of Multiaxial Stress States, Stress Gradients and Elastic Anisotropy on the Evaluation of Residual Stress by X-Rays," *Journal of Applied Crystallography*, Vol. 12, 1979, pp. 489-501.

[11] H. Dölle and J. B. Cohen "Residual Stresses in Ground Steels," *Metallurgical Transactions A*, Vol. 11, No. 1, 1980, pp. 159-164.

[12] B. B. He, K. L. Smith, U. Preckwinkel and W. Schultz, "Micro-Area Residual Stress Measurement Using a Two-Dimensional Detector," *Materials Science Forum*, Vol. 347-349, 2000, pp. 101-106.

[13] B. B. He, U. Preckwinkel and K. Smith, "Advantage of Using 2D Detectors for Residual Stress Measurements," *Advances in X-Ray Analysis*, Vol. 42, 2000, pp. 429-438.

[14] B. B. He, "Introduction to Two-Dimensional X-Ray Diffraction," *Powder Diffraction*, Vol. 18, No. 2, 2003, pp. 71-85.

[15] B. B. He, "Two-Dimensional X-Ray Diffraction," John Wiley & Sons, Inc., New Jersey, 2009, pp. 249-328.

[16] H. Soyama, T. Kikuchi, M. Nishikawa and O. Takakuwa, "Introduction of Compressive Residual Stress into Stainless Steel by Employing a Cavitating Jet in Air," *Surface & Coatings Technology*, Vol. 205, No. 10, 2011, pp. 3167-3174.

[17] O. Takakuwa and H. Soyama, "The Effect of Scanning Pitch of Nozzle for a Cavitating Jet during Overlapping Peening Treatment," *Surface & Coatings Technology*, Vol. 206, No. 23, 2012, pp. 4756-4762.

[18] K. Yukino and R. Uno, "A Method of Evaluating the Dimensional and Orientation Distribution of Crystallites by X-Ray Powder Diffractometer," *Japanese Journal of Applied Physics*, Vol. 25, 1986, pp. 661-666.

[19] T. Kurimura and T. Konishi, "The Optimization of the X-Ray Stress Measurement Condition on Austenitic Stainless Tubes for Electric Power Plants," *Materials Science Research International*, Vol. 8, No. 4, 2002, pp. 175-180.

Single Crystal X-Ray Diffraction Studies on Magnetic $Yb_5Co_4Ge_{10}$*

Udumula Subbarao, Sebastian C. Peter#
New Chemistry Unit, Jawaharlal Nehru Centre for Advanced Scientific Research, Bangalore, India

ABSTRACT

The high quality single crystals of $Yb_5Co_4Ge_{10}$ have been grown by the indium metal flux method and characterized by means of single crystal X-ray diffraction data. $Yb_5Co_4Ge_{10}$ crystallizes in the $Sc_5Co_4Si_{10}$ structure type, tetragonal space group *P*4/*mbm* and lattice constants are $a = b = 12.6369(18)$ Å and $c = 4.1378(8)$ Å. Crystal structure of $Yb_5Co_4Ge_{10}$ composed of three-dimensional $[Co_4Ge_{12}]$ network having five, six and eight membered rings. The three non-equivalent Yb atoms are sandwiched in three different channels created by the $[Co_4Ge_{12}]$ network. Based on the bond length analysis from the crystallographic information, we confirmed that Yb1 and Yb2 atoms are in the trivalent magnetic state and Yb3 is in the divalent non-magnetic state.

Keywords: X-Ray Diffraction; Single Crystal; Intermetallics; Ytterbium

1. Introduction

The ternary intermetallic compounds of general formula $RE_5T_4X_{10}$ (*RE* = rare earths; T = Co, Ni, Rh, Ir and Os and X = Si, Ge, Sn) have been reported mostly for their interesting physical properties [1-22]. These compounds crystallize in the tetragonal $Sc_5Co_4Si_{10}$ structure type having *P*4/*mbm* space group [1]. As per Pearson crystallographic database [23] and inorganic crystal structure database [24], there are 37 compounds reported within the family of $RE_5T_4X_{10}$ compounds and are listed in **Table 1**. Among them only three were studied for the crystal structure from the single crystal XRD data, including our recent discovery of $Yb_5Ni_4Ge_{10}$ mixed valent Yb atoms [2]. However, the majority of these compounds were studied for their physical properties. For examples, the non-magnetic $RE_5T_4X_{10}$ (*RE* = rare earths; T = Co, Rh, Ir, and Os; X = Ge, Si) compounds were reported as superconductors at low temperatures [5,12,17-19]. The $RE_5Rh_4X_{10}$ (*RE* = Gd, Tb, Dy, Ho, Er, Tm; X = Si, Ge, Sn) and $Dy_5Ir_4Si_{10}$ compounds were reported for their interesting magnetic properties and all of them order either ferromagnetically or antiferromagnetically at low temperature [11,18,20,21].

Katoh *et al*. synthesized the $Yb_5Co_4Ge_{10}$ compound in the polycrystalline nature using Bridgman method [22]. They have proposed the mixed valent nature of Yb in $Yb_5Co_4Ge_{10}$ based on the magnetic susceptibility data. In order to understand the structure-property relationship, we need to study the complete crystallography analysis from a high quality single crystal. The chemical bonding nature obtained from the crystal structure can be used to explain the valency of the Yb atoms. Recently, molten metal fluxes have been used as an excellent alternative to the conventional synthetic methods for the exploratory synthesis of new rare earth intermetallic compounds, as well as single crystal growth of already reported compounds [25-39].

In this paper, we report the single crystal growth of $Yb_5Co_4Ge_{10}$ using liquid indium metal flux technique. The crystal structure of this compound has refined from the single crystal XRD data. We also discuss the crystal structure and chemical bonding of $Yb_5Co_4Ge_{10}$ and compared with other compounds.

2. Experimental Section

2.1. Synthesis

The following reagents were used as purchased without further purification: Yb (in the form of metal pieces cut from metal chunk, 99.99% Alfa Aesar), Co (powder,

*Crystallographic information files (CIF).
#Corresponding author.

Table 1. The list of the $RE_5T_4X_{10}$ (*RE* = Rare earth, T = Co, Rh, Ir, Os and Ni; *X* = Si, Ge and Sn) compounds reported.

Compound	*a* (Å)	*c* (Å)	Vol. (Å^3)	Ref.
$Y_5Os_4Ge_{10}$	13.012 (4)	4.2927 (4)	726.8	[3,4]
$Gd_5Rh_4Ge_{10}$	12.987 (8)	4.299 (3)	725.0	[5]
$Dy_5Os_4Ge_{10}$	13.034 (2)	4.299 (3)	725.8	[6]
$Ho_5Os_4Ge_{10}$	12.984 (3)	4.2820 (6)	721.9	[7,8]
$Ho_5Os_4Ge_{10}$	12.984 (3)	4.2820 (6)	721.9	[9]
$Er_5Rh_4Ge_{10}$	12.889 (8)	4.236 (3)	707.7	[5]
$Er_5Ir_4Ge_{10}$	12.865 (9)	4.278 (3)	694.8	[5]
$Tm_5Co_4G_{10}$	12.622 (9)	4.139 (4)	659.4	[5]
$Tm_5Rh_4Ge_{10}$	12.868 (9)	4.230 (3)	700.4	[5]
$Tm_5Ir_4Ge_{10}$	12.846 (8)	4.267 (3)	704.1	[5]
$Yb_5Co_4G_{10}$	12.645 (9)	4.138 (4)	661.6	[5]
$Yb_5Co_4G_{10}$	12.637 (2)	4.1378 (8)	660.7	*
$Yb_5Ni_4G_{10}$	12.672 (2)	4.1598 (8)	667.9	[2]
$Yb_5Rh_4Ge_{10}$	12.886 (9)	4.234 (3)	703.0	[5]
$Yb_5Ir_4Ge_{10}$	12.877 (9)	4.285 (3)	710.5	[5]
$Lu_5Co_4G_{10}$	12.606 (8)	4.125 (3)	655.5	[5]
$Lu_5Rh_4Ge_{10}$	12.850 (80	4.208 (3)	694.8	[5]
$Lu_5Ir_4Ge_{10}$	12.831 (9)	4.252 (3)	700.0	[5]
$Sc_5Co_4Si_{10}$	12.013 (3)	3.936 (2)	568.0	[10]
$Sc_5Co_4Si_{10}$	12.01 (1)	3.936 (5)	567.7	[1]
$Sc_5Rh_4Si_{10}$	12.344 (2)	4.039 (1)	615.4	[11]
$Sc_5Rh_4Si_{10}$	12.330 (3)	4.030 (2)	612.6	[1]
$Sc_5Ir_4Si_{10}$	12.335 (2)	4.083 (1)	621.2	[1,11]
$Y_5Ir_4Si_{10}$	12.599 (2)	4.234 (3)	672.0	[12]
$Dy_5Ir_4Si_{10}$	12.577 (2)	4.234 (1)	670.2	[13,14]
$Ho_5Ir_4Si_{10}$	12.558 (2)	4.218 (1)	665.1	[13]
$Er_5Ir_4Si_{10}$	12.540 (2)	4.208 (1)	661.7	[13]
$Tm_5Ir_4Si_{10}$	12.513 (2)	4.197 (1)	657.1	[13]
$Yb_5Ir_4Si_{10}$	12.503 (3)	4.182 (2)	653.7	[13]
$Lu_5Rh_4Si_{10}$	12.502 (2)	4.137 (2)	646.6	[1]
$Lu_5Ir_4Si_{10}$	12.475 (2)	4.171 (1)	649.1	[1]
$La_5Ir_4Sn_{10}$	14.088 (10)	4.690 (4)	930.8	[15]
$Ce_5Rh_4Sn_{10}$	14.053 (12)	4.621 (5)	912.5	[15]
$Ce_5Ir_4Sn_{10}$	14.027 (15)	4.648 (8)	914.5	[15]
$Pr_5Rh_4Sn_{10}$	14.035 (13)	4.612 (4)	908.4	[15]
$Pr_5Ir_4Sn_{10}$	13.993 (16)	4.632 (6)	906.9	[15]
$Nd_5Rh_4Sn_{10}$	13.992 (17)	4.588 (7)	898.2	[15]
$Sm_5Rh_4Sn_{10}$	13.926 (16)	4.554 (6)	883.1	[15]
$Ho_5Rh_4Ge_{10}$	12.9336 (9)	4.2596 (6)	712.5	[16]

*This work.

99.99% Alfa Aesar), Ge (pieces, 99.999% Alfa Aesar) and In (shots, 99.99% Alfa-Aesar). High quality single crystals of $Yb_5Co_4Ge_{10}$ were obtained by combining ytterbium metal (0.3 g), cobalt powder (0.1 g), germanium (0.3) and indium shots (2 g) in an alumina crucible. The crucible was placed in a 13 mm fused silica tube which was flame sealed under a vacuum of 10^{-5} torr, to prevent oxidation during heating. The reactants were then heated to 1173 K over 8 h, and maintained at that temperature for 2 h to allow proper homogenization, then cooled down to 1073 K in 2 h and kept at this temperature for 72 h. Finally, the sample was allowed to cool slowly to 303 K over 48 hrs. No reactions with the alumina crucible material could be detected. The reaction product was isolated from the excess indium flux by heating at 673 K and subsequently centrifuging through a coarse frit. Any remaining flux was removed by immersion and sonication in glacial acetic acid for 24 h. The final crystalline product was rinsed with water and dried with acetone. A few small gray rod-shaped crystals of $Yb_5Co_4Ge_{10}$ carefully selected for the elemental analysis and single crystal XRD data collection. $Yb_5Co_4Ge_{10}$ is stable in air and no decomposition was observed even after several months.

2.2. Elemental Analysis

Semi-quantitative microanalyses were performed on the single crystals obtained from the flux techniques using a scanning Leica 220i electron microscope (SEM) equipped with Bruker 129 eV energy dispersive X-ray analyzer (EDS). Data were acquired with an accelerating voltage of 20 kV and in 60 s accumulation time. The EDS analysis performed on visibly clean surfaces of the single crystals obtained from the flux method indicated that the atomic composition was close to 5:4:10 in good agreement with the composition obtained from the single crystal data refinement. A typical SEM image of rod-like crystals of $Yb_5Co_4Ge_{10}$ grown from the flux synthesis is shown in **Figure 1**.

2.3. Single Crystal X-Ray Diffraction

Single crystal structural data of $Yb_5Co_4Ge_{10}$ was collected at room temperature on rod shaped single crystals using a Bruker Smart-CCD diffractometer equipped with

Figure 1. Typical SEM image of $Yb_5Co_4Ge_{10}$ single crystal grown from In flux.

a normal focus, 2.4 kW sealed tube X-ray source with graphite monochromatic Mo-Kα radiation (λ = 0.71073 Å) operating at 50 kV and 30 mA, with ω scan mode. A crystal of suitable size (0.5 × 0.05 × 0.05 mm) was cut from a small rod shape crystal and mounted on a thin glass (~0.1 mm) fiber with commercially available super glue. A full sphere of 60 frames was acquired up to 57.87° in 2θ. The individual frames were measured with w steps of 0.50° and an exposure time of 20 s per frame. The programme SAINT [40] was used for integration of diffraction profiles and absorption correction was made with SADABS programme [41]. The systematic absences were leading to the centrosymmetric space group *P4/mbm*. However, the Platon program within WinGx system, ver. 1.80.05 [42] was used to check the additional symmetry, and this was not suggesting any additional symmetry. The structure was solved by SHELXS 97 [43] and refined by a full matrix least-squares method using SHELXL [44] with anisotropic atomic displacement parameters for all atoms. Packing diagrams were generated with Diamond [45]. In order to check for the correct composition, the occupancy parameters were refined in a separate series of least-squares cycles. All bond lengths are within the acceptable range compared to the theoretical values. Details of the crystallographic data are given in **Tables 2** to **5**.

Table 2. Crystal data and structure refinement for $Yb_5Co_4Ge_{10}$.

Formula weight	1826.82
Temperature	293 (2) K
Wavelength	0.71073 Å
Crystal system	Tetragonal
Space group	*P4/mbm*
Unit cell dimensions	a = 12.6369 (18) Å b = 12.6369 (18) Å c = 4.1378 (8) Å
Volume, Z	660.77 (2) Å^3, 2
Density (calculated)	9.18118 g/cm^3
Absorption coefficient	62.152 mm^{-1}
F(000)	1556
Crystal size	0.05 × 0.01 × 0.01 mm^3
θ range	2.28° to 28.93°
Index ranges	$-16 \le h \le 17, -17 \le k \le 17, -5 \le l \le 3$
Reflections collected	6117
Independent reflections	526
Completeness to θ = 28.93°	100%
Refinement method	Full-matrix least-squares on F^2
Data/restraints/parameters	526/0/34
Goodness-of-fit	1.076
Final R indices [>2σ(I)][a]	R_{obs} = 0.0293, wR_{obs} = 0.0671
Extinction coefficient	0.00082 (15)
Largest diff. peak and hole	2.327 and −1.878 e·Å^{-3}

[a] $R = \Sigma||F_o| - |F_c||/\Sigma|F_o|$, $wR = \{\Sigma[w(|F_o|^2 - |F_c|^2)^2]/\Sigma[w(|F_o|^4)]\}^{1/2}$ and calc $w = 1/[\sigma^2(Fo^2) + (0.0318P)^2 + 111.1197P]$ where $P = (Fo^2 + 2Fc^2)/3$.

Table 3. Atomic coordinates (×10^4) and equivalent isotropic displacement parameters (Å^2 × 10^3) for $Yb_5Co_4Ge_{10}$ at 296(2) K with estimated standard deviations in parentheses.

Label	x	y	z	U_{eq}*
Yb1	0	0	0	6 (1)
Yb2	1762 (4)	6762 (4)	5000	6 (1)
Yb3	6113 (4)	1113 (4)	5000	7 (1)
Co	2453 (1)	255 (1)	0	5 (1)
Ge1	688 (1)	5688 (1)	0	6 (1)
Ge2	1554 (1)	1991 (1)	0	8 (1)
Ge3	1640 (1)	56 (1)	5000	8 (1)

*U_{eq} is defined as one third of the trace of the orthogonalized U_{ij} tensor.

Table 4. Anisotropic displacement parameters (Å^2 × 10^3) for $Yb_5Co_4Ge_{10}$ with estimated standard deviations in parentheses.

Label	U_{11}	U_{22}	U_{33}	U_{12}	U_{13}	U_{23}
Yb1	7 (3)	7 (3)	4 (1)	0	0	0
Yb2	7 (3)	7 (3)	3 (1)	−1 (1)	0	0
Yb3	8 (3)	8 (3)	5 (4)	−1 (1)	0	0
Co	6 (1)	6 (1)	4 (1)	−1 (1)	0	0
Ge1	7 (5)	7 (5)	5 (1)	−1 (1)	0	0
Ge2	6 (1)	8 (1)	11 (1)	0	0	0
Ge3	8 (1)	14 (1)	3 (1)	−2 (1)	0	0

The anisotropic displacement factor exponent takes the form: $-2\pi^2[h^2a^{*2}U_{11} + \cdots + 2hka^*b^*U_{12}]$.

Table 5. Selected bond lengths [Å] for $Yb_5Co_4Ge_{10}$ at 296(2) K with estimated standard deviations in parentheses.

Label	Distances	Label	Distances
Yb1-Ge3	2.9300 (9)	Yb3-Co	3.2487 (12)
Yb1-Co	3.1167 (15)	Co-Ge3	2.3232 (9)
Yb2-Ge1	2.8217 (12)	Co-Ge2	2.3603 (19)
Yb2-Ge3	2.9527 (13)	Co-Ge1	2.4112 (17)
Yb2-Ge2	2.9814 (10)	Ge1-Ge1	2.4620 (3)
Yb2-Co	2.9815 (11)	Ge2-Ge2	2.5980 (2)
Yb3-Ge1	3.1228 (8)	Ge2-Ge3	2.9362 (12)
Yb3-Ge3	3.2002 (13)	Ge3-Ge3	2.9341 (17)
Yb3-Ge2	3.2137 (11)		

3. Results and Discussion

3.1. Structure Refinement of $Yb_5Co_4Ge_{10}$

The atomic parameters of $Sc_5Co_4Si_{10}$ [1] were taken as starting parameters and the structure was refined using SHELXL-97 (full-matrix least-squares on F^2) [43] with anisotropic atomic displacement parameters for all atoms. As a check for the correct composition, the occupancy parameters were refined in a separate series of least-squares cycles. Our data confirms the previous X-ray powder data [5], but the atomic positions have been refined with higher precision. Single crystals of $Yb_5Co_4Ge_{10}$

from different synthesis batches were used for the data collection. In the first step of refinement on the single crystal data collected shows that $Yb_5Co_4Ge_{10}$ crystallizes in a primitive tetragonal lattice (*P4/mbm*) within the $Sc_5Co_4Si_{10}$ type structure and lattice constants are $a = b =$ 12.6369 (18) Å and c = 4.1378 (8) Å. This refinement resulted in seven crystallographical positions (three Yb, one Co and three Ge) in the $Yb_5Co_4Ge_{10}$ structure. The data collection and structure refinement for $Yb_5Co_4Ge_{10}$ are listed in **Table 2**.

The standard atomic positions and isotropic atomic displacement parameters of this compound are collected in **Table 3**. The anisotropic displacement parameters and important bond lengths are listed in **Tables 4** and **5**, respectively. Further information on the structure refinements is available from: Fachinformationszentrum Karlsruhe, D-76344 Eggenstein-Leopoldshafen (Germany), by quoting the Registry Nos. CSD-425832.

3.2. Crystal Chemistry of $Yb_5Co_4Ge_{10}$

The crystal structure of $Yb_5Co_4Ge_{10}$ along c-axis is shown in **Figure 2**.

$Yb_5Co_4Ge_{10}$ crystallizes in the primitive tetragonal $Sc_5Co_4Si_{10}$ type structure (space group *P4/mbm*). Crystal structure of $Yb_5Co_4Ge_{10}$ is composed of a complex $[Co_4Ge_{12}]$ polyanion network with three types of one dimensional channels, showed along the c-axis (**Figure 2**), in which Yb atoms are embedded. The cages are shared through the corners of Co and Ge form pentagonal and hexagonal rings and resulted in a three dimensional network. The pentagonal and hexagonal rings are occupied with Yb2 and Yb3 atoms, respectively. These rings are connected through Co and Ge in *ab* plane and resulted in another large 8-membered ring, which is occupied by large Yb1 atoms. A detailed crystal structure description of the compounds crystallizing in the $Sc_5Co_4Si_{10}$ was reported as an example of the $Yb_5Ni_4Ge_{10}$ compound [2].

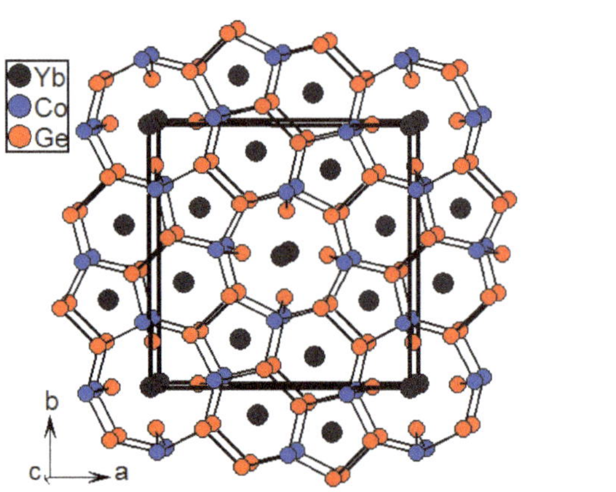

Figure 2. The crystal structure of $Yb_5Co_4Ge_{10}$ as viewed along the *c*-axis; the unit cell is outlined as block solid lines.

Crystallographically there are three different Yb-atoms are present in $Yb_5Co_4Ge_{10}$. All three Yb-coordination environments are significantly different. The Yb1 atom is located at the center of a cage consisting of 12 Ge and 4 Co atoms forming a cage as $Yb[Ge_{12}Co_4]$, Yb2 atoms are surrounded by eight Ge and four Co forming as Yb $[Ge_8Co_4]$ and Yb3 atoms are surrounded by fourteen atoms, $Yb[Ge_{10}Co_4]$. The Yb-Ge and Yb-Co bond lengths in $Yb_5Co_4Ge_{10}$ are in the wide range (2.8218 (9) Å to 3.2138 (10) Å for Yb-Ge and 2.9815 (11) Å to 3.2488 (12) Å for Yb-Co) due to the different size of Yb coordination environments. The average distance obtained for Yb3-Co, 3.2488 (12) Å, is longer than the corresponding distance for Yb1-Co and Yb2-Co with 3.1167 (16) and 2.9815 (11) Å, respectively. Similarly, the average bond distance for the Yb3-Ge is 3.1754 Å, is also longer than the corresponding distance for Yb1-Ge and Yb2-Ge with 3.0175 and 2.9343 Å, respectively. Similar kinds of bonding behavior were observed in the mixed valent $Yb_5Ni_4Ge_{10}$ as well [2]. Since Katoh *et al.* observed the mixed valent Yb in $Yb_5Co_4Ge_{10}$ based on the magnetic susceptibility measurements [22], our crystallographic data suggests the Yb1 and Yb2 atoms are mostly in the trivalent state and Yb3 in the divalent state. Considering the fact that divalent Yb is larger in size (1.86 Å) [46,47] compared to trivalent Yb (1.66 Å) [46,47], Yb3 prefer to be in large cage and others (Yb1 and Yb2) are in the smaller cages.

We also compared the covalent radii and the unit cell volume four Yb based compounds reported in this family, as shown in **Figure 3**. The cell volume of the compounds increases as the covalent radii of the transition metals except in case of Co where there is drop in the cell volume. The drop in the cell volume for Co analogue is due to the major contribution of small size trivalent Yb atoms which was already confirmed in the magnetic susceptibility of two compounds where $Yb_5Ni_4Ge_{10}$ has only 78% Yb^{3+} and $Yb_5Co_4Ge_{10}$ has 94% Yb^{3+}.

The comparison of covalent radii in **Figure 3** also suggests that the metals like Fe, Ru, Os, Pt and Pd are

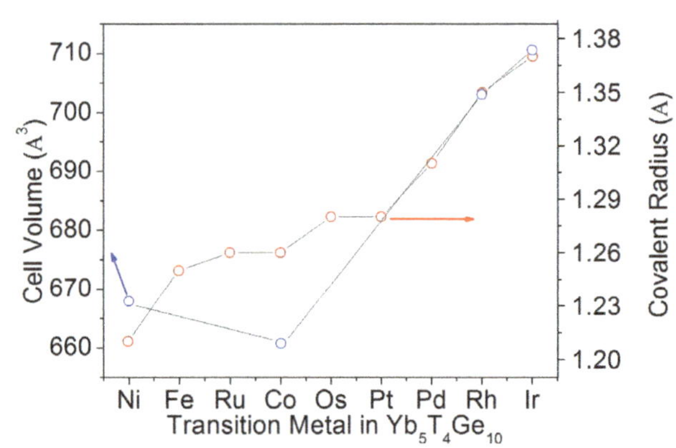

Figure 3. Comparison of the cell volume and covalent radii of transition metals in the $Yb_5T_4G_{e10}$ family.

suitable candidates for the formation $Yb_5T_4Ge_{10}$ compounds because the covalent radii of these metals are within the range of the reported compounds. This assumption can also be supported by the fact that Yb has the ability to show mixed valent compounds to stabilize the compounds.

4. Concluding Remarks

High quality single crystals of $Yb_5Co_4Ge_{10}$ were obtained from the metal flux technique and the crystal structure of $Yb_5Co_4Ge_{10}$ was studied using single crystal X-ray diffraction. Metal flux technique has been proved as vital synthesis method to obtain high quality single crystals. An absolute structure-property relation can be established by understanding the crystal structure from the single crystal X-ray data. Our study opens a lot of scope to find the crystal structure of the compounds which have not been studied as listed in **Table 1** and the synthesis of new compounds as expected based on covalent radii comparison of transition metals.

5. Acknowledgements

We thank Prof. C. N. R. Rao for his support and guidance. Financial support from JNCASR, DST and Sheikh Saqr Laboratory is gratefully acknowledged. Udumula Subbarao thanks CSIR for research fellowship and Sebastian C. Peter thanks DST for Ramanujan Fellowship.

REFERENCES

[1] H. F. Braun, K. Yvon and R. M. Braun, "$Sc_5T_4Si_{10}$ (T = Co, Rh, Ir) and $Y_5T_4Ge_{10}$ (T = Os, Ir) with a New Tetragonal Structure Type," *Acta Crystallographica Section B*, Vol. 36, No. 10, 1980, pp. 2397-2399.

[2] S. C. Peter, S. Rayaprol, M. C. Francisco and M. G. Kanatzidis, "Crystal Structure and Properties of $Yb_5Ni_4Ge_{10}$," *European Journal of inorganic Chemistry*, Vol. 2011, No. 26, 2011, pp. 3963-3968.

[3] K. Ghosh, S. Ramakrishnan, A. D. Chinchure, K. Jonason, V. R. Marathe, G. Chandra and S. Shah, "Heat-Capacity Studies in the $Y_{5-x}Dy_xOs_4Ge_{10}$ System," *Physical Review B*, Vol. 51, No. 17, 1995, pp. 11656-11663.

[4] K. Ghosh, S. Ramakrishnan and G. Chandra, "Magnetism in the $R_5Ir_4Si_{10}$ (R = Ho and Er) Systems," *Physical Review B*, Vol. 48, No. 6, 1993, pp. 4152-4155.

[5] G. Venturini, M. Méot Meyer, E. McRae, J. F. Maréché and B. Roques, "Vingt Nouveaux Germaniures Ternaires $TR_5T_4Ge_{10}$ de Metaux tr des Terres Rares et T = Co, Rh, Ir. Supraconductivite de $Lu_5Rh_4Ge_{10}$ et $Lu_5Ir_4Ge_{10}$," *Materials Research Bulletin*, Vol. 59, No. 12, 1984, pp. 1647-1652.

[6] H. F. Braun, G. Burri and L. Rinderer, "Partial Phase Diagram of the System Sc-Rh-Si," *Journal of the Less Common Metals*, Vol. 68, No. 1, 1979, pp. 1-8.

[7] O. I. Bodak, O. L. Sologub, P. Y. Zavalii and V. E. Zavodnik, "Crystal Structure of the Compound $Ho_5Os_4Ge_{10}$," *Russian Metallurgy* (*Metally*), Vol. 6, 1991, pp. 156-158.

[8] O. I. Bodak and O. L. Sologub, "The Ho-Os-Ge System," *Russian Journal of Inorganic Chemistry*, Vol. 36, 1991, pp. 1353-1355.

[9] O. L. Sologub, K. Hiebl, P. Rogl and H. Noël, "Magnetic Behavior of Holmium-Platinum Group Metal-Germanium Ternary Alloys," *Journal of Alloys and Compounds*, Vol. 245, No. 1-2, 1996, pp. L13-L17.

[10] L. S. Hausermannberg and R. N. Shelton, "Low Temperature Heat Capacity Study of Superconducting Ternary Silicides with the $Sc_5Co_4Si_{10}$-Type Structure," *Physica B+C*, Vol. 135, No. 1-3, 1985, pp. 400-404.

[11] N. G. Patil and S. Ramakrishnan, "Magnetism in the $R_5T_4Sn_{10}$ (R = Ce, Pr, and Nd; T = Rh and Ir) System," *Physical Review B*, Vol. 56, No. 6, 1997, pp. 3360-3371.

[12] H. D. Yang, R. N. Shelton and H. F. Braun, "Superconductivity and Electronic Instability at High Pressure in $Sc_5Co_4Si_{10}$-Type Compounds," *Physical Review B*, Vol. 33, No. 7, 1986, pp. 5062-5065.

[13] H. D. Yang, P. Klavins and R. N. Shelton, "Low-Temperature Physical Properties of $R_5Ir_4Si_{10}$ (R = Dy, Ho, Er, Tm, and Yb) Compounds," *Physical Review B*, Vol. 43, No. 10, 1991, pp. 7688-7694.

[14] S. Ramakrishnan, K. Ghosh and G. Chandra, "Magnetism and Superconductivity in $Sc_{5-x}Dy_xIr_4Si_{10}$ Alloys," *Physical Review B*, Vol. 46, No. 5, 1992, pp. 2958-2963.

[15] G. Venturini, B. Malamana and B. Roques, "New Rare Earth—Transition Metal Stannides with $Sc_5Co_4Si_{10}$ and $TiMnSi_2$—Type Structures. Atomic Size Effects on Their Stability," *Materials Research Bulletin*, Vol. 24, No. 9, 1989, pp. 1135-1139.

[16] R. E. Gladyshevskii, E. Parthé, O. L. Sologub and P. S. Salamakha, "Crystal Structure of Pentaholmium Tetrarhodium Decagermanium, $Ho_5Rh_4Ge_{10}$ with $Sc_5Co_4Si_{10}$ Type," *Zeitschrift für Kristallographie*, Vol. 203, No. 1, 1993, pp. 115-116.

[17] L. S. Hausermannberg and R. N. Shelton, "Low-Temperature Heat-Capacity Study of Superconducting Ternary Silicides and Germanides with the $Sc_5Co_4Si_{10}$-Type Structure," *Physical Review B*: *Condensed Matter*, Vol. 35, No. 13, 1987, pp. 6659-6664.

[18] H. F. Braun and C. U. Segre, "The Superconductivity of $Sc_5T_4Si_{10}$ (T = Co, Rh, Ir) and Isomorphous Compounds," *Solid State Communications*, Vol. 35, No. 10, 1980, pp. 735-738.

[19] N. G. Patil and S. Ramakrishnan, "Magnetism and Superconductivity in $M_5Rh_4Ge_{10}$ (M = Gd, Tb, Dy, Ho, Er,

Tm, Lu, and Y)," *Physical Review B*, Vol. 59, No. 14, 1999, pp. 9581-9589.

[20] M. Kolenda, M. Hofmann, J. Leciejewicz, B. Penc and A. Szytula, "Neutron-Diffraction Studies of $R_5Rh_4Ge_{10}$ (R = Tb, Ho, Er) Compounds," *Applied Physics A*, Vol. 74, No. 1, 2002, pp. S769-S771.

[21] S. Ramakrishnan, K. Ghosh and G. Candra, "Antiferromagnetism in the $Dy_5Ir_4Si_{10}$ System," *Physical Review B*, Vol. 45, No. 18, 1992, pp. 10769-10770.

[22] K. Katoh, T. Tsutsumi, K. Yamada, G. Terui, Y. Niide and A. Ochiai, "Magnetic and Transport Properties of $Yb_5T_4Ge_{10}$ (T = Co, Rh, Ir)," *Physica B*: *Condensed Matter*, Vol. 373, No. 1, 2006, pp. 111-119.

[23] P. Villars and K. Cenzual, "Pearson's Crystal Data-Crystal Structure Database for Inorganic Compounds," ASM International, Materials Park, 2010.

[24] Inorganic Crystal Structure Database, "Fachinformationszentrum Karlsruhe GmbH: Eggenstein-Leopoldshafen," Karlsruhe, 2012.

[25] B. Kindler, D. Finsterbusch, R. Graf and F. Ritter, "Mixed-Valence Transition in $YbInCu_4$," *Physical Review B*, Vol. 50, No. 2, 1994, pp. 704-707.

[26] E. Bauer, "Anomalous Properties of Ce-Cu- and Yb-Cu-Based Compounds," *Advances in Physics*, Vol. 40, No. 4, 1991, pp. 417-534.

[27] P. Wachter, "Handbook on the Physics and Chemistry of Rare Earths," Elsevier Science, Amsterdam, 1994, p. 177.

[28] Y. Matsumoto, S. Nakatsuji, K. Kuga, Y. Karaki, N. Horie, Y. Shimura, T. Sakakibara, A. H. Nevidomskyy and P. Coleman, "Quantum Criticality without Tuning in the Mixed Valence Compound B-$YbAlB_4$," *Science*, Vol. 331, No. 6015, 2011, pp. 316-319.

[29] U. Subbarao and S. C. Peter, "Crystal Structure of $YbCu_6In_6$ and Mixed Valence Behavior of Yb in $YbCu_{6-X}In_{6+X}$ (X = 0, 1, and 2) Solid Solution," *Inorganic Chemistry*, Vol. 51, No. 11, 2012, pp. 6326-6332.

[30] S. Ernst, S. Kirchner, C. Krellner, C. Geibel, G. Zwicknagl, F. Steglich and S. Wirth, "Emerging Local Kondo Screening and Spatial Coherence in the Heavy-Fermion Metal $YbRh_2Si_2$," *Nature*, Vol. 474, 2011, pp. 362-366.

[31] U. Subbarao and S. C. Peter, "Crystal Growth and Properties of $YbCuGa_3$: First Monoclinic System in the *RE*TX_3 Family," *Crystal Growth & Design*, Vol. 13, No. 2, 2013, pp. 953-959.

[32] U. Subbarao, M. Gutmann and S. C. Peter, "New Structure Type in the Mixed-Valent Compound $YbCu_4Ga_8$," *Inorganic Chemistry*, Vol. 52, No. 4, 2013, pp. 2219-2227.

[33] S. C. Peter, S. M. Disseler, J. N. Svensson, P. Carretta and M. J. Graf, "Yb_4LiGe_4—A Yb Mixed Valent Zintl Phase with Strong Electronic Correlations," *Journal of Alloys and Compounds*, Vol. 516, No. 516, 2012, pp. 126-133.

[34] S. C. Peter and M. G. Kanatzidis, "The New Binary Intermetallic $YbGe_{2.83}$," *Journal of Solid State Chemistry*, Vol. 183, No. 9, 2010, pp. 2077-2081.

[35] S. C. Peter, J. Salavador, J. B. Martin, W. Wong-Ng and M. G. Kanatzidis, "New Intermetallics $YbAu_2In_4$ and $Yb_2Au_3In_5$," *Inorganic Chemistry*, Vol. 49, No. 22, 2010, pp. 10468-10474.

[36] M. Chondroudi, S. C. Peter, C. D. Malliakas, M. Balasubramanian, Q. Li and M. G. Kanatzidis, "$Yb_3AuGe_2In_3$: An Ordered Variant of the YbAuIn Structure Exhibiting Mixed-Valent Yb Behavior," *Inorganic Chemistry*, Vol. 50, No. 4, 2011, pp. 1184-1193.

[37] S. C. Peter, S. Rayaprol, M. C. Francisco and M. G. Kanatzidis, "Crystal Structure and Properties of $Yb_5Ni_4Ge_{10}$," *European Journal of Inorganic Chemistry*, Vol. 2011, No. 26, 2011, pp. 3963-3968.

[38] S. C. Peter, M. Chondroudi, C. D. Malliakas, M. Balasubramanian and M. G. Kanatzidis, "Anomalous Thermal Expansion in the Square-Net Compounds RE_4TGe_8 (RE = Yb, Gd; T = Cr-Ni, Ag)," *Journal of the American Chemical Society*, Vol. 133, No. 35, 2011, pp. 13840-13843.

[39] S. C. Peter, S. Sarkar and M. G. Kanatzidis, "Metallic Yb_2AuGe_3: An Ordered Superstructure in the AlB_2-Type Family with Mixed-Valent Yb and a High-Temperature Phase Transition," *Inorganic Chemistry*, Vol. 51, No. 20, 2012, pp. 10793-10799.

[40] SAINT, "Bruker AXS," 6.02 ed., SAINT, Madison, 1999.

[41] G. M. Sheldrick, SADABS, "Empirical Absorption Correction Program," University of Göttingen, Göttingen, 1997.

[42] L. J. Farrugia, "WinGX Suite for Small-Molecule Single-Crystal Crystallography," *Journal of Applied Crystallography*, Vol. 32, No. 4, 1999, pp. 837-838.

[43] G. M. Sheldrick, "A Short History of SHELX," *Acta Crystallographica Section A*, Vol. A64, No. 1, 2008, pp. 112-122.

[44] SHELXTL 5.10, "Bruker Analytical X-Ray Systems," Madison, 1997.

[45] Crystal Impact, "Crystal Impact GbR," Version 3.g, Crystal Impact, Bonn, 2011.

[46] R. T. Sanderson, "Electronegativity and Bond Energy," *Journal of the American Chemical Society*, Vol. 105, No. 8, 1983, pp. 2259-2261.

[47] F. H. Allen, O. Kennard, D. G. Watson, L. Brammer, A. G. Orpen and R. Taylor, "Table of Bond Lengths Determined by X-Ray and Neutron Diffraction," *Journal of the American Chemical Society*, *Perkin Transactions* 2, Vol. 12, 1987, pp. S1-S19.

13

Characterization of Chromium Substituted Cobalt Nano Ferrites Synthesized by Citrate-Gel Auto Combustion Method

M. Raghasudha[1], D. Ravinder[2*], P. Veerasomaiah[3]
[1]Department of Chemistry, Jayaprakash Narayan College of Engineering, Mahabubnagar, India
[2]Department of Physics, Nizam College, Basheerbagh Osmania University, Hyderabad, India
[3]Department of Chemistry, Osmania University, Hyderabad, India

ABSTRACT

Co-Cr nano-ferrites, having the chemical formula $CoCr_xFe_{2-x}O_4$ (where x = 0.0, 0.1, 0.3, 0.5, 0.7, 0.9 and 1.0), were synthesized by the Citrate-gel auto combustion method. Synthesized powders were sintered at 500°C for 4 hours in an air and were characterized by XRD, SEM and EDS. XRD analysis showed cubic spinel structure of the ferrites and the crystallite sizes (D) were found in the range of 6 - 12 nm. The values of lattice parameter (a) decreased and X-ray density (d_x) increased with the increase of Cr content. Scanning Electron Microscopic (SEM) studies revealed nano crystalline nature of the samples. An elemental composition of the samples was studied by Energy Dispersive Spectroscopy (EDS). The observed results can be explained on the basis of composition and crystal size.

Keywords: Co-Cr Nano-Ferrites; Citrate-Gel Auto Combustion; X-Ray Diffraction; SEM; EDS

1. Introduction

Spinel ferrites are commercially important materials due to their excellent magnetic and electrical properties [1]. The interesting physical and chemical properties of the ferrites arise from their ability to distribute the cations among the tetrahedral (A) and octahedral (B) sites [2]. Among the various ferrites, cobalt ferrites have received recent attention as they possess excellent chemical stability, good mechanical hardness, remarkably high electrical resistivity, and large permeability at high frequency and cost effectiveness [3,4]. Due to these properties, cobalt ferrites have been extensively used in electronic devices [3,4], Magnetic Resonance Imaging (MRI), Target drug delivery, Hyperthermia for cancer treatment [5,6], high density storage devices, magnetic fluids [7,8], transformer cores, microwave devices, humidity and gas sensors, etc. Substitution of elements such as Chromium into Cobalt ferrites has been proposed, in order to tailor the magnetic and magneto mechanical properties of these materials [9]. Chromium substituted Cobalt ferrites are of particular interest in which the Cr^{+3} ions are reported to have a strong preference for the B sites (Octahedral) of the spinel structure of the ferrites [10,11]. A few previous investigations regarding Chromium substituted Cobalt ferrites have been reported which mainly concentrated on the Chromium rich side ($1 < x < 2$) [12,13]. It is known that the properties of the ferrites depend strongly on their composition and micro structure which are sensitive to the processing methods used to synthesize them. Several methods have been reported to synthesize chromium substituted Cobalt ferrite nano-materials; these include Powder Ceramic technique [14], Micro-emulsion method [15], Chemical Co-precipitation method [16] and Soft citrate gel method [17]. To the best of my knowledge, a little information is available on nano sized Chromium substituted Cobalt ferrites synthesized by Citrate-gel. Auto combustion method, which is a simple process, speeds up the synthesis and offers a significant saving in time, energy consumption over traditional methods. Hence in the present study we preferred the Citrate Gel auto combustion method for the synthesis of Chromium substituted Cobalt nano-ferrites.

2. Experimental

2.1. Synthesis

A series of Chromium substituted Cobalt nano ferrites

*Corresponding author.

having the chemical formula $CoCr_xFe_{2-x}O_4$ (where x = 0.0, 0.1, 0.3, 0.5, 0.7, 0.9 and 1.0) were prepared by Citrate Gel Auto-combustion method. The starting materials were Cobalt Nitrate ($Co(NO_3)_2{\cdot}6H_2O$), Ferric Nitrate ($Fe(NO_3)_2{\cdot}9H_2O$), Chromium Nitrate ($Cr(NO_3)_2{\cdot}9H_2O$), Citric acid ($C_6H_8O_7{\cdot}H_2O$) and Ammonia (NH_3) all of 99% pure AR grade. Calculated quantities of metal nitrates were dissolved together in a minimum amount of distilled water to get clear solution. An aqueous solution of Citric Acid was then added to the metal nitrate solution. The molar ratio of Citric acid to the total moles of nitrate ions was adjusted to 1:3. The mixture was stirred to obtain a homogeneous solution and then was slowly heated to 80°C at a rate of 5°C/minute using a hot plate magnetic stirrer. P^H of the solution was adjusted to 7 by adding Ammonia (NH_3) solution. A Sol is formed. The resulting solution was evaporated to dryness by heating at about 150°C on a hot plate with continuous stirring. As a result the viscosity rose due to cross linking of carboxylato-metal complexes into a three dimensional structure (pierre *et al.*, 1990; Jang *et al.*, 1995, Narebder abd Messing, 1997) and get started to form a viscous gel (**Figure 1(a)**). When finally all water molecules were removed from the mixture by increasing the temperature to 200°C, the viscous gel began frothing. The gel gave a fast flameless auto combustion reaction with the evolution of large amounts of gases (**Figure 1(b)**). It started in the hottest zones (**Figure 1(c)**) of the beaker and propagated from the bottom to the top like the eruption of a Volcano (**Figure 1(d)**). The reaction was completed in a minute giving rise to dark grey voluminous product with a structure similar to Branched tree (**Figure 1(e)**). Finally the burnt powder was ground and was calcined in air at temperature 500°C for four hours to obtain a spinel phase.

Figure 1. (a) Viscous gel formation; (b) Self ignition; (c) Combustion of the mixture; (d) Propagation of combustion like volcanic eruption; (e) Dark grey voluminous product with a structure similar to a branched tree.

2.2. Characterization

X-ray Diffraction with Cukα (lambda = 1.54 Å) was used to study the single phase nature and nano-phase formation of the Co-Cr ferrite system at room temperature by continuous scanning in the range of 2°C to 85°C.

Micro structural analysis of the prepared samples was carried out by scanning Electron microscopy (SEM) and elemental compositional analysis for all samples was done by Energy Dispersive Spectroscopy (EDS).

3. Results and Discussions

3.1. XRD Analysis

The X-ray diffraction patterns of all the samples were shown in **Figure 2**. XRD patterns and the crystalline phases were identified by comparison with reference data from the ICSD card No. 22-1086 for Cobalt ferrites ($CoFe_2O_4$). The XRD patterns of all the Chromium substituted Cobalt ferrites showed a homogeneous single phased cubic spinel belonging to the space group Fd3m (confirmed by ICSD Ref 22-1086). All the Braggs reflections have been indexed, which confirmed the formation of a well defined single phase cubic spinel structure without any impurity peaks. All the peaks are allowed peaks. The strongest reflection has come from (311) plane that indicates spinel phase.

Crystallite size of each sample was calculated from Scherrers Formula [18] mentioned below by using high intensity peak (311).

$$D = \frac{0.91\lambda}{\beta\cos\theta}$$

where λ = Wavelength of X-ray, β = Full width and Half Maxima in radians, θ = Bragg's angle at the peak position.

Measured crystallite size was in the range of 6 - 12 nm (shown in **Table 1**) for different compositions with a significant increase of 6 nm in response to the increase in Cr substitution. To my knowledge Citrate-gel method is the only method which has resulted in Co-Cr ferrites with such a very small crystallite size.

Lattice parameter "*a*" of individual composition was calculated by using the following formula and values were tabulated in **Table 1**.

$$a = d\sqrt{h^2 + k^2 + l^2}$$

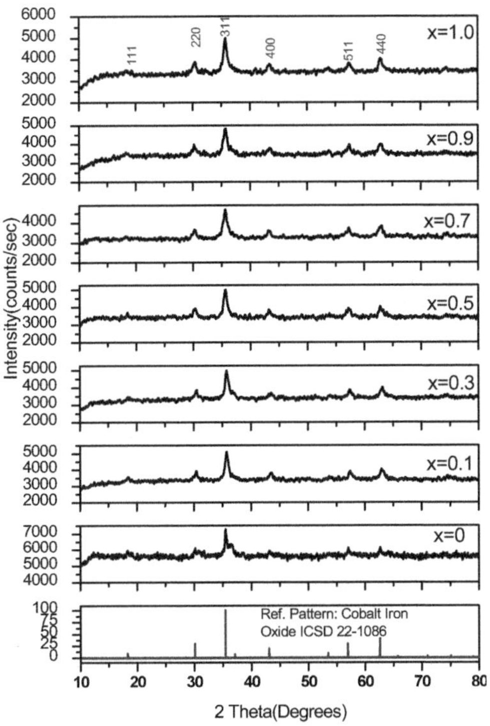

Figure 2. XRD patterns of $CoCr_xFe_{2-x}O_4$ nano ferrite (where x = 0, 0.1, 0.3, 0.5, 0.7, 0.9 and 1.0).

Table 1. Crystallite size (D), Lattice parameter (a) of $CoCr_xFe_{2-x}O_4$ Nano-ferrites with x = 0, 0.1, 0.3, 0.5, 0.7, 0.9 and 1.0.

Ferrite composition	Crystallite size (nm)	Lattice parameter (Å)
$CoFe_2O_4$	6.5	8.390
$CoCr_{0.1}Fe_{1.9}O_4$	12.1	8.380
$CoCr_{0.3}Fe_{1.7}O_4$	12.8	8.367
$CoCr_{0.5}Fe_{1.5}O_4$	11.9	8.356
$CoCr_{0.7}Fe_{1.3}O_4$	10.5	8.346
$CoCr_{0.9}Fe_{1.1}O_4$	10.7	8.335
$CoCrFeO_4$	12.3	8.330

where a = lattice parameter, d = inter planar distance, hkl = miller indices.

The variation of lattice parameter with Cr^{+3} compositions was shown in **Figure 3**. The lattice parameter was found to decrease linearly with increasing Cr concentration. This linear variation indicates that the Co-Cr ferrite system obeys Vegard's law [19].

A similar behavior of lattice parameter with dopant concentration was observed by several investigators in various ferrite systems [20,21]. The decrease in lattice parameter with increase in Cr^{+3} content can be explained on the basis of relative ionic radii of Cr^{+3} and Fe^{+3} ions. As Cr^{+3} ions have small ionic radii (0.63 Å) than that of Fe^{+3}(0.64 Å), a partial replacement of Fe^{+3} ions by Cr^{+3} ions results in a decrease in lattice parameter. Calculated lattice parameter for Cobalt ferrite ($CoFe_2O_4$) was in good agreement with the standard value (8.391 Å) reported from ICSD data.

X-ray density (d_x) for different compositions was calculated using the formula [22] and calculated values were tabulated in **Table 2**.

$$d_x = \frac{ZM}{Na^3}\, gm/cc$$

where Z = Number of molecules per unit cell (8), M = Molecular weight of the sample, N = Avagadro's Number, a = lattice parameter.

The variation of X-ray density with Cr^{+3} compositions was shown in **Figure 4**. X-ray density depends on the lattice parameter and molecular weight of the sample. In the present Co-Cr system with increase in the Cr^{+3} content, it is observed that the molecular weight is decreased (due to greater atomic weight of Fe-55.847 gm/mole and less atomic weight of Cr-51.996 gm/mole) and the lattice

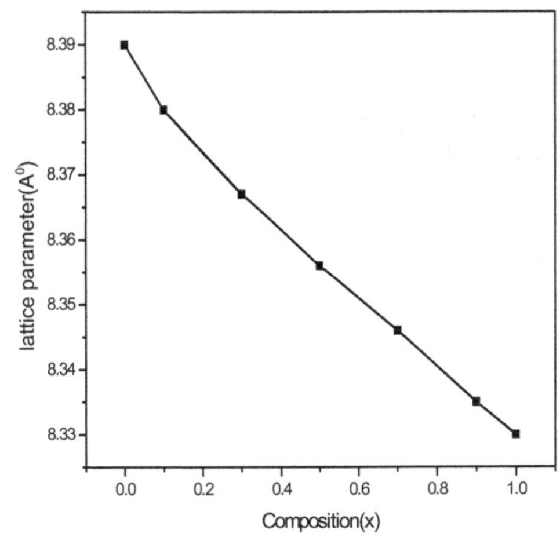

Figure 3. Variation of lattice parameter with Cr composition for Co-Cr nano ferrites.

Table 2. X-ray density (d_x), volume of the unit cell (V) of-$CoCr_xFe_{2-x}O_4$ nano-ferrites with x = 0, 0.1, 0.3, 0.5, 0.7, 0.9 and 1.0.

Ferrite composition	Density d_x (gm/cc)	Volume of unit cell (V)
$CoFe_2O_4$	5.276	590
$CoCr_{0.1}Fe_{1.9}O_4$	5.287	588
$CoCr_{0.3}Fe_{1.7}O_4$	5.294	585
$CoCr_{0.5}Fe_{1.5}O_4$	5.297	583
$CoCr_{0.7}Fe_{1.3}O_4$	5.299	581
$CoCr_{0.9}Fe_{1.1}O_4$	5.302	579
$CoCrFeO_4$	5.303	578

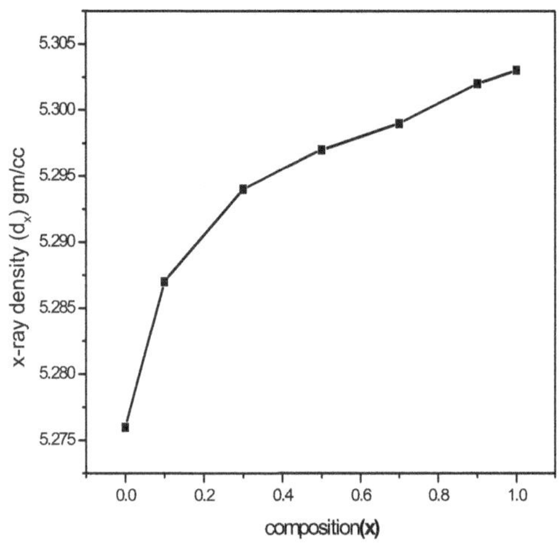

Figure 4. Variation of X-ray density with Cr composition for Co-Cr Nano ferrites.

parameter has also decreased. X-ray density was found to increase with increasing Cr concentration this is due to the fact that decrease in volume (V) of the unit cell overtakes the decrease in mass (M) in the system.

A similar behavior of X-ray density with Cr composition was reported by Md. Javed Iqbal, Mah Rukh Siddiqyah in Co-Cr ferrite system synthesized by Microemulsion method [15].

Volume of unit cell was calculated by using the formula

$$V = a^3 \text{in} \left(\text{Å} \right)^3$$

where 'a' is lattice parameter.

The calculated values were tabulated in **Table 2**. Variation of Volume of unit cell with Cr Composition was shown in **Figure 5**. Volume of unit cell was found to decrease with increase in Cr content, as it depends on lattice parameter which has decreased with increase in Cr content.

X-ray density(d_x), volume of the unit cell (V) of $CoFe_2O_4$ were in good agreement with the standard values (5.274 gm/cc, 590.99$(A^0)^3$) from ICSD data.

The distance between magnetic ions (hopping length) in A site (tetrahedral) and B site (octahedral) were calculated using the relations

$$d_A = 0.25a\sqrt{3}, d_B = 0.25a\sqrt{2}$$

"a" = lattice parameter.

The calculated values of the hopping length for Tetrahedral site(d_A) and octahedral(d_B) of different compositions were tabulated in **Table 3**. The relation between hopping length for Octahedral and Tetrahedral sites as a function of Cr^{+3} content(x) was shown in **Figure 6**. It is observed that the hopping length decreases as the Cr content increases. It may be due to the fact that Cr^{+3} ion has smaller radius (0.63 Å) than Fe^{+3} ion *i.e.* (0.64 Å). This makes the magnetic ions become closer to each other and hopping length decreases.

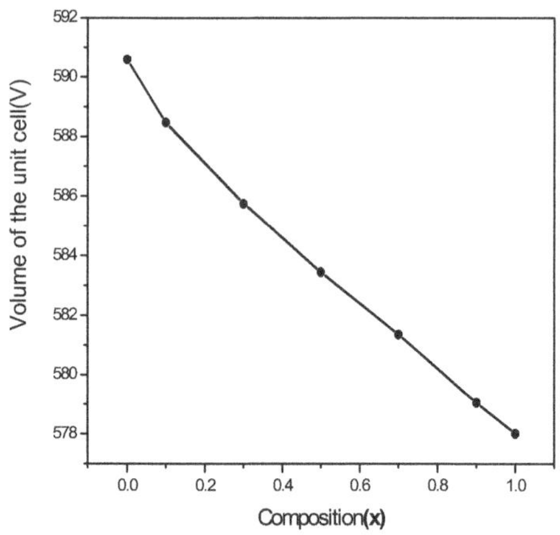

Figure 5. Variation of volume of the unit cell with Cr composition for Co-Cr nano ferrites.

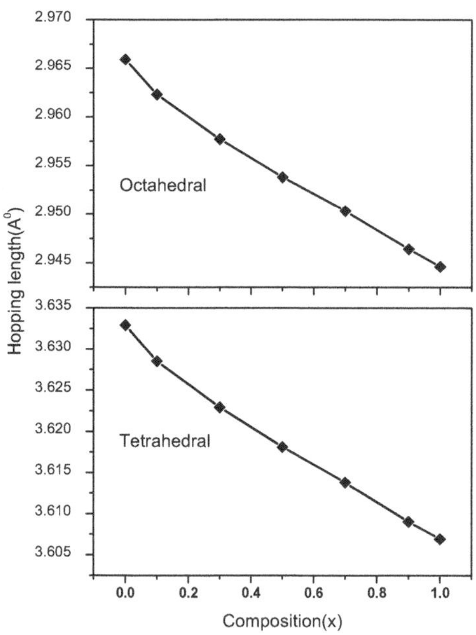

Figure 6. Variation of hopping length d_A and d_B of Co Cr_x $Fe_{2-x}O_4$ with x= 0.0, 0.1, 0.3, 0.5, 0.7, 0.9 and 1.0.

Table 3. Values of hopping length d_A and d_B of Co Cr_x $Fe_{2-x}O_4$ with x = 0.0, 0.1, 0.3, 0.5, 0.7, 0.9 and 1.0.

Ferrite composition	d_A	d_B
$CoFe_2O_4$	3.6329	2.9659
$CoCr_{0.1}Fe_{1.9}O_4$	3.6285	2.9623
$CoCr_{0.3}Fe_{1.7}O_4$	3.6229	2.9577
$CoCr_{0.5}Fe_{1.5}O_4$	3.6181	2.9538
$CoCr_{0.7}Fe_{1.3}O_4$	3.6138	2.9503
$CoCr_{0.9}Fe_{1.1}O_4$	3.6090	2.9464
$CoCrFeO_4$	3.6069	2.9446

Similar behavior of hopping length with Cr^{+3} composition was observed for the Mg-Cr system synthesized by double sintering technique reported by V.B. Kawade, G.K. Bichile, K.M. Jadhav [21].

3.2. Morphology by SEM

Morphology of the prepared samples by Citrate-gel method was studied using scanning electron microscope (SEM) where the secondary electron images were taken at different magnifications to study the morphology. The scanning electron microscopic images of all the synthesized samples were shown in **Figure 7**.

The images show that the particles have an almost homogeneous distribution, and some of them are in agglomerated form. It is evidenced by SEM images that the aggregation of particles lies in nanometric region. The particles were observed as uniform grains (in different SEM images) confirming the crystalline structure of Co-Cr ferrites which were detected by XRD studies. The formation of Fe_2O_4 was chemically favored by heating during the synthesis where as final reaction was completed during the sintering where the pores between the particles were removed combined with growth and strong bonds by agglomeration. It is also observed that agglomeration of the compound was increased with change in composition of the sample from x = 0 to x = 1 which may be due to the Cr^{+3} ions. SEM images reveal that with increase in Cr composition the grain size has increased (x = 0 to x = 1), hence is an evidence for XRD analysis.

3.3. Elemental Analysis by EDS

The elemental analysis of all the Co-Cr nano ferrite samples with different compositions was analyzed by Energy Dispersive Spectrometer (EDS) and the elemental % and atomic % of different elements in the were shown in the **Table 4**. The EDS pattern for samples with x = 0.0, 0.1 and 0.5 were shown in **Figure 8** which indicates the elemental and atomic composition in the sample. The compounds show the presence of Co, Cr, Fe and O without precipitating cations.

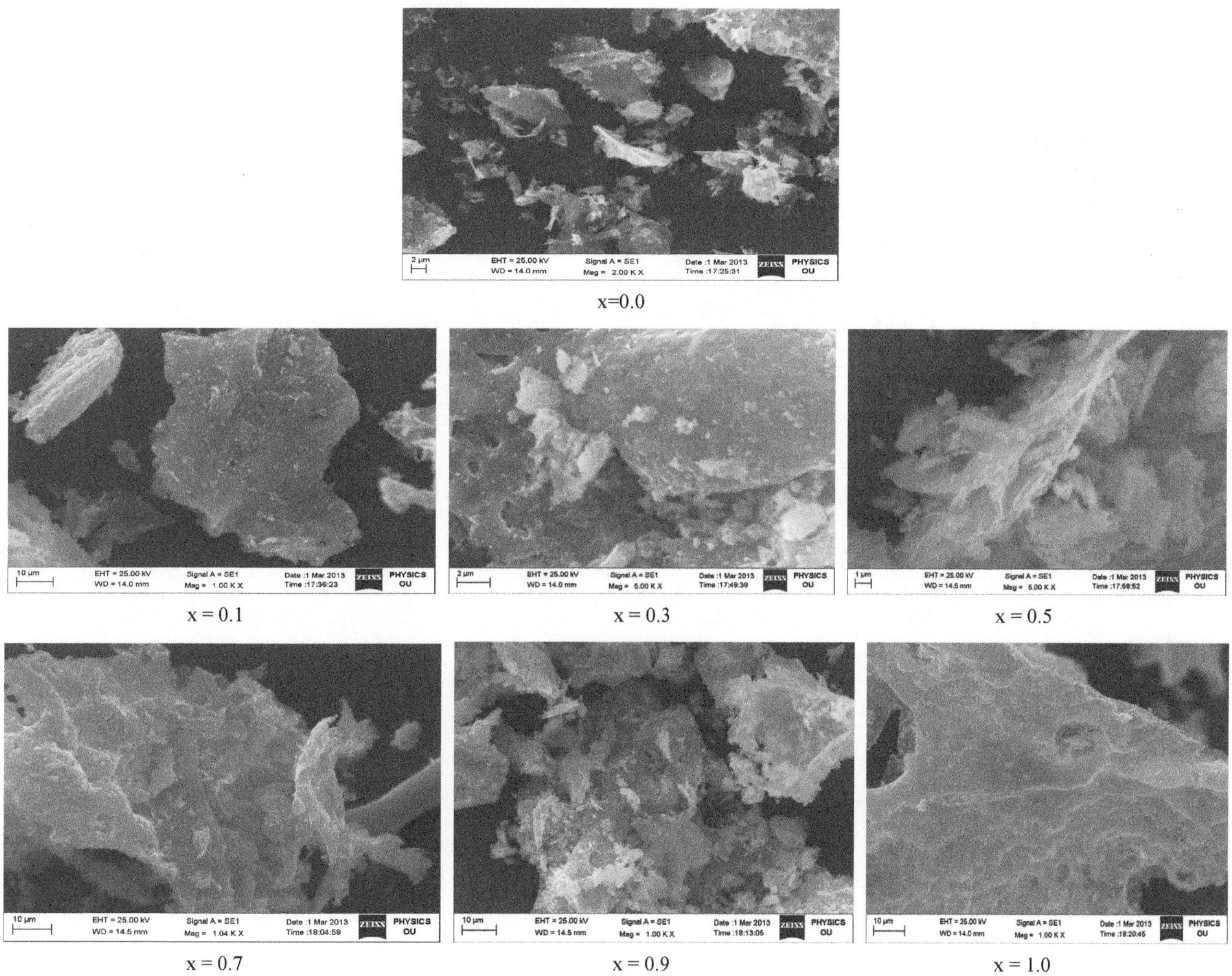

Figure 7. SEM images of Co Cr_x $Fe_{2-x}O_4$ with x = 0.0, 0.1, 0.3, 0.5, 0.7, 0.9 and 1.0.

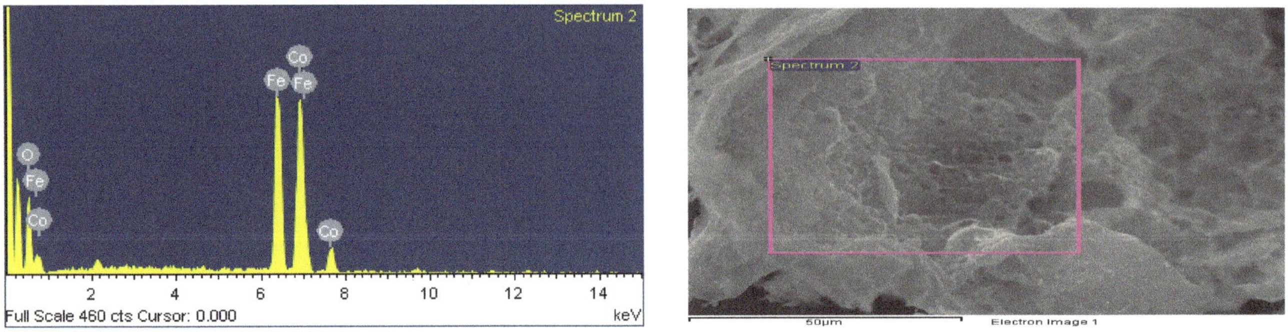

x = 0.0

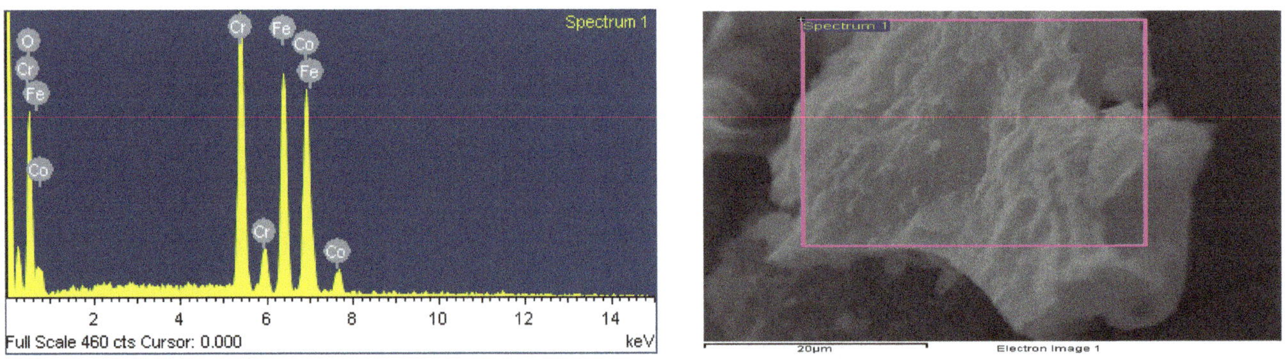

x = 0.1

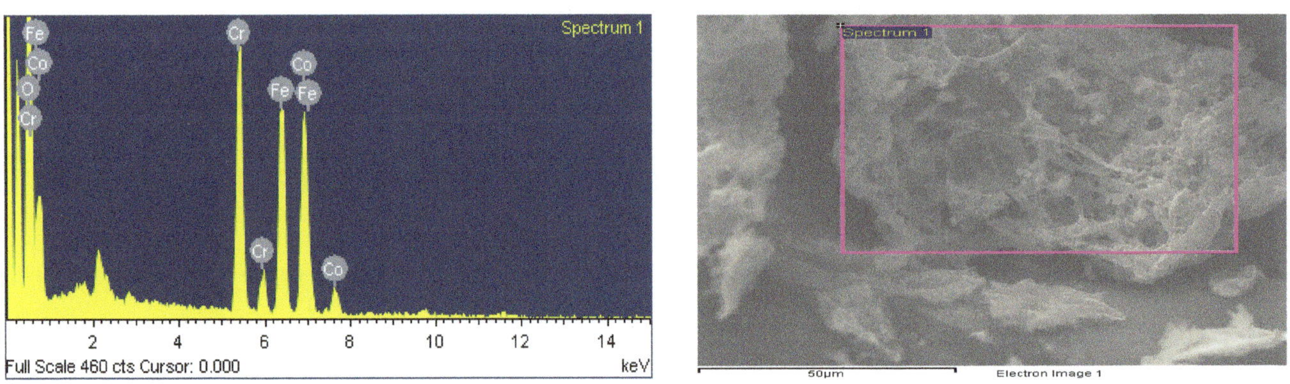

x = 0.5

Figure 8. EDS patterns of Co Cr_x $Fe_{2-x}O_4$ (for x = 0.0, 0.1, 0.5).

Table 4. Elements of each sample composition analyzed by (% weight) obtained by EDS.

Element	O		Fe		Co		Cr	
Ferrite composition	Element%	Atomic %	Element %	Atomic %	Element %	Atomic %	Element %	Atomic %
$CoFe_2O_4$	19.93	47.18	37.89	25.7	42.18	27.12	-	-
$CoCr_{0.1}Fe_{1.9}O_4$	44.25	73.44	18.91	8.99	20.61	9.29	16.23	8.29
$CoCr_{0.3}Fe_{1.7}O_4$	34.99	65.22	20.84	11.13	24.87	12.58	19.3	11.07
$CoCr_{0.5}Fe_{1.5}O_4$	22.44	50.17	25.97	16.63	28.12	17.06	23.47	16.14
$CoCr_{0.7}Fe_{1.3}O_4$	27.02	56.34	23.51	14.04	27.9	15.79	21.56	13.83
$CoCr_{0.9}Fe_{1.1}O_4$	13.07	34.31	26.86	20.2	31.95	22.77	28.12	22.72
$CoCrFeO_4$	34.16	64.36	21.3	11.5	24.63	12.6	19.91	11.54

4. Conclusions

- Citrate Gel auto combustion technique is a convenient way for obtaining a homogeneous nano sized mixed Co-Cr ferrites. The process involves no impurity pickup and material loss. It is a very simple and economical method where no specific heating or cooling rate is required. It is a low temperature processing technique and requires shorter sintering duration.
- X-ray diffraction pattern confirms the formation of cubic spinel structure in single phase without any impurity peak. It is in good agreement with the standard data from ICSD.
- The crystallite size of the various Co-Cr ferrites was in the range of 6 - 12 nm.
- The lattice parameter has decreased with the increase of Cr substitution in Co-Cr ferrites which indicate that the mixed Co-Cr ferrite system obeys the Vegard's law.
- SEM micrographs of various compositions indicate the morphology of the particles is similar. They reveal largely agglomerated, well defined nano particles of the sample powder with inhomogeneous broader grain size distribution.
- EDS data give the elemental % and atomic % in the mixed Co-Cr ferrites and it shows the presence of Co, Cr, Fe and O without precipitating cations.

5. Acknowledgements

One of the authors, M. Raghasudha, is thankful to the chairman K. S. Ravi Kumar, Jayaprakash Narayan college of Engineering for his support in establishing the lab for the synthesis of samples. One of the authors (D.R) is grateful to Prof. T. L. N. Swamy, Principal Nizam college for his encouragement to carry out this research work. The authors are grateful to Prof. C. Gyana Kumari, Head, Department of Chemistry for her encouragement in carrying out research activity. The authors are also thankful to Prof. M. Vithal, Department of Chemistry, O.U. for his support in characterization of samples.

REFERENCES

[1] P. V. Reddy and T. S. Rao, "Dielectric Behavior of Mixed Li-Ni Ferrites at Low Frequencies," *Journal of Less Common Metals*, Vol. 86, 1982, pp. 255-261.

[2] G. Blasse, *Philips Research Reports*, Suppl. 3, 1964.

[3] J. Smit and H. P. J. Wijn, "Ferrites," Philips Technical Library, Eindhoven, 1959.

[4] K. Ishino and Y. Narumiya, "Development of Magnetic Ferrites: Control and Application of Losses," *American Ceramic Society Bulletin*, Vol. 66, No. 10, 1987, pp. 1469-1474.

[5] D. H. Kim, S. H. Lee, K. N. Kim, K. M. Kim, I. B. Shin and Y. K. Lee, "Cytotoxicity of Ferrite Particles by MTT and Agar Diffusion Methods for Hyperthermic Application," *Journal of Magnetism and Magnetic Materials*, Vol. 293, No. 1, 2005, pp. 287-292.

[6] S. Somiya, S. Aldinger, N. Claussen, R. M. Uchino, R. M. Uchino, K. Koumto and K. Kanenoin, "Handbook of Advanced Ceramics, Vol. II, Processing and Their Applications," Elsevier Academic Press, Waltham, 2003, p. 394.

[7] F. Zhang, S. Kantake, Y. Kitamoto and M. Abe, "Spin-Spray Ferrite-Plated co Ferrite Films with High Coercivity for Perpendicular Magnetic Recording Media," *IEEE Transactions on Magnetics*, Vol. 35, No. 5, 1999, pp. 2751-2753.

[8] R. Arulmurugan, G. Vaidyanathan, S. Sendhilnathan and B. Jeyadavan, "Mn Zn Ferrite Nanoparticles for Ferrofluid Preparation: Study on Thermal-Magnetic Properties," *Journal of Magnetism and Magnetic Materials*, Vol. 298, No. 2, 2006, pp. 83-94.

[9] J. A. Paulsen, A. P. Ring, C. C. HLO, J. E. Snyder and D. C. Jiles, "Manganese-Substituted Cobalt Ferrite Magnetostrictive Materials for Magnetic Stress Sensor Applications," *Journal of Applied Physics*, Vol. 97, No. 4, 2005, Article ID: 044552.

[10] R. C. O. Handley, "Modern Magnetic Materials Principles and Applications," Wiley, New York, 2000.

[11] A. Hauet, J. Teillet, B. Hannoyer and M. Lenglet, "Mössbauer Study of Co and Ni Ferrichromites," *Physics Status Solidi* A, Vol. 103, No. 1, 1987, pp. 257-261.

[12] L. G. Antoshina, A. N. Goryaga and R. R. Annarv, "Magnetization Processes of the Sample $CoFeCrO_4$ with Frustrated Magnetic Structure," *Journal of Magnetism and Magnetic Materials*, Vol. 195, No. 3, 1999, pp. 608-610.

[13] T. Tsushima, "Magnetic Properties of Ferrite-Chromite Series of Nickel and Cobalt," *Journal of Physics Society of Japan*, Vol. 18, 1963, pp. 1162-1166.

[14] K. Krieble, C. C. H. Lo, Y. Melikhov and J. E. Snyder, "Investigation of Cr Substitution in Co Ferrite ($CoCr_xFe_{2-x}O_4$) Using Mossbauer Spectroscopy," *Journal of Applied Physics*, Vol. 99, No. 8, 2006, Article ID: 08M912.

[15] M. J. Iqbal and M. R. Siddiquah, "Electrical and Magnetic Properties of Chromium Substituted Cobalt Ferrite Nano Materials," *Journal of Alloys and Compounds*, Vol. 453, No. 1-2, 2008, pp. 513-518.

[16] M. Anis-ur-Rehman, M. Ansari, Z. N. Mughal, M. S. Awan and A. Maqsood, "Synthesis and Thermoelectric Studies in Cr doped Cobalt Ferrite Nano Paticles," *Key Engineering Materials, Advanced Materials XII*, Vol. 510-511, 2012, pp. 171-176.

[17] P. P. Hankare, U. B. Sankpal, R. P. Patel, P. D. Lokhande and R. Sasikala, "Synthesis, Characterization and Catalytic Activity of Chromium Substituted Cobalt Ferro

Spinels," *Materials Science and Engineering*: *B*, Vol. 176, No. 2, 2011, pp. 103-109.

[18] B. D. Cullity, "Elements of XR-Diffraction," Addison Weseley Publishing, Reading, 1959, p. 132.

[19] L. Vegard, "The Constitution of Mixed Crystals and the Space Occupied by Atoms," *Zeitsch rift fiir Physics*, Vol. 5, No. 17, 1921, pp.17-23.

[20] V. B. Kawade, G. K. Bichile and K. M. Jadhav, "X-Ray and Infrared Studies of Chromium Substituted Magnesium Ferrite," *Material Letters*, Vol. 42, No. 1-2, 2000, pp. 33-37.

[21] R. Arulmurugan, B. Jeyadevan, G. Vaidyanathan and S. Sendhilnathan, "Effect of Zinc Substitution on Co-Zn and Mn-Zn Ferrite Nanoparticles Prepared by Coprecipitation," *Journal of Magnetism and Magnetic Materials*, Vol. 288, 2005, pp. 470-477.

[22] R. C. kumbale, P. A. sheikh, S. S. Kamble and Y. D. kolekar, "Effect of Cobalt Substitution on Structural Magnetic and Electric Properties of Nickel Ferrite," *Journal of Alloys and Compounds*, Vol. 478, No. 1-2, 2009, pp. 599-603.

14

High Resolution X-Ray Diffraction Analyses of (La,Sr)MnO_3/ZnO/Sapphire(0001) Double Heteroepitaxial Films

Katsuhiko Inaba[1], Shintaro Kobayashi[1], Kenichi Uehara[2,3], Akira Okada[2], Sanapa Lakshmi Reddy[4], Tamio Endo[2]

[1]X-Ray Research Laboratory, Rigaku Corporation, Tokyo, Japan
[2]Graduate School of Engineering, Mie University, Mie, Japan
[3]Adwel R&D Co., Ltd., Shiga, Japan
[4]Department of Physics, Sri Venkateswara Degree College, Kadapa, India

ABSTRACT

Thin films of (La,Sr)MnO_3 (LSMO) were deposited on ZnO underlayer on sapphire (Sap) substrate by ion beam sputtering at the growth temperatures from 650˚C to 750˚C with supply of either plasma oxygen or molecular oxygen. The epitaxial relationships of lattices and crystalline qualities of LSMO/ZnO/Sap double-hetero systems were extensively analyzed by means of X-ray diffraction techniques with a modern high resolution XRD system. The epitaxial growth of out-of-plane (110) phase of the LSMO overlayer on the (0001) ZnO underlayer on the (0001) sapphire substrate is promoted at the higher temperatures accompanying the suppression of other competitive growth of the LSMO (001) and (111) phases. The plasma oxygen supply more enhances this suppression of the LSMO (111) phase at 750˚C. The complex in-plane orientational relationships of these three phases were clarified by the precise analyses based on various measurements of high resolution out-of-plane XRD and in-plane XRD, pole figure, and reciprocal space mappings. They are summarized as follows.

$\mathbf{[1\bar{1}0](110)LSMO//[1\bar{1}00](0001)ZnO//[11\bar{2}0](0001)Sap}$,

$\mathbf{[110](001)LSMO//[1\bar{1}00](0001)ZnO//[11\bar{2}0](0001)Sap}$, and

$\mathbf{[11\bar{2}](111)LSMO//[1\bar{1}00](0001)ZnO//[11\bar{2}0](0001)Sap}$ (μ-mode), where the notations such as $\mathbf{[1\bar{1}0]}$ indicate the in-plane crystalline directions. The two possible configurations of LSMO (111) phase were discussed. These relationships were confirmed also by the data analyses of out-of-plane wide-range reciprocal space mappings using the 2-dimensional X-ray detector.

Keywords: X-Ray Diffraction; Heteroepitaxy; LSMO/ZnO; Orientational Relations; In-Plane XRD; Reciprocal Space Mapping

1. Introduction

Various combinations of oxide thin films have been studied for fabrications of new functional devices. Perovskite (La,Sr)MnO_3 (LSMO) manganate shows versatile electrical and magnetic properties such as colossal magnetoresistance [1] and is a p-type semiconductor at higher temperatures. While ZnO is an n-type widegap semiconductor which is studied as transparent conductive oxide for thin film transistors and display applications, since it is transparent for visible light but absorbs UV light. We are trying to obtain a novel p-n heterojunction of LSMO/ZnO system whose characteristics can be modified by magnetic field, temperature and UV light [2]. For these purposes, the controls of the crystalline orientations of these materials are crucial to obtain superior and controllable junction characteristics.

Bhosle and Narayan [3] investigated an epitaxial growth of an (110) out-of-plane oriented LSMO directly on a c-plane sapphire (Sap), and reported the epitaxial orientational relationship as

$\mathbf{\left[1\bar{1}0\right](110)LSMO//\left[1\bar{1}00\right](0001)Sap}$. In this article, we define (110) and (0001) as the surface planes which show the out-of-plane crystalline orientations of the grown thin film, and $[\mathbf{1\bar{1}0}]$ and $[\mathbf{1\bar{1}00}]$ as the crystalline axes parallel to the surface planes which show the in-plane crystalline orientations of the films, respectively, hereafter. The lattice matching between the LSMO film and the sapphire substrate for the case of Bhosle and Narayan is expected to be good since $d_{LSMO,2\bar{2}0}$ = 0.13753 nm (a_{LSMO} = 0.3890 nm) is very close to $d_{Sap,3\bar{3}00}$ = 0.13746 nm (a_{Sap} = 0.4762 nm). However, the lattice mismatch in the perpendicular direction within the surface plane is as high as 19%. They reported that this relatively large mismatch was accommodated by matching of four (200) planes of the LSMO films with five $(\bar{2}110)$ planes of sapphire.

In our case, the lattice mismatch should be considered between LSMO films and ZnO underlayers. The interplanar distances of major lattice planes perpendicular to the surface plane are $d_{LSMO,1\bar{1}0}$ = 0.27506 nm and $d_{LSMO,002}$ = 0.19450 nm for the LSMO films, and $d_{ZnO,11\bar{2}0}$ = 0.16250 nm and $d_{ZnO,1\bar{1}00}$ = 0.28146 nm (a_{ZnO} = 0.3250 nm) for the ZnO underlayer. A small lattice mismatch of −2.2% is expected by the coincidence of $d_{LSMO,1\bar{1}0}$ and $d_{ZnO,1\bar{1}00}$, and also a lattice mismatch of 20% is expected by the coincidence of $d_{LSMO,002}$ and $d_{ZnO,11\bar{2}0}$. Thus, the epitaxial growth of the LSMO film on the ZnO underlayer with the epitaxial orientation of $[1\bar{1}0](110)$LSMO//$[1\bar{1}00](0001)$ZnO is expected. A relatively large mismatch along LSMO [001] direction would be accommodated by matching of four (200) planes of the LSMO films with five $(\bar{2}110)$ planes of ZnO underlayers. Thus some improvement of the crystalline quality of LSMO epitaxial layer can be expected. Further, we propose that the in-plane orientation of LSMO against the sapphire substrate can be controlled by insertion of intermediate ZnO layer.

In this study, we try to extensively analyze lattice relationships of epitaxial growth and crystalline qualities in the LSMO/ZnO/sapphire double-hetero system. We use here X-ray Diffraction (XRD) technique with the modern high resolution XRD systems, such as out-of-plane 2θ-θ XRD scans, ω rocking curves, in-plane XRD scans, ϕ rocking curves, pole figure measurements, and reciprocal space mappings.

2. Experimental

2.1. Thin Film Fabrications

The thin films of (La,Sr)MnO_3 (LSMO) overlayer were epitaxially grown on the ZnO underlayer on the (0001) sapphire (Sap) substrate by ion beam sputtering technique. Molecular oxygen (ML) or plasma oxygen (PL) was supplied during the growth. Deposition conditions for the LSMO overlayer were varied as a function of the substrate temperature (Ts: 650˚C - 750˚C) and the oxygen partial pressure (Po). Details of the film fabrication can be referred in the previous papers [1,4].

Four LSMO/ZnO/Sap double-hetero epitaxial films were mainly studied in this article. Two of them are termed as PL-LT and PL-HT in which the LSMO films were grown at the low Ts (650˚C) and high Ts (750˚C) with the supply of plasma oxygen, respectively. The other two are termed as ML-LT and ML-HT in which the LSMO films were grown at the low Ts (650˚C) and high Ts (700˚C) with the supply of molecular oxygen, respectively. Many other double-layer films grown at different deposition conditions were also characterized to give deliberate and correct discussions, while the experimental results are not shown for these samples in this paper.

2.2. XRD Analysis

All the XRD measurements were performed with the "SmartLab" diffractometer system (Rigaku Corp.) [5], equipped with high power 9 kW Cu rotating anode generator (**Figure 1(a)**). The sample is manipulated with 3-circles (ω, χ, ϕ) and the detector is moved with 2-circles (2θ and $2\theta\chi$). The goniometer configuration is schematically shown in **Figure 1(b)**. The samples of LSMO/ZnO/Sap films were extensively analyzed with this XRD system in terms of their crystalline qualities (mosaic spreading; tilt and twist), lattice strains, and orientational relationships between LSMO and ZnO/sapphire substrate. We employed various measurement techniques, such as out-of plane XRD measurements and X-ray Rocking Curves (XRCs), in-plane XRD measurements and XRCs, Pole Figure (PF) measurements, and out-of-plane wide-range Reciprocal Space Mapping (RSM) using the 2-dimensional (2D) X-ray detector. These techniques should be briefly mentioned in this section.

2.2.1. Out-of-Plane XRD Measurements: 2θ-θ Scans and ω-Scans

For the high resolution XRD analysis with double axis geometry, a Ge220-2bounce channelcut monochromating collimator was employed in the incident optics to use the Cu$K\alpha_1$ radiation (λ = 0.154059 nm) [6,7]. A resolution of optics for XRC was confirmed to be better than 0.004 deg by the rocking curve measurement for the sapphire 0006 reflection. The 2θ-θ scans were performed to detect diffraction signals arising from the lattice planes parallel to the sample surface plane as shown in **Figure 2**, which give the information of preferential growth axes (lattice planes) of the LSMO overlayer (and ZnO underlayer). The ω-scan rocking curve measurements were performed on the LSMO peaks to evaluate the tilt mosaic spreading of LSMO overlayer.

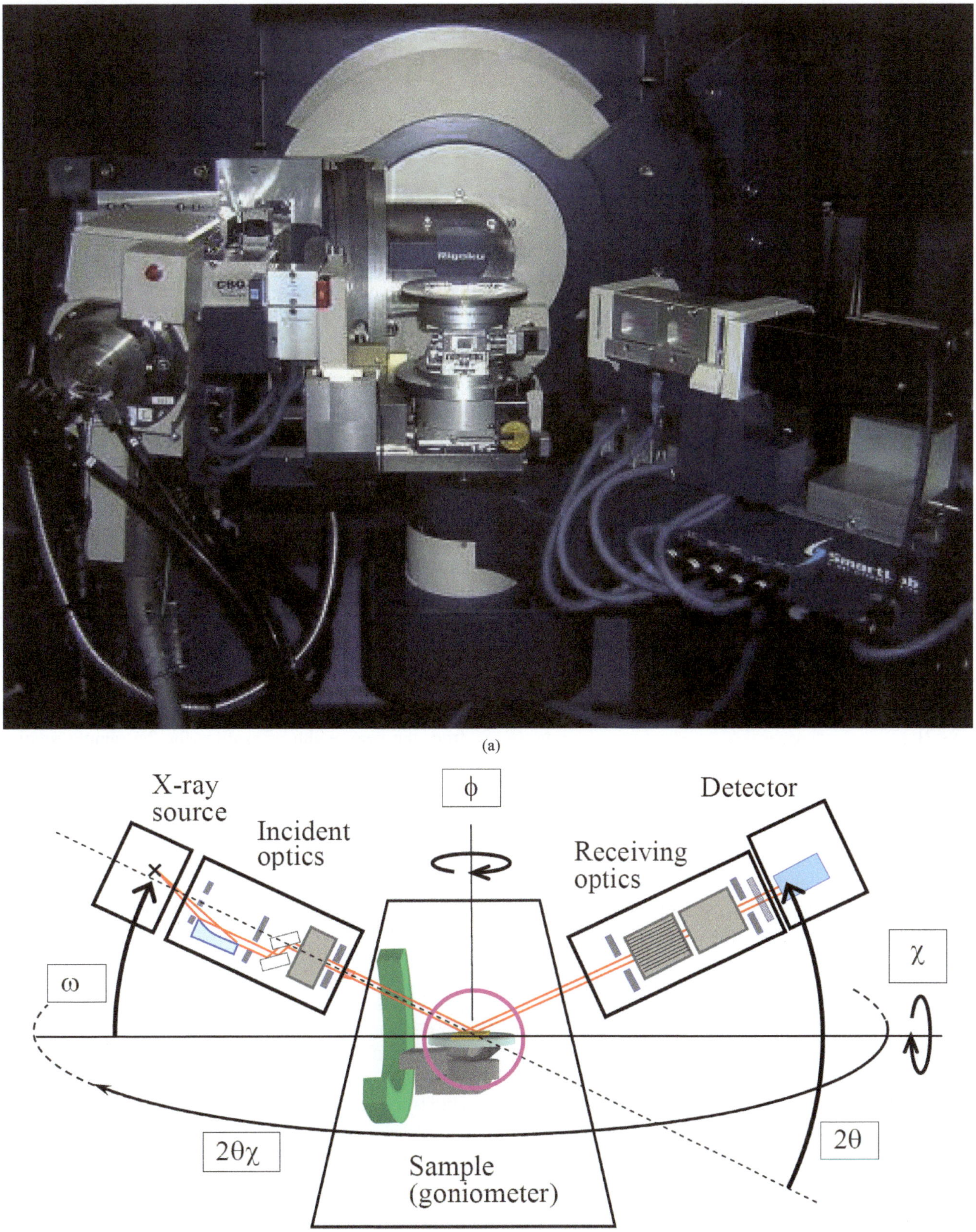

Figure 1. (a) An over view of the 5-circle thin film X-ray diffractometer equipped with the $2\theta\chi$ axis (SmartLab); (b) A schematic illustration of 5-circle goniometer system.

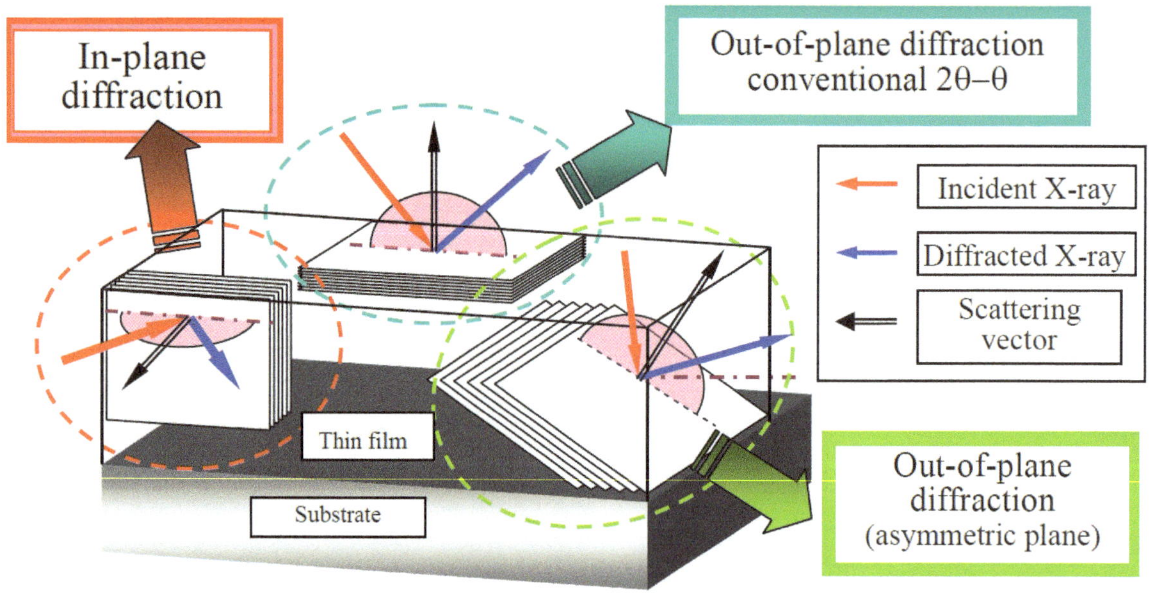

Figure 2. A schematic diagram of the geometries of thin film X-ray diffraction measurement, showing relations between the lattice planes of a thin film sample and X-ray geometries.

2.2.2. In-Plane XRD Measurements: $2\theta\chi$-ϕ Scans, ϕ-Scans, and In-Plane Reciprocal Space Mappings

The in-plane XRD is a powerful technique for thin film characterizations, since this technique is geometrically sensitive to film surface layers. Further we can directly access to the lattice planes perpendicular to the film surface which is illustrated in **Figure 3** as well as **Figure 2** [8-12].

Regarding the surface sensitivity, the divergence of the incident X-ray beam in the normal direction to the sample surface greatly affects the sensitivity, which is indicated by vertical arrows in **Figure 3(b)**. Therefore, the in-plane XRD measurements with SmartLab were performed by controlling the incident angle (ω) to the sample surface plane in the accuracy of around 0.01 deg. There we utilized the well shaped parallel incident X-ray beam formed by the parabolic multilayered mirror, using CuKα radiation (λ = 0.154186 nm as the weighted average of CuKα_1 and CuKα_2). This measurement with the grazing incidence beam geometry was performed by the sample rotation with ϕ-axis and detector motion of 2χ, while the sample was kept on the horizontal plane without tilting the sample by χ-axis (**Figure 1(b)**). In this article, the incident angle to the sample surface in the in-plane XRD measurements was set to be around 0.32 deg which is close to the angle for the total external reflection of LSMO [8,11].

A resolution of the in-plane XRD profiles should depend on the optical elements employed, such as the Parallel Slit Collimator (PSC) in the incident side, and the Parallel Slit Analyzer (PSA) in the receiving side (**Figure 3**). The horizontal resolution of 0.50 deg was set in this study. The direct accessibility of this technique to the lattice planes perpendicular to the sample surface plane enables us to analyze the in-plane lattice constants directly by the coupling motion ($2\theta\chi$-ϕ) of detector ($2\theta\chi$) and sample rotation (ϕ). Moreover, the directional distributions of perpendicular crystalline planes near the surface plane (*i.e.*, twist mosaic spreading) can be evaluated from the peak width of profiles for the sample rotation (ϕ), while the detector is positioned at fixed $2\theta\chi$ angle. As illustrated in **Figure 3**, the in-plane reciprocal space mapping (in-plane RSM) measurements were performed by the iterative motions of incremental step of $\Delta\phi$ and $2\theta\chi$-ϕ scans [11].

2.2.3. Pole Figure Measurements Using the In-Plane Axis [13]

The pole figure (PF) measurement is an indispensable technique for the characterization of orientational relationships in the complex heteroepitaxial thin film systems. A sample is generally manipulated with tilting and rotated with ϕ-axis while keeping the 2θ-θ configuration. This geometry is called as a skew geometry, where the angles of the incident X-rays and exit X-rays to the sample surfaces should be kept identical. The χ-axis is commonly used for tilting a sample in usual PF measurements in a case of conventional XRD systems. However, in the present PF measurement using the in-plane axis, the lattice planes inclined to the surface plane can be accessed by the tilting motion through the combining motions of 2θ and $2\theta\chi$, without tilting the sample. As can be easily seen from the schematic illustration of the goniometer

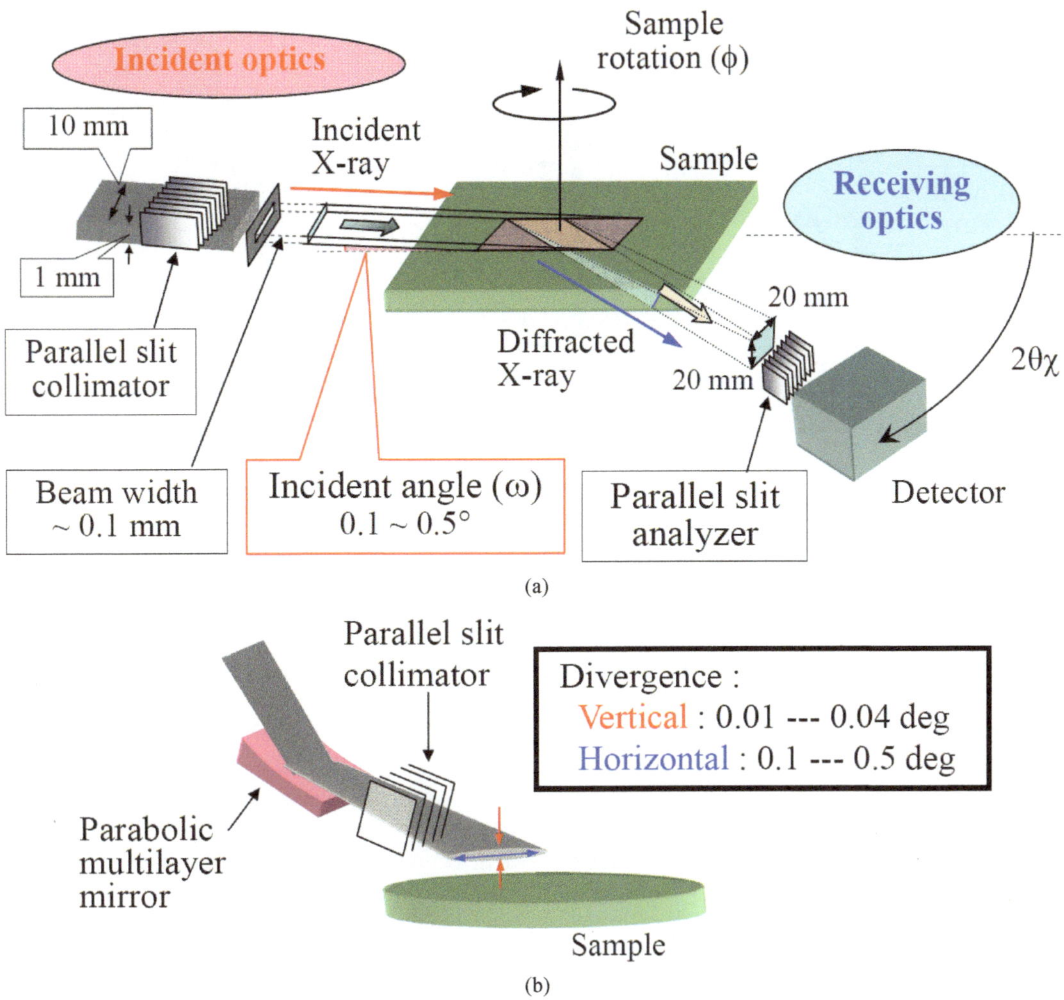

Figure 3. (a) A schematic illustration of goniometer setup and motion for in-plane XRD measurements; (b)) A schematic illustration of the incident optics setup for in-plane XRD measurements using a parabolic multilayer X-ray mirror.

motion in **Figure 4**, a whole hemisphere pole figure measurement is accomplished with this PF measurement using the in-plane axis. The azimuthal angle (β) is denoted for the relative angle of ϕ from the referential azimuthal direction of a sample, for example, $[11\bar{2}0]$ axis of the sapphire substrate. The same configuration of the optics is employed with that for the in-plane XRD measurement (**Figure 3**). This is the strong merit of this system we improved for precise in-plane measurements. We measured the PF on the present samples using this novel in-plane system.

2.2.4. Out-of-Plane Wide-Range Reciprocal Space Mapping Using the 2D X-Ray Detector [14]

The RSM measurements were complementarily done to cover the PF measurements for the characterization of orientational relationships in the complex heteroepitaxial thin film systems. The RSMs of wide angular ranges can be measured using the 2D detector and point-shaped incident X-ray beam as shown in **Figure 5(a)**. The PILATUS 100K/R 2-dimensional solid-state semiconductor detector is employed in this measurement, instead of a 0-dimensional scintillation detector. The CBO-*f* device [14] is employed in the incident optics, which acts as a converter of the line-shaped X-ray beam to the point -shaped X-ray beam (diameter of around 0.4 mmϕ). The photograph of this system is shown in **Figure 5(a)**.

Both of the RSM and the PF measurements should access to the lattice plane inclined with ψ with respect to the sample surface as illustrated in **Figure 5(b)**. Due to the geometrical restriction to keep the skew geometry (2θ-θ geometry), the sample is tilted with the χ-axis, while the 2D detector was scanned with time-delayed integration (TDI) mode (**Figure 5(b)**). The area of RSM measured by this mode is painted in red where the diffraction vector is marked by a red arrow in **Figure 5(b)**. For a comparison, the area of RSM measured by the conventional mode of ω-shift and 2θ-ω scan is shown as a sector colored in blue. In this mode, the sample is tilted with the ω-axis, instead of the χ-axis, which is a so-called co-planar ge-

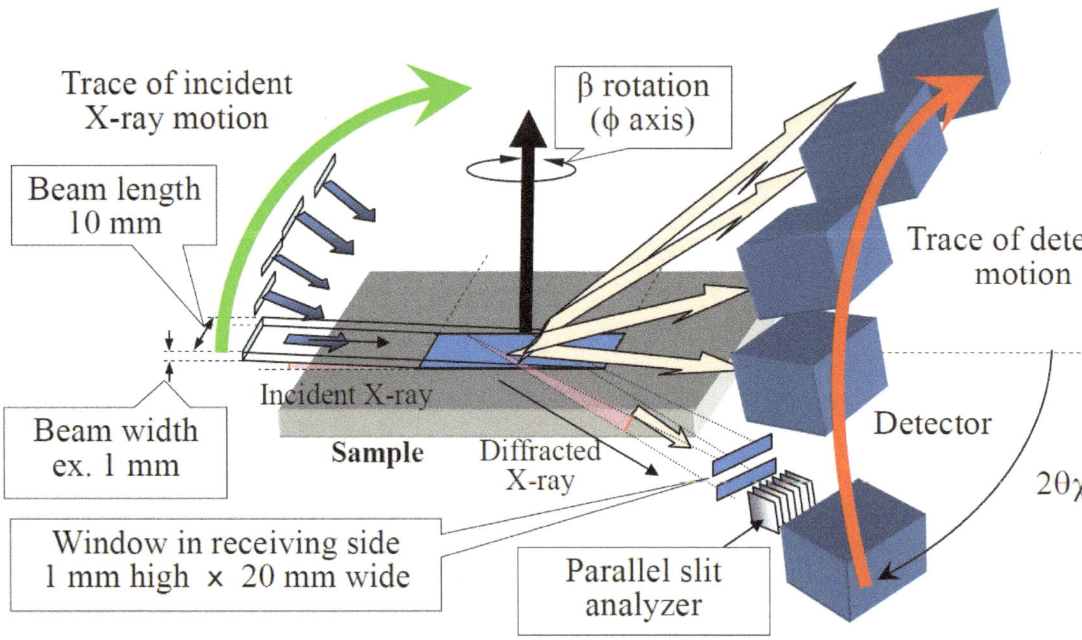

Figure 4. A schematic illustration of the goniometer motion for PF measurement using the in-plane axis.

ometry (**Figure 5(b)**). An optical resolution to the tilting direction in this mode is defined with the divergence of the incident optics to the omega direction, and the higher resolution is expected when the monochomating collimator is employed. The area of in-plane RSM is shown as a section colored in green. The capture angles in the χ tilting direction for a single 2θ-ω scan with the 2D detector was set as ±7.5 deg with a sample-detector distance of 150 mm.

The wide range RSM data can be obtained by iterative motions of the goniometer for small χ steps and $2\theta/\omega$ scans. The 2D data should be shown either in the goniometer coordinates (the two coordinate axes are tilting angles (χ) and 2θ angles) or in the reciprocal space coordinates. The two orthogonal axes are employed in a scale of the reciprocal unit (1/Å is used in this study), those are the direction along the surface plane and the direction normal to the surface plane.

3. Results and Discussion

3.1. Out-of-Plane XRD Measurements

Figure 6 shows the out-of-plane 2θ-θ scan profiles for the four samples. The LSMO overlayer peaks of 110, 111, 002, 220, 222 reflection can be clearly observed as well as 0002 and 0004 reflections from the ZnO underlayer, and 0006 reflections from the sapphire substrate. A sharp peak at $2\theta = 64.5$ deg is the 0009 reflections from sapphire substrates, which is intrinsically the forbidden reflections but actually observed due to the effect of multiple reflections [15]. From these data, it is assumed that phases with three different preferred orientation textures competitively exist in the LSMO overlayer grown on the ZnO underlayer on the (0001) sapphire substrate. These are, (110) growth phase, (111) growth phase and (001) growth phase (represented by 002 reflection peak). Hereafter we denote these out-of-plane oriented LSMO phases as the (110) phase, (111) phase and (001) phase. Extended profiles around the peaks of LSMO 110, 111 and 002 reflections are shown in **Figures 7(a)-(c)**, respectively. The peak intensities of LSMO 110 of PL-HT and ML-HT are several times stronger than those of PL-LT and ML-LT. This indicates that the LSMO (110) growth is greatly promoted at the higher Ts (700°C and 750°C) than at the lower Ts (650°C). Looking at the 111 peaks, the intensity for ML-HT is much stronger than that for ML-LT and PL-LT. Conspicuously the 111 peak for PL-HT cannot be observed. The 002 peak intensity for ML-HT is weaker than that for PL-LT and ML-LT. The 002 peak for PL-HT cannot be observed again.

It is quite clear that the (110) phase growth is much promoted at the higher temperature than at the lower temperature, contrary the (001) phase growth is promoted at the lower temperatures than at the higher temperatures. This must be resulted from a different lattice matching between the LSMO overlayer and ZnO underlayer. Probably the LSMO (110) phase has the better lattice matching at the higher temperatures, while the LSMO (001) phase has the better lattice matching at the lower temperature due to different thermal expansion rates of these lattices. It was partially reported by our previous paper [4]. However, the LSMO (111) phase behaves curiously. It is promoted at the higher temperatures up to 700°C, but it cannot be grown at 750°C. Exactly speaking, at 750°C

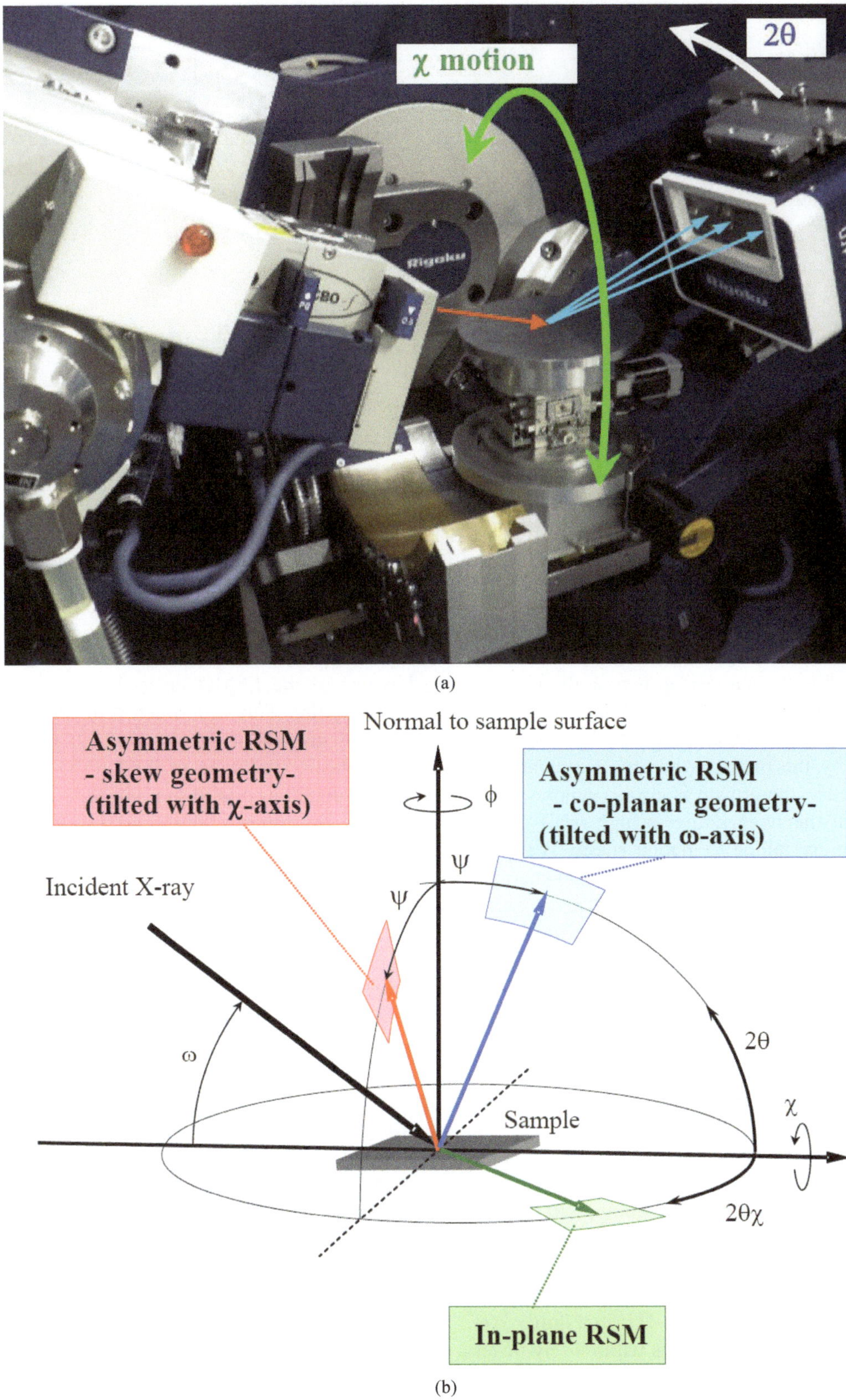

Figure 5. (a) A photograph of SmartLab system equipped with the 2D detector for the wide-range reciprocal space mapping measurement. The incident X-ray beam is shown by a red arrow, and the diffracted X-ray beams are shown by blue arrows. Tilting motion for the sample with χ-axis is shown by a curved green arrow; (b) Geometry of the reciprocal space mapping measurement.

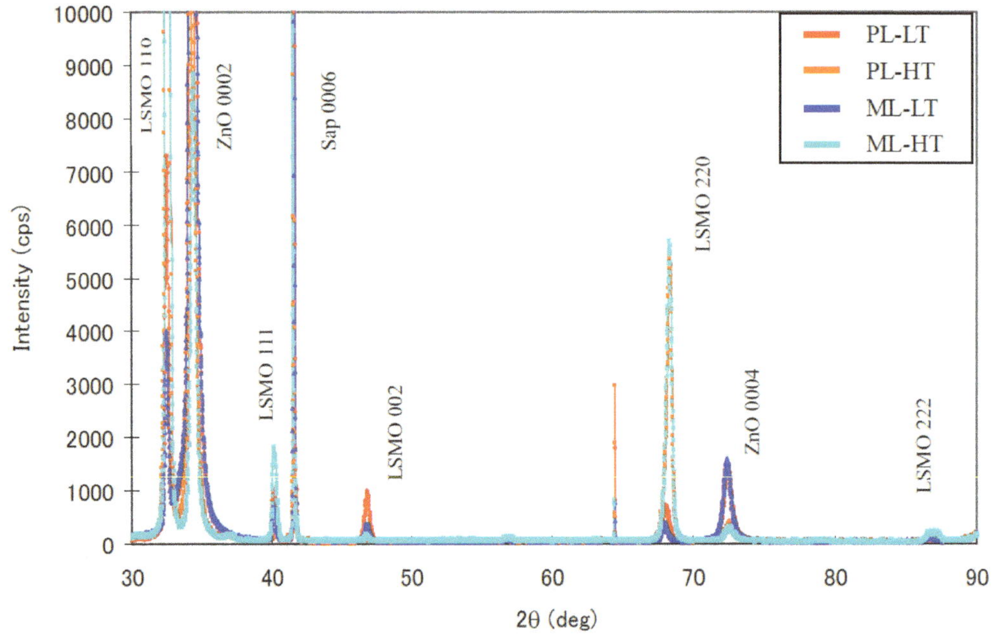

Figure 6. Out-of-plane XRD: 2θ-θ scan profiles for the four samples. The horizontal axis is shown in 2θ angle (deg), and the vertical axis is the intensity in counts per second (cps) unit with linear scale.

LSMO (111) phase can be grown faintly under the molecular supply, while it cannot be grown at all under the plasma supply. We can recognize an outstanding rule of the plasma effect for the three phases. The growth is always enhanced by the plasma when the temperature is low, on the contrary, the growth is always suppressed by the plasma when the temperature is high. This clearly indicates that the crystalline formation energy is supplemented by the plasma energy when the thermal energy is insufficient. The plasma energy, however, gives excess energy when the thermal energy is sufficient, leading to decomposition of LSMO crystal structure. This must be the reason why the LSMO (111) and (001) phases cannot be grown at the higher temperature of 750°C under the plasma supply.

It should be noted that as shown in **Figure 7(d)** asymmetric tailing of the peaks to the lower angles in 2θ can be observed for the 220 reflections of PL-HT and ML-HT grown at the higher temperatures. We decomposed the (220) peak of ML-HT as schematically shown by two broken curves. This tail in the lower 2θ angle is found to be overlapping with the peaks of the other samples of ML-LT and PL-LT grown at the lower temperature. Further, a doublet peak is clearly observed in the higher order 222 reflection of ML-HT in **Figure 7(e)**, which apparently implies two phases with closely different lattice constants exist in the LSMO overlayer. Estimated lattice constants from these peaks are **a** = 0.3890 (±0.0003) nm for PL-HT and **a** = 0.3875 (±0.0003) nm for ML-HT after the deconvolution analysis of the 220 reflections. The value of **a** = 0.3875 nm is smaller than the bulk value, which is expected to arise from near-interface domains of the LSMO layers being constrained by the ZnO/sapphire. This is resulted from the lattice matching epitaxial growth of promoted (110) phase. This might cause elastic shrinkage of the LSMO lattices along the growth direction due to the tensile stress near the interface. On the contrary, the larger value of **a** = 0.3890 nm is close to that estimated from the peaks for PL-LT and ML-LT, and this value coincides with that of the bulk LSMO.

The orientational distribution of the crystallite axis of the (110) phase can be analyzed from the tilt spreading in the ω rocking curve profiles for the LSMO 110 reflections shown in **Figure 8**. Full Width at Half Maximum (FWHM) values are around 2.0 deg for the four samples, which are matched to the FWHM values for ZnO 0002 reflections (not shown here). Thus it is supposed that the tilting of the overlayer LSMO (110) phase is grossly limited by that of the ZnO underlayer. Looking the curves carefully, we notice the same rule with the above. The FWHM value is smaller for the higher temperatures than for the lower temperature samples, implying that the LSMO with better mosaic spreading can be grown at the higher temperatures. At the higher temperature, the LSMO with better crystalline quality can be grown by the molecular supply than by the plasma supply, implying that the plasma gives the excess energy for the crystal growth at the higher temperature. On the contrary, the better mosaic crystalline LSMO can be grown by the plasma supply than by the molecular supply at the low temperature. It implies that the plasma energy can supplement the

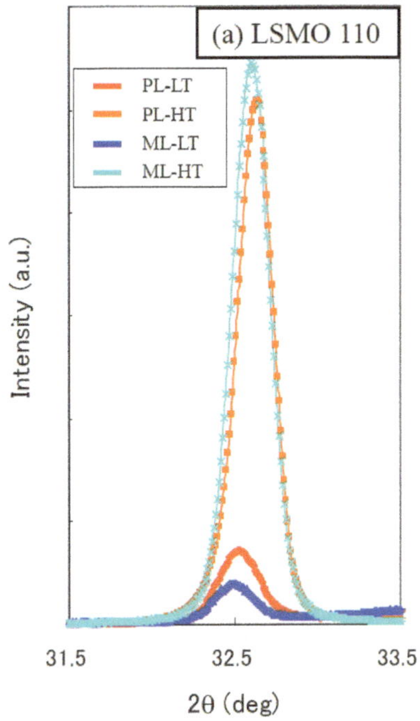

(a) LSMO 110 reflections

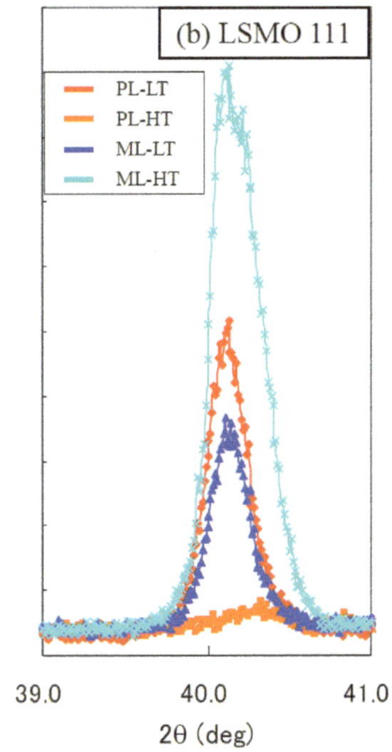

(b) LSMO 111 reflections

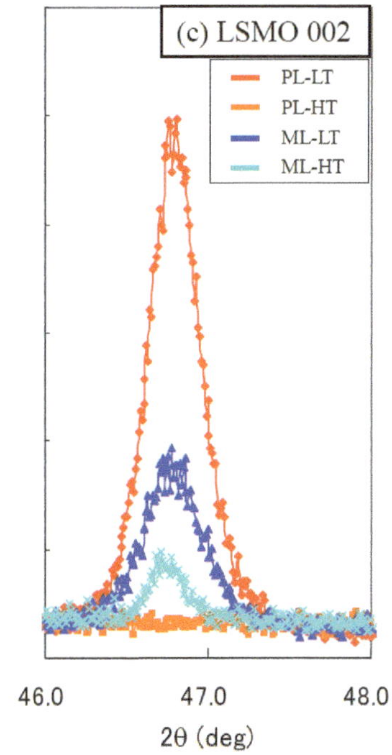

(c) LSMO 002 reflections

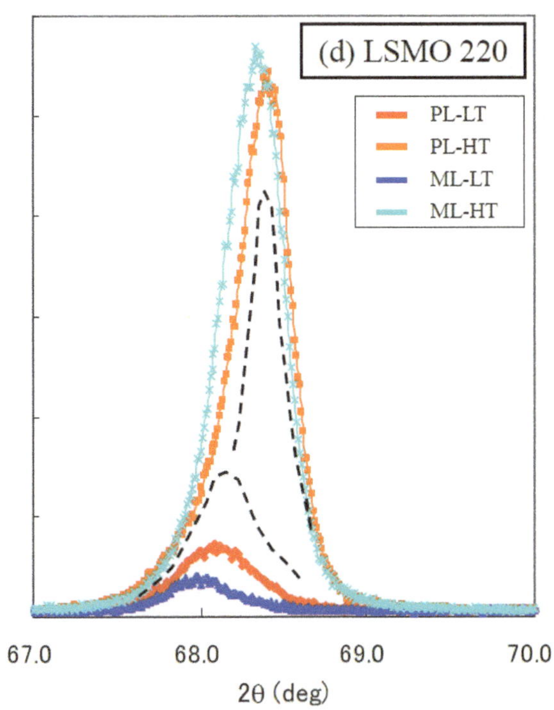

(d) LSMO 220 reflections

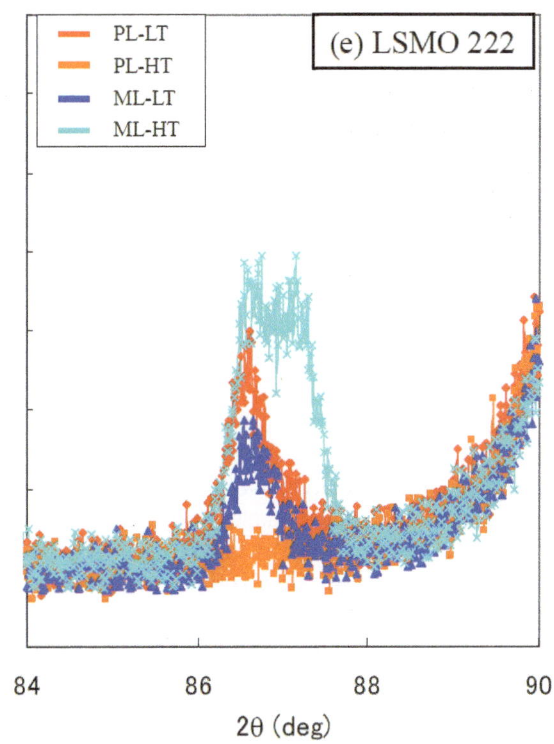

(e) LSMO 222 reflections

Figure 7. Extended 2*θ*-*θ* profiles of the LSMO (110), (111) and (001) phases in Figure 6. (a) LSMO 110 reflections; (b) LSMO 111 reflections; (c) LSMO 002 reflections; (d) LSMO 220 reflections; (e) LSMO 222 reflections. Two black broken curves in Figure 7(d) are schematically shown as the deconvoluted peaks of the sample ML-HT for eye guide.

shortage of the thermal energy when the thermal energy is insufficient, leading to the better mosaic crystallinity.

An inset in **Figure 8** is a rocking curve profile for the sapphire 0006 refection. The FWHM value of this peak is 0.0035 deg, which corresponds to the resolution of the present optics.

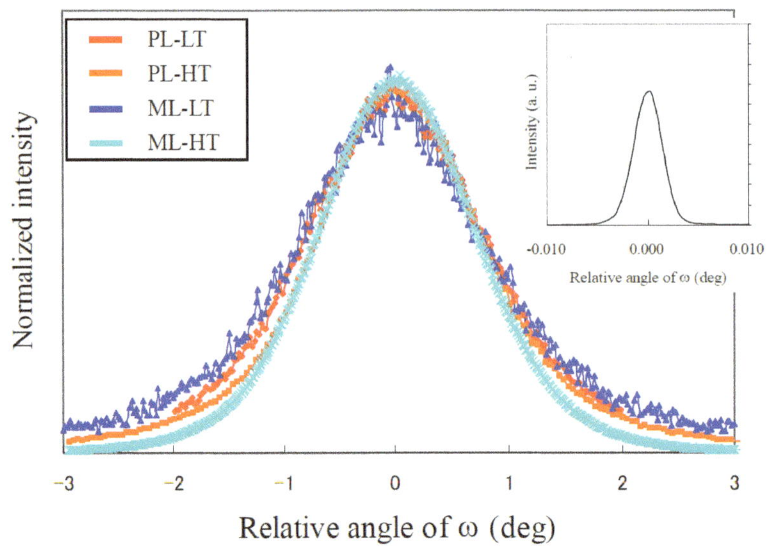

Figure 8. High resolution ω rocking curve profiles for the LSMO 110 reflections for the four samples. Horizontal axes are shown in the relative angles of ω in degree. Vertical axes are shown in the normalized intensity with the peak top intensity. The inset shows a rocking curve profile for the sapphire 0006 refection. This gives a resolution of the employed optics as around 0.0035 deg in FWHM.

3.2. In-Plane XRD Measurements

In-plane Reciprocal Space Mapping (RSM) data [11,16, 17] for the four samples are shown in **Figures 9(a)-(d)** with the reciprocal space coordinates in the unit of 1/Å. In these figures, the horizontal axis is aligned to $\langle \mathbf{1\bar{1}00} \rangle$ axis of the sapphire substrate and also to $\langle \mathbf{11\bar{2}0} \rangle$ axis of the ZnO underlayer, and the vertical axis is aligned to $\langle \mathbf{11\bar{2}0} \rangle$ axis of the sapphire substrate and also to $\langle \mathbf{1\bar{1}00} \rangle$ axis of the ZnO underlayer.

The $2\theta\chi$-ϕ coupled scans, which should correspond to the 2θ-θ scans for the out-of-plane XRD, sweep the directions in the reciprocal space radially from the origin of reciprocal space. That is, we sweep by the iterative motions of incremental step of $\Delta\phi$ and $2\theta\chi$-ϕ coupled scans. (the $2\theta\chi$ scan under various fixed ϕ with infinitesimally small interval). The $2\theta\chi$ corresponds to the distance between the vertical crystalline plane, and the ϕ corresponds to the in-plane direction of this crystalline plane. An example of the traces of $2\theta\chi$-ϕ coupled scans is demonstrated by a red broken arrow in **Figure 9(a)**. The circumferential position denotes the azimuthal angle (β), which is traced with an arc of yellow solid curve in **Figure 9(a)**. The orientational distribution of the vertical lattice plane receives the circumferential direction in this figure, that is, the distribution traces an arc with a fixed equidistance (corresponding to a d-value) from the origin.

Here we mention first on the mosaic spread of twisting, that can be evaluated by the in-plane rocking curve (ϕ-scan) profiles in this measurement system. The in-plane rocking curve profiles traced from these 2D figures in **Figures 9(a)-(d)** for the 110 reflections of LSMO are shown in **Figure 10** for the four samples. The FWHM values for the samples of PL-LT and ML-LT grown at the lower temperature are around 5.0 deg. These peaks with rather wide width indicate that the vertical (110) lattice plane is basically oriented in certain directions (exactly mentioned later) but should have some distribution in the film plane. In other word, the crystallites (having the out-of-plane orientations of (001), (110) and (111)) are basically in-plane aligned but exhibit slight dispersion within surface planes at some extent. We can recognize faint rings in **Figures 9(a)** and **(c)** for the samples of PL-LT and ML-LT. These rings imply an existence of small amount of randomly distributed components of the grains. Then it should be noted that the films grown at the lower temperature have both of the fundamentally aligned major grains and the randomly distributed minor grains.

In contrast to the wide spread of FWHM observed in the samples of PL-LT and ML-LT, the FWHM values in the samples of PL-HT and ML-HT grown at the higher temperatures are rather small as around 2.0 deg, and it is close to the value for $\mathbf{1\bar{1}00}$ reflections of the ZnO underlayer (not shown here). It is supposed from these facts that the twisting of LSMO film follows that of ZnO underlayer during the film growth of LSMO at the higher temperatures. However, such constraint of the LSMO lattice from the ZnO lattice is not so strong at the lower temperature. Due to this relaxation, the LSMO films grown at the lower temperature have the larger spread of twisting. This gives a very important knowledge concerning the in-plane orientations in the epitaxial growth.

The lattice constant of LSMO along the in-plane direction can be determined as **a** = 0.389 - 0.390 nm from the 110 reflections. These values are almost the same with

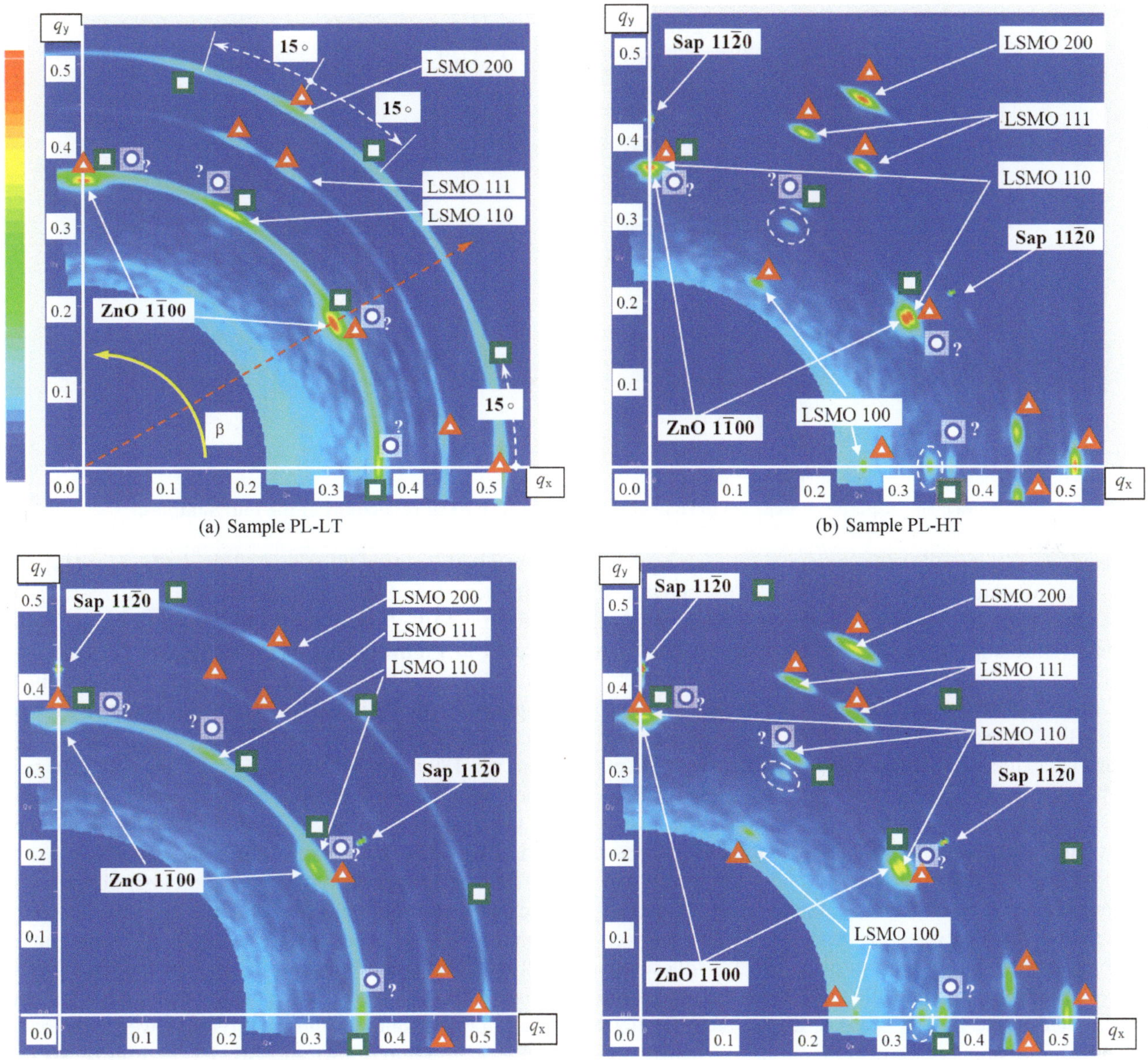

(a) Sample PL-LT (b) Sample PL-HT

(c) Sample ML-LT (d) Sample ML-HT

Figure 9. In-plane reciprocal space mappings (RSM) for the four samples. (a) Sample PL-LT; (b) Sample PL-HT; (c) Sample ML-LT; (d) Sample ML-HT. Horizontal axis corresponds to sapphire $\langle 1\bar{1}00\rangle$ and ZnO $\langle 11\bar{2}0\rangle$, and vertical axis corresponds to sapphire $\langle 11\bar{2}0\rangle$ and ZnO $\langle 1\bar{1}00\rangle$. The contributions of the (001), (110), and (111) phases are shown with the marks of □, △, and ○, respectively. The faint spots marked with white broken circles are resulted from $2\bar{2}0$ reflections of $ZnAl_2O_4$ phases. (see text).

those estimated by the out-of-plane 2θ-θ measurement mentioned above. We would like to mention here, the crystallinity (*i.e.*, twist) of the LSMO overlayer can be improved by inserting the ZnO underlayer with the smaller lattice mismatch between the LSMO overlayer and the sapphire substrate with the larger lattice mismatch, compared with the case of direct growth of LSMO on the sapphire substrate [3].

Next, we mention on the in-plane orientations of the out-of-plane (001), (110) and (111) oriented phase grains, which can be deduced from the results shown in **Figures 9(a)-(d)**. Before going into the details, we need fundamental background knowledge which assists our easy understanding. These grains are denoted by (□) for the (001) phase, by (△) for the (110) phase and by (○) for the (111) phase, respectively in **Figures 9(a)-(d)**. We use the diffracting vertical planes of (200), (110) and (111) included in these three grains. The diffracting planes are

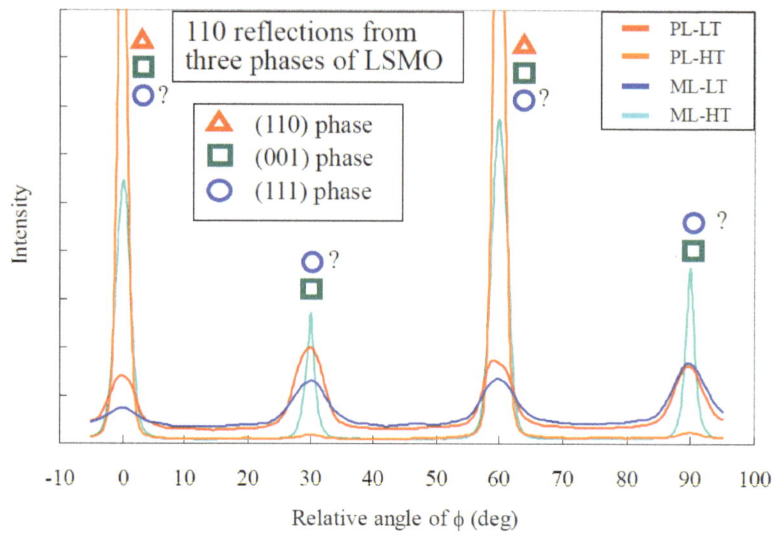

Figure 10. In-plane ϕ-rocking curve profiles for the LSMO 110 reflections of the four samples.

indicated in **Figures 9(a)-(d)** appearing on each equidistant circumference. However, it should be noted that each grain does not always have all these three vertical planes, for example, the (001) grain does not have the vertical (111) plane. The LSMO has the cubic symmetry and the (001) phase should show four-fold symmetry within the surface plane, basically. However, as mentioned later, the LSMO has three equivalent alignments on the (0001) ZnO with the hexagonal symmetry showing six-fold symmetry. Then the LSMO does not always show the exact four-fold symmetry, but shows apparent six-fold like symmetry.

Assisted by such background knowledge, we can digest the relations between the grains and diffracting vertical planes observable in the in-plane XRD as follows. A) Only the reflections from 110 and 001 vertical planes can be detected in the (001) grain (□); B) All the reflections from 001, 110 and 111 vertical planes can be detected in the (110) grain (▲); C) Only the reflection from 110 vertical plane can be detected in the (111) grain (○).

The 111 reflection is only related to the (110) grain, then it is efficient to analyze the in-plane orientational relationships using the 111 reflection first. The azimuthal angles β corresponding to the diffraction spots from the cubic LSMO, should appear at $\beta = \mathbf{m} \times \mathbf{180}^\circ + \mathbf{n} \times \mathbf{60}^\circ \pm \mathbf{35.3}^\circ + \beta_0$, where the first term of $\mathbf{m} \times \mathbf{180}^\circ$ comes from the azimuthal frequency following the two-fold symmetry of the in-plane appearance of the rectangular plane surface lattice of the LSMO (110) phase as illustrated in **Figure 11(a)**. Whereas the second term of $\mathbf{n} \times \mathbf{60}^\circ$ actually comes from the LSMO (110) phase, but it is affected by the azimuthal frequency following the six-fold symmetry of the in-plane hexagonal (0001) ZnO underlayer, because the rectangular (110) LSMO has the three equivalent arrangements on the hexagonal ZnO as illustrated also in **Figure 11(a)**. The third term of $\pm\mathbf{35.3}^\circ$ comes from the intersecting angles of $[\mathbf{1\bar{1}0}]$ axis and $[\mathbf{\bar{1}\bar{1}1}]$ or $[\mathbf{1\bar{1}1}]$ axis as illustrated in **Figure 11(a)**. Similarly, it can be understood from **Figure 11(a)** that the 110 reflection spots should appear at $\beta = \mathbf{m} \times \mathbf{180}^\circ + \mathbf{n} \times \mathbf{60}^\circ + \beta_0$, and the 100 reflection spots should appear at $\beta = \mathbf{m} \times \mathbf{180}^\circ + \mathbf{n} \times \mathbf{60}^\circ \pm \mathbf{90}^\circ + \beta_0$. Thus, the 110 spots and 100 spots appear with the regular interval of 60 deg in the β position. There is the shift of 30 deg in the β position between the 110 spots and the 100 spots. All the diffraction spots originated from the (110) phase are indicated by the symbol ▲ in **Figure 9**. These diffraction spots for the samples grown at the higher temperatures (PL-HT in **Figure 9(b)** and ML-HT in **Figure 9(d)**) are observed more clearly due to the stronger peak intensities than for the samples grown at the lower temperature (PL-LT in **Figure 9(a)** and ML-LT in **Figure 9(c)**) as expected from **Figure 10**. This is also resulted from the fact that the (110) phase growth is more promoted at the higher temperatures as pointed out by the results of 2θ-θ scan (see **Figure 7**). We can summarize the in-plane orientations of LSMO (110) phase against the (0001) oriented ZnO underlayer and the (0001) oriented sapphire substrate as $[\mathbf{1\bar{1}0}](\mathbf{110})\mathbf{LSMO}//[\mathbf{1\bar{1}00}](\mathbf{0001})\mathbf{ZnO}//[\mathbf{11\bar{2}0}](\mathbf{000\bar{1}})\mathbf{Sap}$.

Bhosle and Narayan [3] reported as $[\mathbf{1\bar{1}0}](\mathbf{110})\mathbf{LSMO}//[\mathbf{1\bar{1}00}](\mathbf{0001})\mathbf{Sap}$. The in-plane relation of the lattice orientation of our $[\mathbf{1\bar{1}0}](\mathbf{110})\mathbf{LSMO}//[\mathbf{1\bar{1}00}](\mathbf{0001})\mathbf{ZnO}$ is the same with that of their [1-10](110)LSMO//[1-100](0001)Sap in the system of rectangular on hexagonal lattices. However, the relation between LSMO and sapphire substrate is different. Then it is important that we can control the orientation of LSMO against the sapphire substrate by

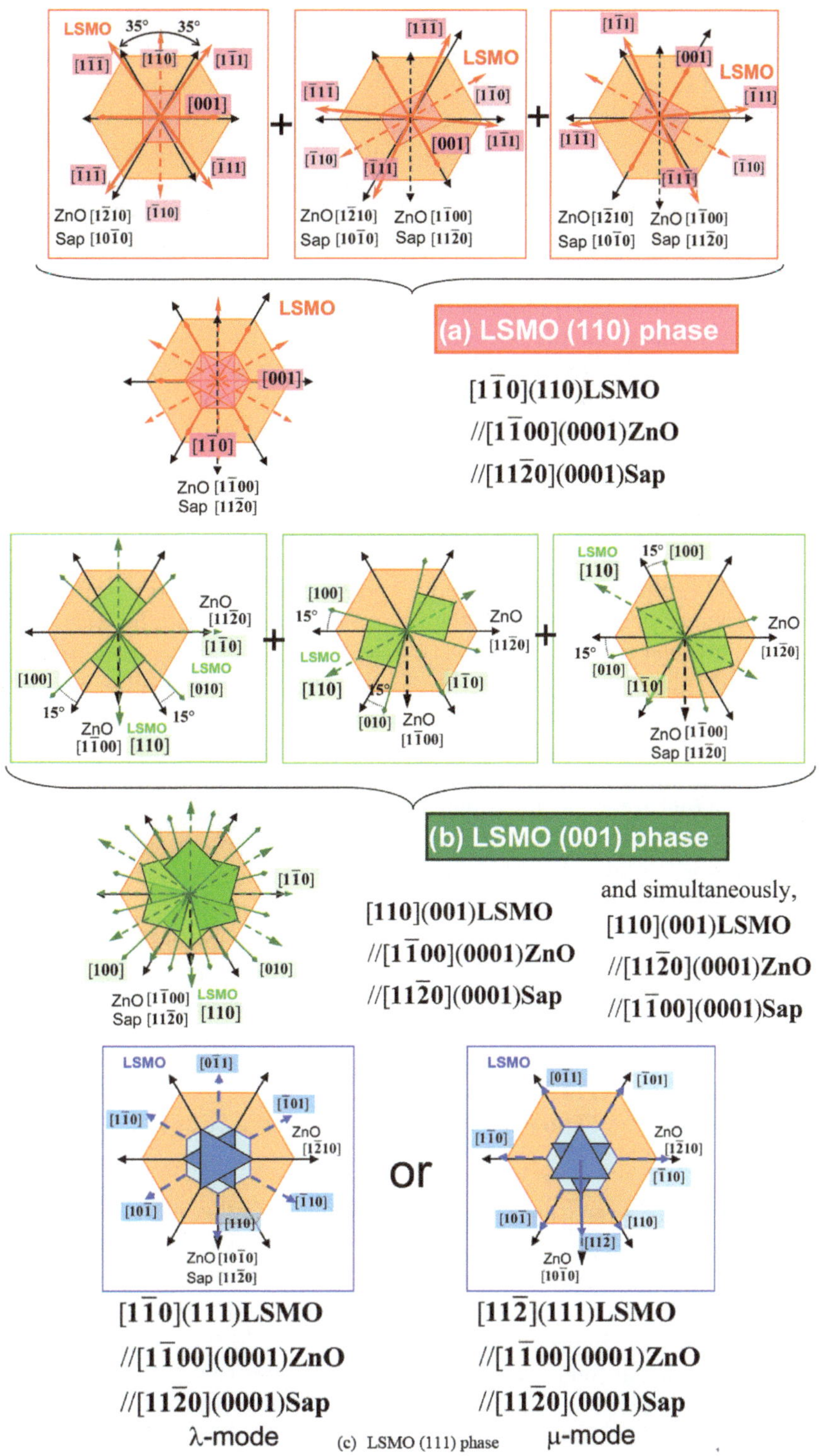

Figure 11. Schematic illustrations of the in-plane axes alignments of LSMO lattices of the three phases against the ZnO and sapphire lattices. The ZnO hexagonal lattices are shown in orange hexagons. (a) LSMO (110) phase: the LSMO lattices are shown in red rectangles, principle axes of LSMO are shown in red arrows; (b) LSMO (100) phase: the LSMO lattices are shown in green squares, principle axes of LSMO are shown in green arrows; (c) LSMO (111) phase: The overlayer LSMO lattices are shown in dark blue triangles, and principle axes of LSMO are shown in blue arrows. The two apparent triangles are shown for the 3-dimensional unit cell of LSMO (111) phase, because when the (111) grain is viewed from the top, we see one triangle, but when it is viewed from the bottom, we see the other triangle with 60 deg rotation. There are two possible arrangements, λ-mode and μ-mode, of the (111) phase as shown. The λ-mode is expected by the results of in-plane XRD shown in Figure 9, but we cannot exclude the possibility of μ-mode.

inserting the intermediate buffer layer.

The 110 and 200 diffraction spots originated from the (001) phase (marked by □ in **Figures 9(a)-(d)**) appear with the regular interval of 30 deg in the β position as illustrated in **Figure 11(b)**. The 110 diffraction spots should appear at $\beta = \mathbf{m} \times \mathbf{180}\circ + \mathbf{n} \times \mathbf{60}\circ + \beta_1$ and the 200 diffraction spots should appear at $\beta = \mathbf{m} \times \mathbf{180}\circ + \mathbf{n} \times \mathbf{60}\circ \pm \mathbf{45}\circ + \beta_1$. There is the shift of 15 deg in the β position between the 110 and 200 spots due to the four-fold symmetry of the LSMO (001) phase and to their three equivalent arrangements on the six-fold symmetry of ZnO (0001) plane. The 200 diffraction spots from the (001) phase (□) are observed with the 15 deg deviation in the β position from the 200 spots from the (110) phase (△). This result suggests that there should be overlapping of the 110 spots from the (110) phase and that from the (001) phase. However, the regular interval in the β position for the 110 spots from the (001) phase is 30 deg, while those from the (110) phase is 60 deg. Thus, the 110 spots from the (001) phase can be independently observed as seen in **Figures 9(a)** and **(c)**. The 110 spots and the 200 spots from the (001) phase are very weak for the ML-HT (**Figure 9(d)**), and the 200 spots from the (001) phase are not detected in the PL-HT (**Figure 9(b)**). These results suggest again that the growth of LSMO (001) phase is suppressed at the higher temperatures, and the plasma oxygen supply is more effective to suppress this growth. This is the same suggestion given by the above result of out-of-plane XRD (**Figure 7(c)**). We can summarize the in-plane orientations of LSMO (001) phase against the (0001) oriented ZnO underlayer and the (0001) oriented sapphire substrate as $[\mathbf{110}](\mathbf{001})\mathbf{LSMO}//\left[\mathbf{1\bar{1}00}\right](\mathbf{0001})\mathbf{ZnO}//\left[\mathbf{11\bar{2}0}\right](\mathbf{0001})\mathbf{Sap}$. The four-fold symmetry of LSMO (001) phase causes the axial alignment of LSMO $\left[\mathbf{1\bar{1}0}\right]$ axis to ZnO $\left[\mathbf{1\bar{2}10}\right]$ axis and Sap $\left[\mathbf{1\bar{1}00}\right]$ axis, simultaneously (see **Figure 11(b)**). This orientational relationship can be alternatively noted as $\left[\mathbf{1\bar{1}0}\right](\mathbf{00\bar{1}})\mathbf{LSMO}//\left[\mathbf{11\bar{2}0}\right](\mathbf{0001})\mathbf{ZnO}//\left[\mathbf{1\bar{1}00}\right](\mathbf{0001})\mathbf{Sap}$

The 110 diffraction spots from the ($1\bar{1}1$) phase (○) show 60° intervals (**Figure 11(c)**). If we make an assumption that they are on the same β-positions with the spots of ZnO $\mathbf{1\bar{1}00}$, then we can summarize the in-plane orientations of LSMO (111) phase against the (0001) oriented ZnO underlayer and the (0001) oriented sapphire substrate as $\left[\mathbf{1\bar{1}0}\right](\mathbf{111})\mathbf{LSMO}//\left[\mathbf{1\bar{1}00}\right](\mathbf{0001})\mathbf{ZnO}//\left[\mathbf{11\bar{2}0}\right](\mathbf{000\bar{1}})\mathbf{Sap}$. We termed the LSMO grown with this orientational relationship as "λ–mode" [4]. However, the other possibility of the orientational relationship for the (111) phase cannot be excluded, which is denoted as $\left[\mathbf{11\bar{2}}\right](\mathbf{111})\mathbf{LSMO}//\left[\mathbf{1\bar{1}00}\right](\mathbf{0001})\mathbf{ZnO}//\left[\mathbf{11\bar{2}0}\right](\mathbf{0001})\mathbf{Sap}$. This orientation is termed as the μ-mode [4], shown on the right hand side of **Figure 11(c)**. A difference between these two possible epitaxial orientations is the in-plane axial alignment, *i.e.*, whether $\left[\mathbf{1\bar{1}0}\right]\mathbf{LSMO}$ or $\left[\mathbf{11\bar{2}}\right]\mathbf{LSMO}$ is aligned with $[\mathbf{1\bar{1}00}]$ $\mathbf{ZnO}$ and $\left[\mathbf{11\bar{2}0}\right]\mathbf{Sap}$. The diffraction spots from the LSMO ($11\bar{1}$) phase (○) that could be observed in **Figures 9(a)** to **(d)** were limited for 110 spots, and these spots might be overlapped with other 110 spots from the (110) or (001) phase. From this reason, we cannot exclude the occurrence of LSMO (111) phase with the μ-mode. Since we cannot determine the position of 110 spots caused from the LSMO (111) phase, a mark of " ? " is attached on the blue circles (○) indicating the (111) phase in **Figures 9(a)-(d)** and **Figure 10**. However, this ambiguity can be clarified by utilizing the pole figure measurement as mentioned next.

For our visual comprehension, we show in **Figure 12** the 3D pictures for the out-of-plane and in-plane epitaxial orientations of three LSMO phases over the ZnO underlayer on the sapphire substrate. Regarding the (111) phase, only the possible μ-mode is shown here.

Faint spots marked by white broken circles in **Figures 9(b)** and **(d)** can be identified as the $\mathbf{2\bar{2}0}$ diffraction spots from $ZnAl_2O_4$ spinel phase. Dovienko *et al.*, [18] and Wessler, *et al.*, [19] reported the epitaxial growth of this phase in the vicinity between ZnO epitaxial film on the c-sapphire substrate, with the epitaxial orientation of $\left[\mathbf{11\bar{2}0}\right](\mathbf{0001})\mathbf{ZnO}//\left[\mathbf{1\bar{1}0}\right](\mathbf{111})\mathbf{ZnAL_2O_4}//\left[\mathbf{10\bar{1}0}\right](\mathbf{0001})\mathbf{Sap}$. In our study, the same in-plane orientation with these reports was observed in the two samples of PL-HT and ML-HT grown at the higher temperatures. This $ZnAl_2O_4$ spinel phase is probably formed at the interface of ZnO underlayer and sapphire substrate as a very thin layer. The detection of this signal verifies the excellent in-plane sensitivity of the in-plane XRD technique.

3.3. Pole Figure Measurements Using the In-Plane Axis

In order to further clarify the epitaxial orientations of the (111) phase of LSMO overlayer, we performed the pole figure (PF) measurements for the LSMO 111 reflections. We should remind here that the (111) phase does not have the vertical (111) plane in its grain, then we cannot adopt the in-plane measurement. That is a reason why we adopt the pole figure measurement to detect the 111 reflection from the (111) phase. The choice of 111 reflections is based on two conditions. One reason is that the orientational relationships can be easily deduced from their diffraction spot positions in the PF data with respect to the lattices of ZnO underlayer and sapphire lattices. The other reason is that there are no reflections in the vicinity of 2θ value for this reflection (around 40 deg). The only exception is the sapphire 0006 reflection whose 2θ value is 41.68 deg. But its position to be appeared in

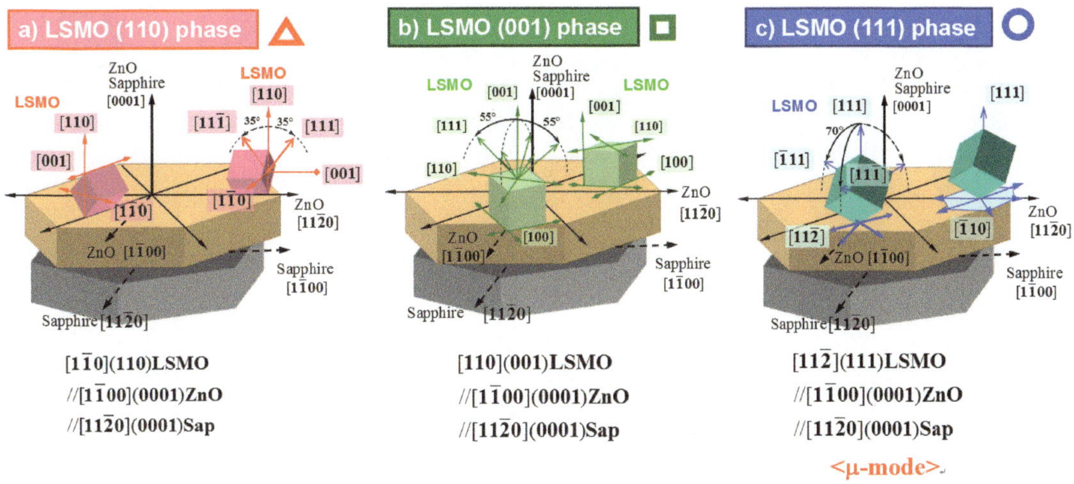

Figure 12. Three dimensional views of the orientational relationships of lattices for the three phases of LSMO on the (0001) ZnO on the (0001) sapphire substrate. Only the μ-mode configuration is shown for the (111) phase.

the PF data is already known well (at the center position of the PF data), and this might not be an obstacle for the orientational analysis of the LSMO overlayer.

We focused on comparison of the PF measurements for the samples of ML-LT and ML-HT to analyze this problem. The experimental results of diffraction spots are shown in **Figures 13(a) and (b)** for the 111 reflections from the (111) phase of ML-LT and ML-HT, respectively. These 2D data are shown in polar coordinates [20], where the tilting angles (radial direction: α) and the azimuthal angles (circumferential direction: β) are swept. If we define the tilting angle of a lattice plane from the sample surface plane as ψ (see **Figure 5(b)**), ψ = 0 deg corresponds to lattice planes parallel to the surface plane, and ψ = 90 deg corresponds to lattice planes perpendicular to the surface plane. **Figures 13(a)** and **(b)** are shown with polar coordinates with α and β [12], where, α is defined as $\alpha = 90 - \psi$, and β is the azimuthal angle from the direction where one of **11$\bar{2}$0** sapphire reflections is observed. We should note that the diffraction signals can be clearly found in the whole range of α, that is, α = 0 deg (corresponding to the in-plane XRD geometry) to α = 90 deg (corresponding to the 2θ-θ scan geometry). This means that the signal covers the whole hemisphere of a diffraction space.

The symbols used in these figures are the same as shown, for example, in **Figure 12**. Red arrows indicate the directions of $\langle \mathbf{11\bar{2}0} \rangle$ axes of sapphire and $\langle \mathbf{1\bar{1}00} \rangle$ axes of the ZnO underlayer. A circle in red indicates the tilting of 70 deg (α = 20 deg), and a dotted circle in white indicates the tilting of 35 deg (α = 55 deg). Peaks at α = 0 deg (corresponding to the in-plane geometry) and those at α = 55 deg can be attributed to the LSMO (110) phase (see **Figure 12**). Peaks at α = 20 deg are originated from the LSMO (111) phase. The latter peaks are observed at the same β positions with those for the **11$\bar{2}$0** reflections of sapphire. This result reveals that the $[\mathbf{11\bar{2}}]$ axis of LSMO (111) phase is aligned to the $[\mathbf{11\bar{2}0}]$ axis of sapphire substrate but not the $[\mathbf{1\bar{1}0}]$ axis. Consequently, the epitaxial orientation for the (111) LSMO (111) phase can be determined as $[\mathbf{11\bar{2}}](\mathbf{111})$ $[\mathbf{1\bar{1}00}](\mathbf{0001})$ **LSMO // ZnO //** $[\mathbf{11\bar{2}0}](\mathbf{0001})$**Sap**. This is termed as the μ-mode and shown in **Figure 11(c)** which is equivalently noted as $[\mathbf{1\bar{1}0}](\mathbf{111})$ **LSMO //** $[\mathbf{11\bar{2}0}](\mathbf{0001})$**ZnO //** $[\mathbf{1\bar{1}00}](\mathbf{0001})$**Sap**.

The peaks of the 111 reflections in the (110) phase (▲) of "ML-HT" (**Figure 13(a)**) are observed with higher intensity than for those of "ML-HT" (**Figure 13(b)**), suggesting again the growth promotion of the (110) phase at the higher temperatures. It should be noted that the peaks marked with ▲ were observed at the outer end circle in **Figure 13(b)**, where the measuring geometry corresponds to the in-plane XRD. It can be easily confirmed that the distribution pattern for these peaks (around 30 deg and 90 deg in β) is the same as that for the 111 reflections in the in-plane RSM data in **Figure 9(d)** (around 0 deg and 60 deg in β). The peaks of the 111 reflections in the (001) phase should be observed at α = 35 deg, but these signals are infinitesimally faint since the volumetric contribution of this (001) phase is very small.

3.4. Out-of-Plane Wide-Range Reciprocal Space Mapping Using the 2-Dimensional X-Ray Detector

Lastly we tried to confirm the validity of the epitaxial orientational relationships mentioned above by an alternative measurement. That is, out-of-plane wide-range RSM measurements were performed using the 2D detector for the samples of ML-LT and ML-HT. The results are shown in **Figures 14(a)** and **(b)** for the ML-LT and

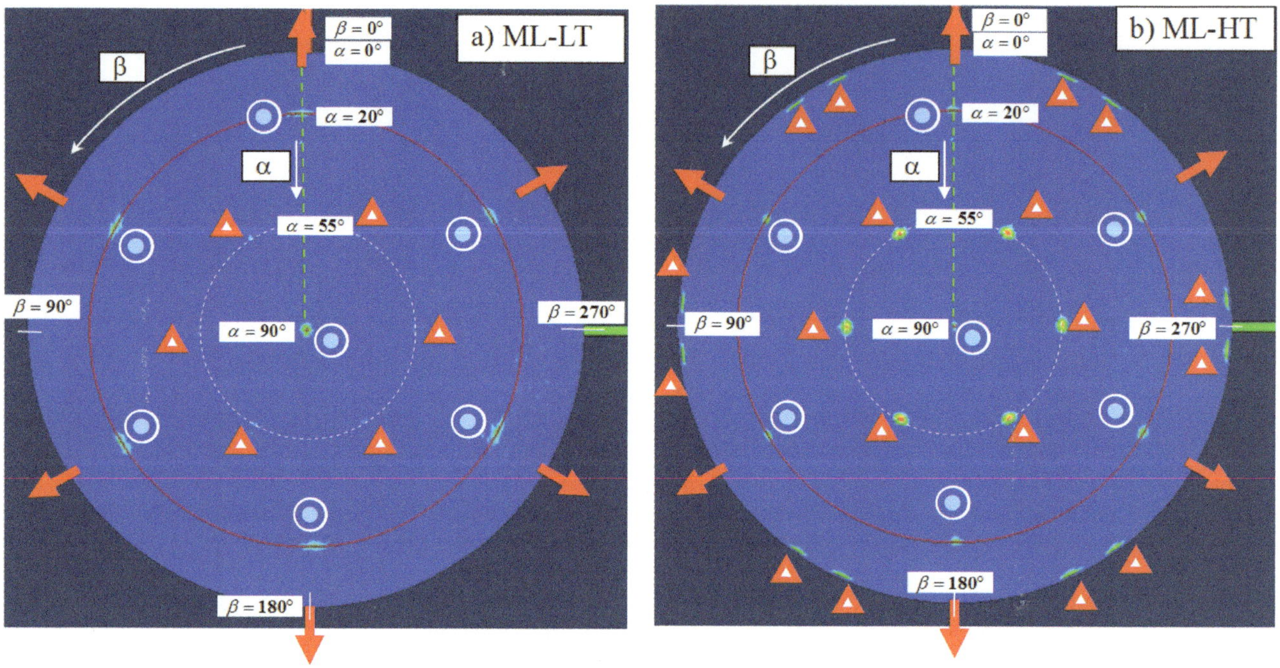

Figure 13. The pole figures obtained from the measurements using the in-plane axis of LSMO 111 reflections for the samples (a) ML-LT and (b) ML-HT. The red arrows indicate the directions of $\langle 11\bar{2}0\rangle$ axes of sapphire substrate and $\langle 1\bar{1}00\rangle$ axes of the ZnO underlayer. The detected intensity is scaled in log scale. A red circle indicates the tilting of 70 deg from the surface plane, and a broken circle indicates the tilting of 35 deg from the surface plane. The center of the figure corresponds to the measurement of 2θ-θ scans, where lattice planes are parallel to the sample surface plane. The outer circle corresponds to the vertical plane diffraction obtained by the in-plane XRD measurement.

ML-HT, respectively. The horizontal axisqxcorresponds to the directions of $\langle \mathbf{11\bar{2}0}\rangle$ axis of sapphire substrate and $\langle \mathbf{1\bar{1}00}\rangle$ axis of the ZnO underlayer, and the vertical axis qzcorresponds to the directions of $[\mathbf{0001}]$ axis of sapphire and ZnO. The detected intensity is scaled in log scale.All the diffraction peaks in **Figures 14(a)** and **(b)** can be indexed by the above mentioned epitaxial orientational relationships as shown in the figures.The positions of all the peaks are in good coincidence with the peak positions determined above.Consequently this measurement provides the validity of the above measurements and analyses..

All the peaks in **Figures 14(a)** and **(b)** can be indexed with a present model of epitaxial orientational relationships, and the positions of all these peaks are in good coincidence with peak positions estimated from the model, proving the validity of this model.

4. Summary

The LSMO overlayer was grown on the (0001) ZnO underlayer on the (0001) sapphire substrate by the ion beam sputtering at the growth temperatures from 650°C to 750°C with the supply of molecular oxygen or plasma oxygen. The out-of-plane and in-plane epitaxial orientational relationships of the lattices and crystalline qualities were analyzed using various X-ray diffraction techniques for the LSMO/ZnO/Sap double-hetero epitaxial systems. The growth of out-of-plane (110) oriented phase of LSMO is promoted at the higher temperatures, accompanying the suppression of other competitive growth of the LSMO (001) and (111) phases. The growth of LSMO (111) phase is remarkably suppressed by supplying the plasma oxygen at the higher temperature due to the excess energy. The slight contraction of lattice constants along the growth direction was found in the samples grown at the higher temperature. The tilting of lattices of the LSMO (110) phase was evaluated by the peak breadth of ω rocking curves as around 2.0 deg. This value is almost the same for the all samples investigated. The twisting of LSMO lattices was evaluated by the in-plane ϕ rocking curves, indicating the improvement of crystalline quality for the samples grown at the higher temperatures.

The complex epitaxial orientaional relationships of the three LSMO phases were clarified by the precise analyses using the high resolution measurements of the out-of-plane XRD, the in-plane XRD, the in-plane RSM, and PF. They are summarized as

$[1\bar{1}0](110)$LSMO / /$[1\bar{1}00](0001)$ZnO / /$[11\bar{2}0](0001)$Sap,

$[110](001)$LSMO / /$[1\bar{1}00](0001)$ZnO / /$[11\bar{2}0](0001)$Sap,

$[11\bar{2}](111)$LSMO / /$[1\bar{1}00](0001)$ZnO / /$[11\bar{2}0](0001)$Sap (μ-mode).

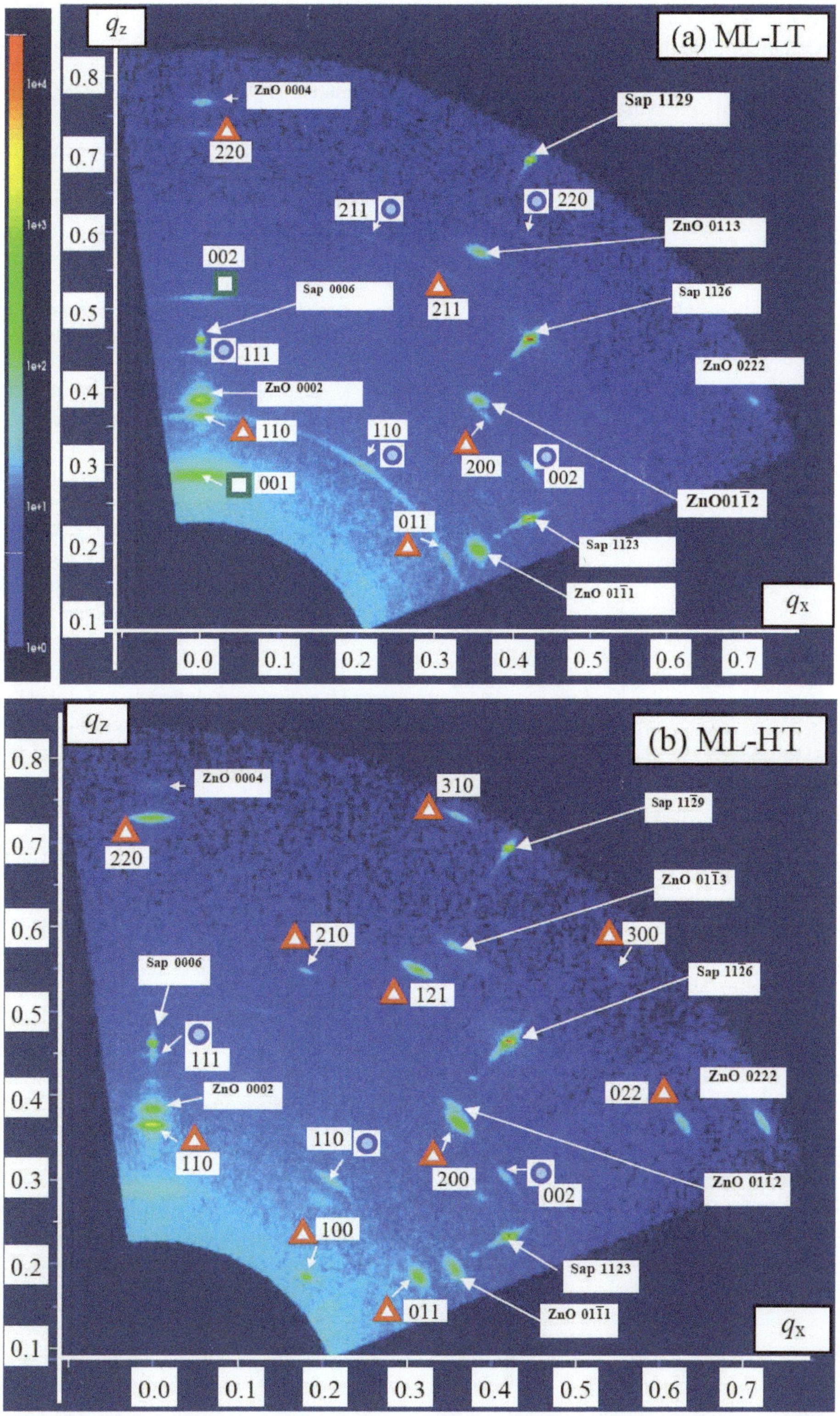

Figure 14. The Out-of-plane wide-range reciprocal space mappings obtained by using the 2D X-ray detector for the samples (a) ML-LT and (b) ML-HT. The horizontal axis corresponds to the directions of $\langle 1\bar{1}00 \rangle$ axis of the ZnO underlayer and $\langle 11\bar{2}0 \rangle$ axis of the sapphire substrate. The vertical axis corresponds to the directions of [0001] axis of the ZnO underlayer and sapphire substrate. The detected intensity is scaled in log scale. The index numbers indicate the reflecting planes. The LSMO phases are indicated by the same symbols with Figure 12.

The validity of these analyses for the epitaxial orientations of LSMO/ZnO/Sap double heteroepitaxial layers were confirmed by the alternative analysis of the out-of-plane wide-range reciprocal space mappings using the 2D X-ray detector.

REFERENCES

[1] T. Endo, K. Uehara, T. Yoshii, M. Yokura, H. Zhu, J. Nogues, J. Colino and K. Endo, "Peculiar Electric and Magnetic Properties of La(Ba)MnO_3 Thin Films," *Transactions of Materials Research Society of Japan, 20th Anniversary Special Issue*, 2012, pp. 65-76.

[2] K. Lord, D. Hunter, T. M. Williams and A. K. Paradhan, "Photocarrier Injection Effect and p-n Junction Characteristics of $La_{0.7}Sr_{0.3}MnO_3$/ZnO and Si Heterostructures," *Applied Physics Letters*, Vol. 89, No. 5, 2008, Article ID: 052116.

[3] V. Bhosle and J. Narayan, "Epitaxial Growth and Magnetic Properties of $La_{0.7}Sr_{0.3}MnO_3$ Films on (0001) Sapphire," *Applied Physics Letters*, Vol. 90, 2007, Article ID: 101903.

[4] K. Uehara, A. Okada, A. Okamoto, M. Yokura, S. L. Reddy, S. Kobayashi, K. Inaba, N. Iwata, R. Philip, H. Kezuka, M. Matsui and T. Endo, "Hetero-Epitaxial Growth of Cubic La(Sr)MnO_3 on Hexagonal ZnO, In-Plane Orientations of La(Sr)MnO_3 (001), (110), and (111) Phases," *Japanese Journal of Applied Physics*, Vol. 51, 2012, Artile ID: 11PG07.

[5] K. Inaba, "X-Ray Thin-Film Measurement Techniques I. Overview," *The Rigaku Journal*, Vol. 24, No. 1, 2008, pp. 10-15.

[6] P. F. Fewster, "X-Ray Scattering from Semiconductors," Imperial College Press, 2000.

[7] T. Konya, "X-Ray Thin-Film Measurement Techniques III. High Resolution X-Ray Diffractometry," *The Rigaku Journal*, Vol. 25, No. 2, 2009, pp. 1-8.

[8] K. Omote and J. Harada, "Grazing-Incidence X-Ray Diffractometer for Determining In-Plane Structure of Thin Films," *Advances in X-Ray Analysis*, Vol. 43, 2000, pp. 192-200.

[9] S. Matsuno, M. Kuba, T. Nayuki, S. Soga and P. W. T. Yuen, "Microstructural Characterization of Thin Films and Surfaces by a New Grazing Incident X-Ray Diffractometer," *The Rigaku Journal*, Vol. 17, No. 2, 2000, pp. 36-44.

[10] Y. F. Chen, S. K. Hong, H. J. Ko, V. Kirshner, H. Wenish, T. Yao, K. Inaba and Y. Segawa, "Effects of an Extremely Thin Buffer on Heteroepitaxy with Large Lattice Mismatch," *Applied Physics Letters*, Vol. 78, No. 21, 2001, pp. 3352-3354.

[11] S. Kobayashi, "X-Ray Thin-Film Measurement Techniques IV. In-Plane Diffraction Measurements," *Rigaku Journal*, Vol. 26, No. 1, 2010, pp. 3-11.

[12] S. K. Han, D. C. Oh, J. H. Song, K. Inaba, T. Yao and S. K. Hong, "Lattice Deformation in a-Plane ZnO Films Grown on r-Plane Al_2O_3 Substrates Grown by Plasma-Assisted Molecular-Beam Epitaxy," *Applied Physics Express*, Vol. 5, 2012, Article ID: 081101.

[13] K. Nagao and E. Kagami, "X-Ray Thin-Film Measurement Techniques VII. Pole Figure Measurement," *Rigaku Journal*, Vol. 27, No. 2, 2011, pp. 6-14.

[14] S. Kobayashi and K. Inaba, "X-Ray Thin-Film Measurement Techniques VIII. Detectors and Series Summary," *Rigaku Journal*, Vol. 28, No. 1, 2012, pp. 8-13.

[15] M. Renninger, "Beitrag zur Kenntnis der Roentgenorgraphischen Unterschiede Zwischen den Beiden Diamant-Typen,"*Acta Crystallographica*, Vol. 8, 1955, pp. 606-610.

[16] M. Seki, T. Konya, K. Inaba and H. Tabata, "Epitaxial Thin Films of $InFe_2O_4$ and $InFeO_3$ with Two-Dimensional Triangular Lattice Structures Grown by Pulsed Laser Deposition," *Applied Physics Express*, Vol. 3, 2010, Article ID: 105801.

[17] G. Saint-Girons, C. Priester, P. Regreny, G. Patriarche, L. Largeau, V. Favre-Nicolin, G. Xu, Y. Robach, M. Gendry and G. Hollinger, "Spontaneous Compliance of the InP/$SrTiO_3$ Heterointerface," *Applied Physics Letters*, Vol. 92, No. 24, 2008, Article ID: 241907.

[18] K. Dovidenko, S. Oktyabrsky, A. K. Sharma and J. Narayan, "TEM Characterization of ZnO and AlN/ZnO Thin Films Grown on Sapphire," *Materials Research Society Symposium Proceedings*, Vol. 526, 1998, pp. 311-316.

[19] B. Wessler, F. F. Lange and W. Mader, "Textured ZnO Thin Films on (0001) Sapphire Produced by Chemical Solution Deposition," *Journal of Materials Research*, Vol. 17, No. 7, 2002, pp. 1644-1650.

[20] M. Birkholz, "Thin Film Analysis by X-Ray Scattering," Wiley-VCH Verlag GmbH & Co., 2006.

Effect of Pressure on the Superconducting and Mechanical Properties of $Bi_{1.6}Pb_{0.4}Sr_{1.8}Ba_{0.2}Ca_2Cu_{2.2}Ni_{0.8}O_{10+\delta}$ System

Ghazala Y. Hermiz[*], Bushra A. Aljurani, Hassan A. Thabit
Physics Department, College of Science, University of Baghdad, Baghdad, Iraq

ABSTRACT

Solid state reaction method was used to prepare samples of $Bi_{1.6}Pb_{0.4}Sr_{1.8}Ba_{0.2}Ca_2Cu_{2.2}Ni_{0.8}O_{10+\delta}$. Superconducting and mechanical properties by Vickers microhardness measurements have been carried out to examine the effects of pressure. Electrical measurement demonstrates the best transition temperature was under 0.7 GPa. An improvement of mechanical properties was found when the pressure increases from 0.3 GPa to 0.9 GPa. On the other side an increases of the pressure to 1.1 GPa decrease in the micro hardness, Young modulus and yield strength.

Keywords: Pressure Effect; Superconductor; Mechanical Properties

1. Introduction

Considerable efforts have been made in the synthesis, processing, and characterization of BiSrCaCuO superconducting ceramics because these materials have a higher T_c and good chemical durability (resistance to moisture), and also they contain no rare earth elements [1-3]. However, the bulks sometimes fracture due to the thermal stress on the cooling process or the electromagnetic force during the magnetization. Thus, it has been recognized that improvement and understanding of the mechanical properties of bulks are indispensable for practical application.

Ahn [4] prepared superconducting BiSrCaCu(Ni)O ceramics by the gel method using an aqueous solution containing a tartaric acid. Their ceramic microstructure were studied using IR, TGA, XRD, resistance measurements, and SEM. They found that the nickel dopant substituting for Cu gives rise to the gradual decrease of the T_c and the pure phase 2212 ceramics were obtained at sintering temperature 1098 K for 24 hs. SEM pictures showed that liquid phase was formed when the samples were sintered temperatures higher than 1098 K.

The effect of silver additions on the mechanical and superconducting properties of sintered bulk YBCO, BSCCO-2212, and BSCCO-2223 has been evaluated by Joo *et al.* [5] their results show that strength and fracture toughness of YBCO and BSCCO bars increased with increasing Ag content up to 30 vol% Ag. These improvements in strength and fracture toughness are due to the presence of Ag particles that may induce compressive stresses in the superconducting matrix and resist crack propagation by pinning the propagating cracks. On the other hand, the hardness of YBCO and BSCCO decreased with increasing Ag contents because of the lower hardness of Ag. Addition of Ag showed no adverse effects on superconducting properties of YBCO or BSCCO superconductors.

Katagiri *et al.* [6] summarize the evaluation process of the mechanical properties for various high-temperature superconducting bulks. The Young's modulus, Poisson's ratio, fracture strength, fracture toughness and hardness are evaluated by tensile, bending, compression and hardness tests. The mechanical properties are anisotropic, mainly due to the crystallographic structure and pre-existing micro-cracks associated with it.

Madre *et al.* [7] measured the flexural strength of 1 wt% Ag-doped $Bi_2Sr_2CaCu_2O_{8+\delta}$ thin rods textured by a laser heated floating zone as a function of the environmental conditions (air versus water) at room temperature. Loading rates spanning three orders of magnitude (1, 10 and 100 μm/min) were used to explore their susceptibility to the environmental conditions. These mechanical tests were completed with electrical characterization (critical current at 77 K and resistivity from 77 to 300 K) of samples submerged in distilled water for different time lengths (0, 12 and 120 h). While $Bi_2Sr_2CaCu_2O_{8+\delta}$ has been shown to be unstable during contact with water molecules, the Ag-doped Bi-2212 textured rods tested

[*]Corresponding author.

are very inert to the water environment, with respect to their mechanical and electrical properties, due to the presence of a narrow (≈150 μm) low textured outer ring formed in the growth process.

In the previous work [8], mechanical properties of $Bi_{1.6}Pb_{0.4}Sr_{1.8}Ba_{0.2}Ca_2Cu_{2.2}Ni_{0.8}O_{10+\delta}$ superconducting system pressed under 0.5 GPa were investigated. In this research, effect of pressure on the critical temperature (T_c), oxygen content (δ) and mechanical properties of the superconducting system have been studied.

2. Experimental Part

$Bi_{1.6}Pb_{0.4}Sr_{1.8}Ba_{0.2}Ca_2Cu_{2.2}Ni_{0.8}O_{10+\delta}$ samples were prepared by conventional solid-state reaction method. Appropriate amounts of Bi_2O_3(99.9%), Pb_3O_4(99.9%), $SrCO_3$ (99.9%), $BaCO_3$(99.9%), CaO(99.9%), CuO(99.9%) and Ni_2O_3(99.9%) powders were used as starting mate- rials. The powder of precursor was mixed together by using agate mortar. The mixture homogenization takes place by adding a sufficient quantity of 2 - propanol to form a past during the process of grinding from about (1 h). In the second step, the materials were grounded to a fine powder and then calcined in air at 1073 K for 24 hs, the mixture was then pressed into pellets (1.3 cm) in dia- meter and (0.2) cm thick, using hydraulic type (SPECAC), under pressure of 0.3, 0.5, 0.7, 0.9 and 1.1 GPa. The pellets were sintered in air at 1123 K for 140 hs.

Iodometric titration was used to find the oxygen content (δ) in the samples. Four probe dc method at temperature range (77 - 300) K was used to measure the resistivity (ρ) and to determine the critical temperature (T_c).

The mechanical properties, such as the Vickers microhardness (H_V), Young modulus (E) and yield strength (Y) of superconductor samples were calculated as referred in the previous paper [8].

3. Results and Discussion

In order to examine the effect of the pressure on critical temperature T_c, samples of $Bi_{1.6}Pb_{0.4}Sr_{1.8}Ba_{0.2}Ca_2Cu_{2.2}Ni_{0.8}O_{10+\delta}$ were sintered at 1123 K for 140 hs under different pressures (0.3, 0.5, 0.7, 0.9 and 1.1) GPa. The resistivity behavior as a function of temperature for $Bi_{1.6}Pb_{0.4}Sr_{1.8}Ba_{0.2}Ca_2Cu_{2.2}Ni_{0.8}O_{10+\delta}$ at different pressures are shown in **Figure 1**. From this figure we can see the increase of the critical temperature T_c from 113 K to 117 K with increasing pressure from 0.5 GPa and 0.7 GPa, while an increase of the pressure to 1.1 GPa causes a reduction of transition temperature to 93 K. It is believed that the increase of the critical temperature with increasing pressure from 0.3 to 0.7 GPa, as listed in **Table 1** may be due to the increase of the carrier concentration n_h in the CuO_2 planes, the change of n_h within the unit cell resulting in the improvement of the critical temperature [9]. Also Vonsovsky *et al.* [10] indicate that the deviation from stoichioctry and disordering of these compound lead to decrease the value of T_c with the pressure.

It is found from **Table 1** that the oxygen content (δ) of the samples increases as the pressure increases. This is clearly shown for samples under 0.3 GPa, 0.5 GPa and 0.7 GPa, the oxygen content (δ) is 0.191, 0.246 and 0.368 respectively. The enhancement of the pressure is going to enhance the hole concentration in the Cu-O_2 layers, also enhancing the critical temperature T_c from nearly 110, 113 K to 117 K respectively, while increases the pressure to 1.1 GPa decreases the oxygen content (δ) and the critical temperature T_c as shown in **Figure 2**. This indicates that the increase in the critical temperature T_c and the oxygen content (δ) values is attributed to the increase of oxygen absorption during crystallization process of the superconductors. From the results we can conclude that the oxygen content (δ) equal 0.368 is optimum

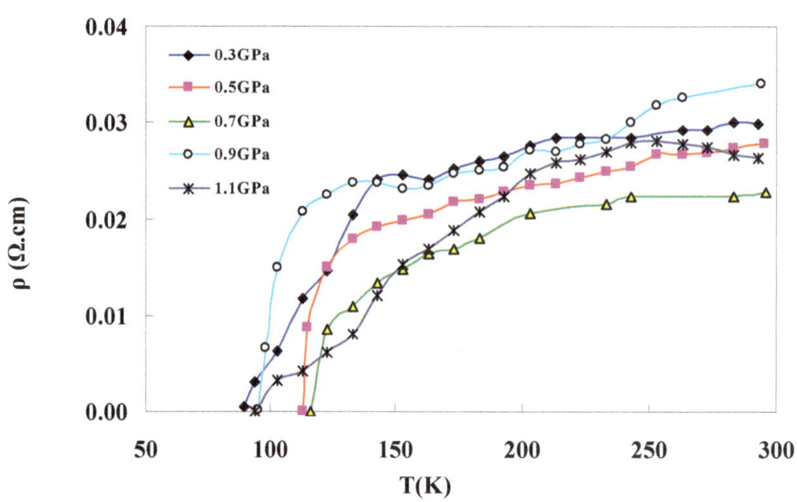

Figure 1. Temperature dependence of resistivity for $Bi_{1.6}Pb_{0.4}Sr_{1.8}Ba_{0.2}Ca_2Cu_{2.2}Ni_{0.8}O_{10+\delta}$ under different pressure.

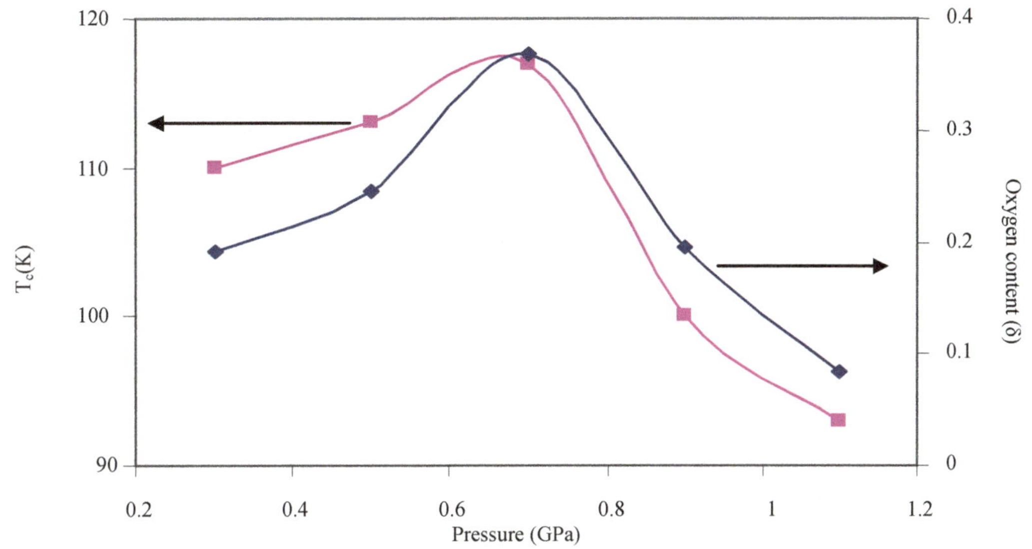

Figure 2. Transition temperature and oxygen content for $Bi_{1.6}Pb_{0.4}Sr_{1.8}Ba_{0.2}Ca_2Cu_{2.2}Ni_{0.8}O_{10+\delta}$ under different pressure.

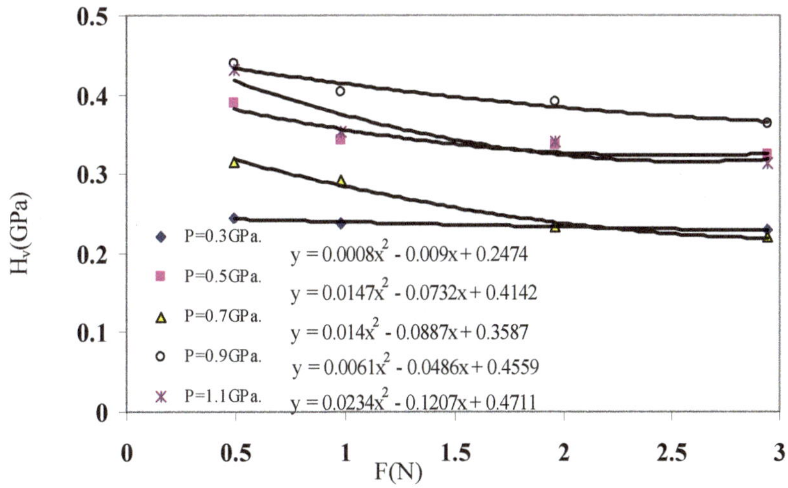

Figure 3. Variation of microhardness with applied load for different pressures.

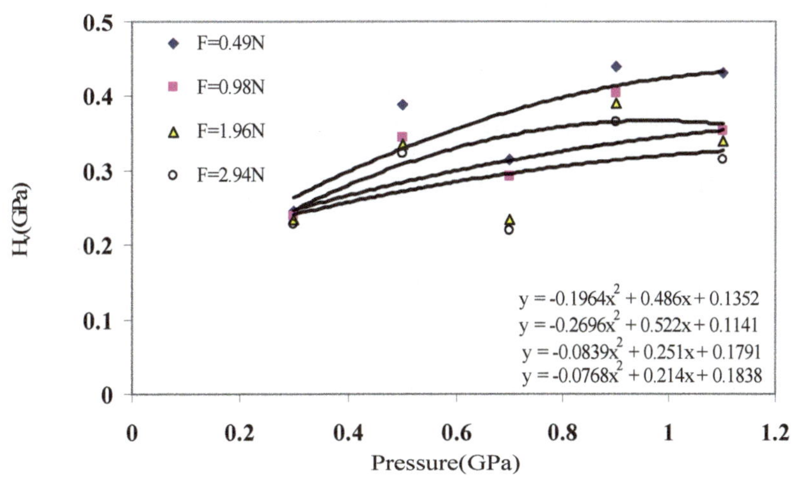

Figure 4. The variations of microhardness (H_v) as a function of pressure for different loads.

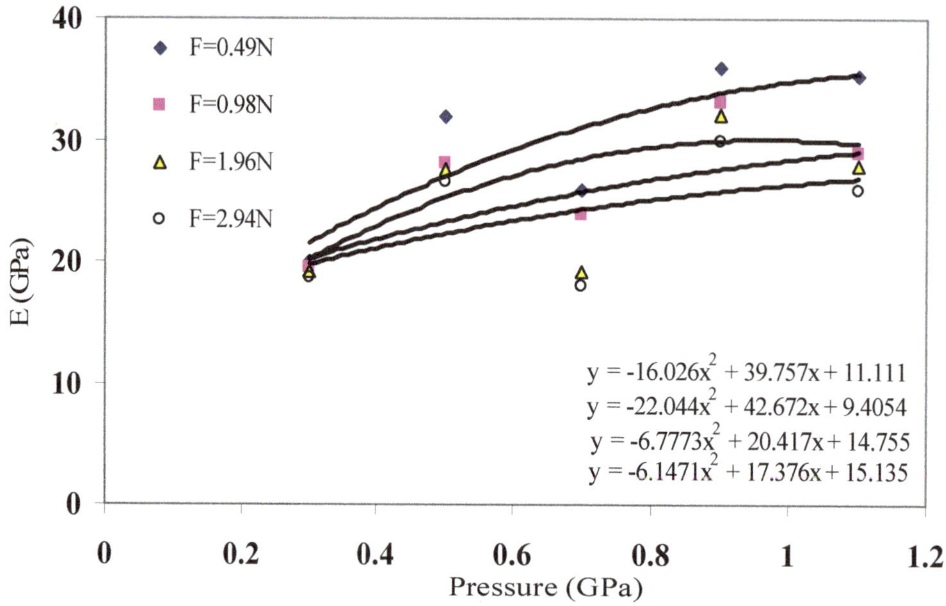

Figure 5. Variations of Young's modulus (E) as a function of pressure for different loads.

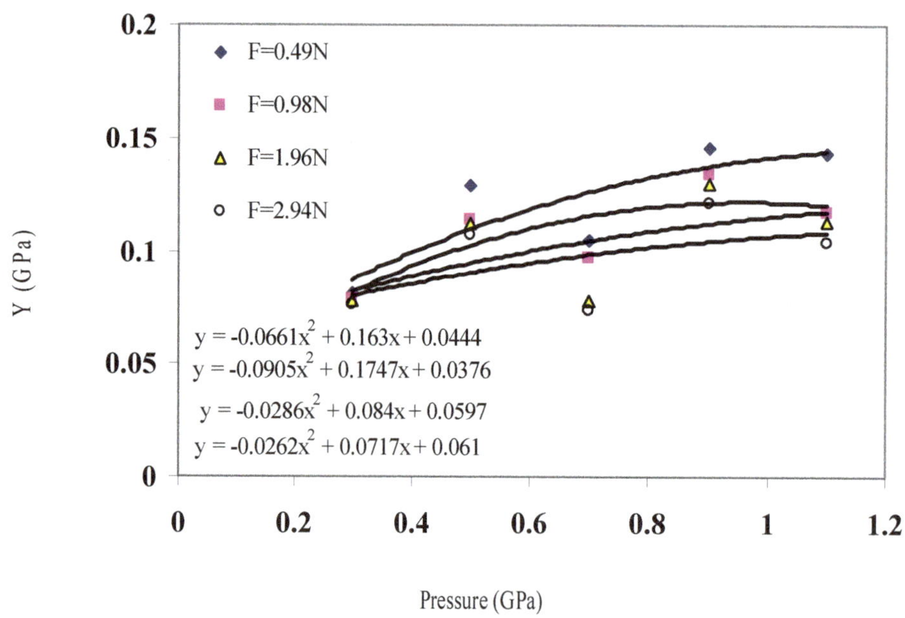

Figure 6. Variations of yield strength (Y) as a function of pressure for different loads.

Table 1. Oxygen content and critical temperature (T_c) for $(Bi_{0.8}Pb_{0.2})_2(Sr_{0.9}Ba_{0.1})_2Ca_2Cu_{2.2}Ni_{0.8}O_{10+\delta}$ under different pressures.

Pressure (GPa)	δ	T_c (K)
0.3	0.191	110
0.5	0.246	113
0.7	0.368	117
0.9	0.196	100
1.1	0.083	93

Table 2. Microhardness (H_v), Young modulus (E) and yield strength (Y) for different load for $Bi_{1.6}Pb_{0.4}Sr_{1.8}Ba_{0.2}Ca_2Cu_{2.2}Ni_{0.8}O_{10+\delta}$ under different pressure and different loads.

Pressure (GPa)	F (N)	Hv (GPa)	E (GPa)	Y (GPa)
0.3		0.244	20.01	0.081
0.5		0.389	31.91	0.129
0.7	0.49	0.315	25.82	0.105
0.9		0.439	36.01	0.146
1.1		0.430	35.28	0.143
0.3		0.238	19.53	0.079
0.5		0.343	28.13	0.114
0.7	0.98	0.292	23.93	0.0973
0.9		0.404	33.11	0.134
1.1		0.352	28.85	0.117
0.3		0.234	19.211	0.078
0.5		0.336	27.57	0.112
0.7	1.96	0.234	19.18	0.078
0.9		0.391	32.07	0.130
1.1		0.340	27.89	0.113
0.3		0.228	18.72	0.076
0.5		0.323	26.52	0.107
0.7	2.94	0.22	18.03	0.0733
0.9		0.364	29.90	0.121
1.1		0.314	25.80	0.104

value for the critical temperature T_c 117 K under pressure 0.7 GPa.

The effects of pressure on the mechanical properties of $Bi_{1.6}Pb_{0.4}Sr_{1.8}Ba_{0.2}Ca_2Cu_{2.2}Ni_{0.8}O_{10+\delta}$ samples were studied. The samples were pressed under 0.3, 0.5, 0.7, 0.9 and 1.1 GPa. It is noticed from **Figure 3** a decreases of the microhardness with increases of load for samples under different pressures. This behavior may attribute to the presence of weak grain boundaries of the superconducting ceramics.

Figures 4-6 and **Table 2** shows the variation of Vickers microhardness, Young modulus and yield strength, as a function of pressure for different loads. An increase of microhardness, Young's modulus and yield strength was observed with increasing pressure from 0.3 GPa to 0.9 GPa. This result indicates that an enhancement of the pressure work on article contacts which get rearranged to fill the voids and pores resulting in more densification of the specimen [11], which leads to a remarkable increase in the mechanical resistance of the samples. On the other side an increase of pressure to 1.1 GPa reduces the microhardness, Young's modulus and yield strength. This decrease of mechanical properties is due to the decrease in strength of bonding between the grains and consequently causes the degradation [12].

4. Conclusions

1) The highest critical temperature was 117 K under 0.7 GPa.

2) A remarkable increase, in the mechanical resistance of the samples was observed with the increase of the pressure from 0.3 GPa to 0.9 GPa. On the other hand an increase of the pressure to 1.1 GPa reduces these properties.

REFERENCES

[1] T. D. Xiao, K. E. Gonsalves and P. R. Strutt, "Synthesis of Aluminum Nitride/Boron Nitride Composite Materials," *Journal of the American Ceramic Society*, Vol. 76, No. 4, 1993, pp. 987-992.

[2] P. L. Chen and I. W. Chen, "Sintering of Fine Oxide Powders: I, Microstructural Evolution," *Journal of the American Ceramic Society*, Vol. 79, No. 12, 1996, pp. 3129-3135.

[3] D. R.Yang, D. S. Tsai and H. C. Liu, "Raising Pyrolysis Yield of Preceramic Polymers of Silicon Carbonitride," *Journal of Material Science*, Vol. 30, No. 17, 1995, pp. 4463-4468.

[4] B. S. Ahn, "Synthesis of BiSrCaCu(Ni)O Ceramics from the Gel Precursors and the Effect of Ni Substitution," *Bulletin Korean Chemical Society*, Vol. 23, No. 9, 2002, pp. 1304-1308.

[5] J. Joo, J. P. Singh, T. Warzynski, A. Grow and R. B. Poeppel, "Role of Silver Addition on Mechanical and Superconducting Properties of High-T_c Superconductors" *Applied Superconductivity*, Vol. 2, No. 6, 1994, pp. 401-410.

[6] K. Katagiri, A. Murakami and K. Kasaba, "Evaluation of Mechanical Properties of High-Temperature Superconducting Bulks Fabricated by a Melt-Processing," *Cryogenics*, Vol. 48, No. 3-4, 2008, pp. 87-94.

[7] M. A. Madre1, Sh. Rasekh, J. C. Diez and A. Sotelo, "Dynamic Fatigue Behaviour of Ag-Doped Bi-2212 Textured Thin Rods," *Journal of Physics: Conference Series*, Vol. 153, No. 1, 2009, Article ID: 012021.

[8] G. Y. Hermiz, B. A. Aljurani and H. A. Thabit, "Mechanical Properties of $Bi_{1.6}Pb_{0.4}Sr_{1.8}Ba_{0.2}Ca_2Cu_{3-x}Ni_xO_{10+\delta}$ Superconducting System," *Journal of Superconductivity and Novel Magnetism*, Vol. 25, No. 6, 2012, pp. 1629-1634.

[9] X. J. Chen, H. Q. Lin and C. D. Gong, "Pressure Dependence of T(c) in Y-Ba-Cu-O Superconductors," *Physical Review Letter*, Vol. 85, No. 10, 2000, pp. 2180-2183.

[10] S. V. Vonsovsk, Yu. A. Lzumov and E. Z. Kurmaev, "Superconductivity of Transition Metals," Springer-Verlag, Berlin, 1982.

[11] V. M. Anandakumar and M. A. Khadar, "Microhardness Studies of Nanocrystalline Calcium Tungstate," *Crystal Research and Technology*, Vol. 43, No. 2, 2008, pp. 193-199.

[12] O. Gorur, M. Nursoy, C. Terzioglu, A. Varilci and I. Belenli, "Influence of Annealing Temperature on the Microstructure and Mechanical Properties of MgB_2," *Journal of Physics: Conference Series*, Vol. 153, No. 1, 2009, Article ID: 012012.

A Study of γ-Fe_2O_3 Nanoparticles Modified Using $ZnCl_2$ during Synthesis*

Longlong Chen, Jian Li#
School of Physical Science & Technology, Southwest University, Chongqing, China

ABSTRACT

Surface modification of γ-Fe_2O_3 nanoparticles synthesized by a chemically induced transition has been attempted by adding $ZnCl_2$ during synthesis. The structure of the modified particles was studied using X-ray diffraction (XRD) and transmission electron microscopy (TEM). The experimental results showed that $ZnFe_2O_4$ grew epitaxially on the γ-Fe_2O_3 crystallites to form composite nanoparticles with the spinel structure, on which $FeCl_3$ might be adsorbed. The apparent grain size d_c estimated by XRD using the Scherrer equation was larger rather than smaller than the average particle size $\langle d \rangle$ measured by TEM. This paradox can be explained by the effect of absorption in the coating heterolayer.

Keywords: Composite Nanoparticle; Spinel Structure; X-Ray Diffraction; Scherrer Equation

1. Introduction

The interest in nanomaterials has grown because of their distinct optical, magnetic, electronic, mechanical, and chemical properties compared with those of the bulk materials. Magnetic nanoparticles with a size less than 100 nm represent an important class of artificial-nanostructure materials; they allow investigations of fundamental aspects of magnetic ordering phenomena and could lead to new technological applications [1]. Studies of magnetic nanoparticles have focused on the development of novel synthetic methods [2]. In previous work, we presented a method to prepare magnetic nanoparticles using a chemically induced transition, which involved treating a hydroxide precursor with an $FeCl_2$ solution [3-5].

A nanocomposite is a material composed of two or more phases, one of which has a grain size of less than 100 nm. The resulting combination of different physical or chemical properties may give rise to completely new materials [6,7]. Magnetic nanocomposites have applications ranging from ferrofluids to separation science and technology [8]. Much research has shown that composite nanoparticles can be prepared by surface modification of particles, either during synthesis or in a postsynthesis step [9]. Such composite nanoparticles can be described in terms of a core-shell model [10].

X-ray diffraction (XRD) is a useful technique for characterizing the physicochemical makeup of unknown solids. For ferrite nanoparticles with the spinel structure, the average size d_c of the grains in a sample can be estimated from the (311) peak using the Scherrer equation as

$$d_c = \frac{k\lambda}{\beta \cos\theta} \tag{1}$$

where k is a constant (equal to 0.89), λ is the wavelength of the X-rays, θ is the Bragg diffraction angle of the (311) plane, and β is the full width at half-maximum (FWHM) of the (311) diffraction peak [11,12]. Generally, the average grain size given by XRD is less than the average size obtained from transmission electron microscopy (TEM), which is termed the physical size. However, oriented attachment may occur, in which a relatively large crystalline structure is formed by the attachment of crystalline primary particles [13], so that there may be some agglomeration of the nanoparticles in such a way that the particle size obtained from XRD is larger than the physical size obtained from TEM [12]. In the work presented here, γ-Fe_2O_3/$ZnFe_2O_4$ magnetic composite nanoparticles prepared by surface modification were studied, and a new explanation is proposed for the fact that the particle size estimated using XRD is larger than the physical size measured by TEM.

*Financial support for this work was provided by the National Science Foundation of China (No. 11074205).
#Corresponding author.

2. Experimental

Nanoparticles were synthesized by a method based on a chemically induced transition. The method involved two steps.

1) *Preparation of the precursor*. An aqueous solution of $FeCl_3$ (40 mL, 1 M) and $Mg(NO_3)_2$ (10 mL, 2 M, in 0.05 M HCl) was added to an NaOH solution (500 mL, 0.7 M). The mixture was heated and then boiled for 5 min with stirring. The red-brown precursor precipitated gradually after the heating was stopped.

2) *Synthesis of the nanoparticles*.

a) *Unmodified nanoparticles*. The precursor was added to a $FeCl_2$ solution (400 mL, 0.25 M), which was heated at boiling point for 30 min. After the heating was stopped, the product separated gradually from solution. It was then dehydrated by aceton and dried naturely to obtain γ-Fe_2O_3 nanoparticles [3], which we refer to as the unmodified sample.

b) *Modified nanoparticles*. During the above process of nanoparticle synthesis, after the $FeCl_2$ solution with added precursor had been heated at boiling point for 20 min, $ZnCl_2$ solution (50 mL, 2 M) was added to perform modification. The mixture was then boiled continuously for 10 min. Particles were precipitated after the heating was stopped. Finally, nanoparticles modified by the addition of $ZnCl_2$ were obtained after dehydrating and drying; these are referred to as the modified sample.

To characterize the modified sample, the morphology of the particles was observed using TEM at 200 kV, and the crystal structure was analyzed using XRD with Cu Kα radiation. For comparison, the unmodified particles ware also characterized by XRD.

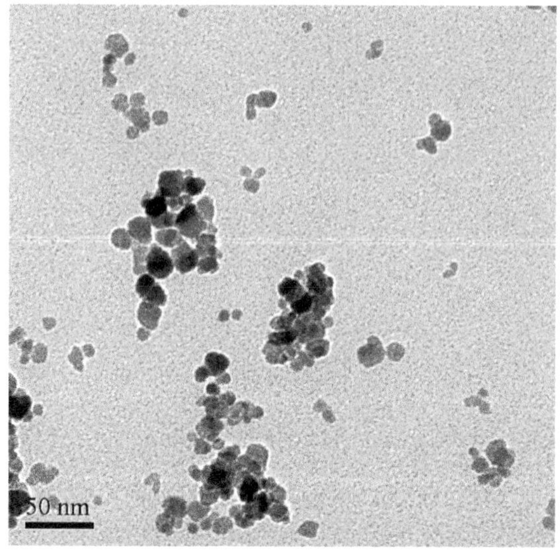

Figure 1. Typical TEM image of the modified sample.

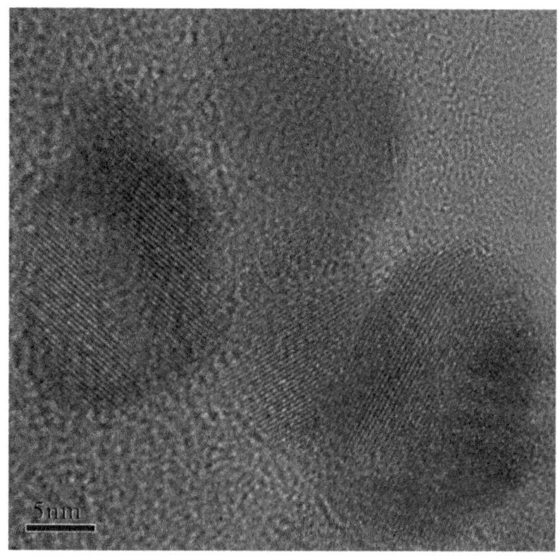

Figure 2. HRTEM image of an agglomeration in the modified sample.

3. Results and Analysis

TEM observation indicated that the modified particles were nearly spherical nanoparticles, as shown in **Figure 1**. Statistical analysis showed that the sizes of the particles fitted a log-normal distribution; the median diameter d_g and standard deviation $\ln \sigma_g$ were 10.47 nm and 0.27, respectively. For spherical particles with a log-normal size distribution, the average diameter $\langle d \rangle$, which is just so-called physical size measured by TEM, can be calculated from [14]

$$\langle d \rangle = \exp\left[\ln d_g + 0.5\ln^2 \sigma_g\right]. \quad (2)$$

The average diameter of the particles was given by this equation as 10.86 nm. The results of high-resolution TEM (HRTEM) revealed that the nanoparticles were highly crystalline, and in the aggregations of the nanoparticles, the particles exhibited different crystalline orientations, as shown in **Figure 2**.

XRD patterns of an unmodified and a modified sample are shown in **Figure 3**. It is evident that the unmodified particles have a ferrite-like spinel structure with the features of magnetite (PDF #39-1346), whose lattice constant is 0.8352 nm. For the modified sample, there were also clear diffraction peaks from the spinel structure, but there were no diffraction peaks corresponding to the (110) and (210) planes of γ-Fe_2O_3, and the peak for the (310) plane was weaker, although other peaks were stronger than in the unmodified sample. Therefore, we judged that in addition to γ-Fe_2O_3, there could be another phase with the spinel structure in the modified sample, for which the diffraction from the (110), (210), and (310) planes is apparently too weak to produce peaks in the XRD pattern. As a consequence, it can be inferred from the XRD results that the nanoparticles in the modified sample possess mainly the spinel structure. The diameter of the grains in the modified sample could be obtained from the

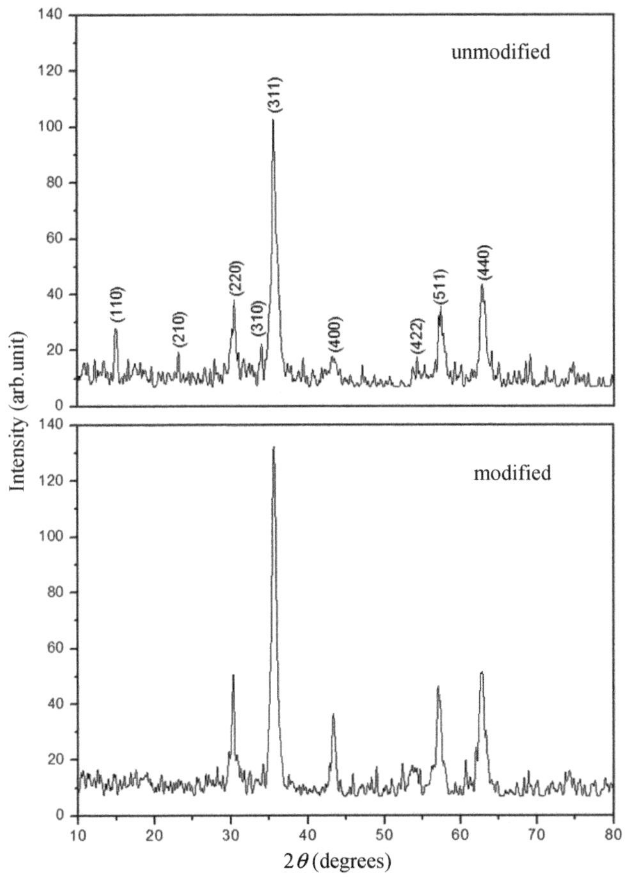

Figure 3. XRD patterns of unmodified and modified sample.

(311) peak using the Scherrer equation as $d_c = 13.43$ nm.

4. Discussion

The XRD analysis indicates that the unmodified sample consisted of γ-Fe_2O_3 nanocrystallites, which could be coated with adsorbed $FeCl_3$ [15]. During synthesis of the modified sample with $ZnCl_2$ addition, we infer that γ-Fe_2O_3 crystallites were formed first. This is attributed to the precursor transforming in the $FeCl_2$ solution, by the same process as in the formation of the unmodified nanoparticles. After $ZnCl_2$ was added to the solution, another crystal layer with a spinel structure similar to the γ-Fe_2O_3 phase grew epitaxially on the γ-Fe_2O_3 crystallites. Unlike the unmodified sample, the modified sample did not show diffraction peaks from the (110) and (210) planes, and the peak from the (310) plane was weaker in the modified sample. So, it can be judged that the additional crystal layer is the $ZnFe_2O_4$ phase, for which diffraction lines from the (110), (210) and (310) planes are not present (PDF #22-1024), and the lattice constant is 0.8441 nm. The standard intensity data *I(f)* for both γ-Fe_2O_3 and $ZnFe_2O_4$ are listed in **Table 1**.

In addition, it was noticed that for the modified sample, the average size d_c obtained by XRD analysis (=13.43 nm) is larger than the average size $\langle d \rangle$ obtained from TEM (=10.47 nm). This is a similar result to that in [12], in which it was attributed to the oriented attachment of crystalline primary particles. However, the HRTEM results indicate that for the modified sample, the crystalline orientations of the nanoparticles in an agglomeration are different. Moreover, if instrumental broadening is taken into account [16], the apparent grain size becomes higher than the size d_c given above. Instead, the paradox of the difference between d_c and $\langle d \rangle$ could be related to an absorption effect. We believe that the modified particles may be coated with adsorbed $FeCl_3$. For particles coated with a different material, the X-rays diffracted by the interior can be absorbed by the coating layer, so that the diffraction peaks become weaker. For example, when α-Al_2O_3 nanoparticles were coated with $La(OH)_3$, the diffraction peaks of the α-Al_2O_3 were very weak, and the main diffraction peaks agreed with an $La(OH)_3$ phase [17]. This absorption effect could result in the FWHM β of the diffraction peak becoming less, so that the apparent d_c of the spinel structure would become larger for γ-Fe_2O_3/$ZnFe_2O_4$ nanoparticles coated with $FeCl_3$. This can be explained as follows.

In practical applications, the FWHM β can be calculated as the ratio between the integrated intensity and the maximum intensity [18], *i.e.*

$$\beta = \frac{\int I \, d(2\varepsilon)}{I_m} \tag{3}$$

where I is the normal diffraction intensity, I_m is the maximum of the intensity, and 2ε is the deviation of 2θ from the diffraction condition, as shown in **Figure 4**. This value of β is generally termed the integral width. When the diffraction from the grains is absorbed by a coating layer, the FWHM is written as

$$\beta' = \frac{\int I' \, d(2\varepsilon)}{I'_m} \tag{4}$$

where I' is the curve of the diffraction intensity and I'_m is the maximum of the intensity with absorption effect from the coating layer. Generally, after the X-ray with intensity I_0 passing through a matter of thick t, the intensity becomes $I_t = I_0 e^{-\mu t}$, where μ is coefficient of linear absorption. Obviously, the longer the path of the X-ray going is, the weaker the I_t is. Thus, while absorption effect of the coating layer is considered, the curve of diffraction intensity can be described as

$$I' = A(2\varepsilon) I \tag{5}$$

where I is the curve of diffraction intensity without the absorption effect and $A(2\varepsilon)$ is absorbing factor in relation to 2ε. Since 2ε is very small, the $A(2\varepsilon)$ can be looked as a

Table 1. Standard intensity data *I(f)* for γ-Fe_2O_3 (PDF #39-1346) and $ZnFe_2O_4$ (PDF #22-1012).

hkl	110	210	220	310	311	400	422	511	440
γ-Fe_2O_3	5	5	35	2	100	16	10	24	34
$ZnFe_2O_4$	-	-	35	-	100	17	12	30	35

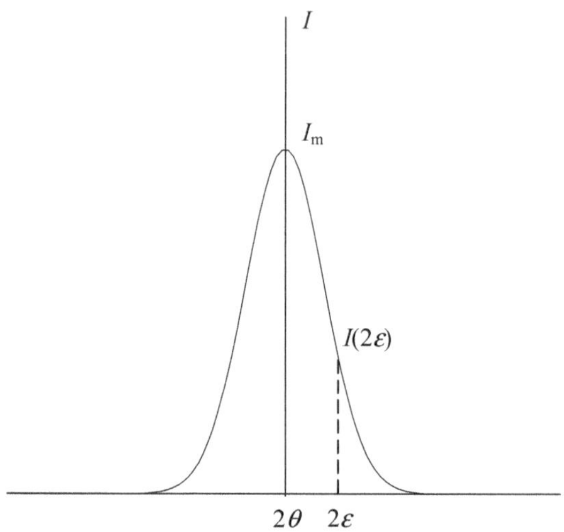

Figure 4. Schematic curve of X-ray diffraction intensity.

constant A less than one. Consequently, the FWHM β' of the diffraction intensity with the absorption effect can be written as

$$\beta' = A\frac{\int I\,d(2\varepsilon)}{I'_m} \tag{6}$$

where I'_m is the maximum of the diffraction intensity in the presence of the absorption effect. Obviously, if I'_m is regarded as I_m, $\beta' = A\beta$. Since $A < 1$, $\beta' < \beta$. This means that in an actual XRD measurement, the curve of the diffraction intensity in the presence of an absorption effect is narrower than that for normal diffraction without an absorption effect, as shown in **Figure 5**. Hence, it can be said in summary that because of the effect of absorption, the grain size d_c estimated using the Scherrer equation can be not only larger than the actual grain size but also larger than the physical size measured by TEM.

Also, when the volume fraction of diffracting units (in a mixture typically containing different crystallites) is less than one, the maximum of the diffraction intensity can be described by $I''_m = BI_m$ and the curve of the intensity is given by $I'' = BI$, where B is defined as the content factor and is less than one. So, the FWHM β'' of the diffraction intensity with a volume fraction of diffracting units less than one is

$$\beta'' = \frac{\int I''\,d(2\varepsilon)}{I''_m} = \frac{\int I\,d(2\varepsilon)}{I_m} = \beta \tag{7}$$

Equation (7) shows that the FWHM of the diffraction curve remains unaffected; it has a constant value β. In other word, the ratio between the integral intensity and maximum intensity is unchanged for both the diffraction with insufficient diffraction unit and normal diffraction.

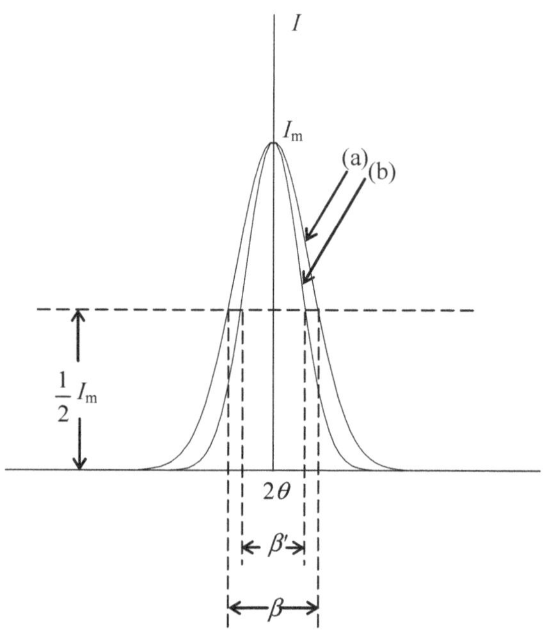

Figure 5. Comparison between intensity curves for (a) normal diffraction and (b) diffraction with absorption effect.

5. Conclusion

X-ray diffraction is an important tool for identifying the structure of nanoparticles from the data on interplanar spacings and relative intensities that it provides. In the work presented here, the XRD results confirmed that during synthesis, $ZnFe_2O_4$ grew epitaxially on γ-Fe_2O_3 crystallites to form composite nanoparticles, which were then possibly coated with an $FeCl_3$ layer. The size of the nanoparticles can be estimated from the XRD results using the Scherrer equation, and should be less than the physical size measured by TEM. However, our work shown that for nanoparticles with a coating heterolayer, the effect of absorption by the coating could weaken the XRD intensity from the interior of the particle compared with that for a bare particle without a coating layer, so that the FWHM of the diffraction peak would be reduced, making the apparent average size of the grains measured by XRD larger than the actual size. Consequently, it is possible that the apparent size of the grains obtained by XRD may be larger than the physical size given by TEM. When the volume fraction of diffracting units is less than one, both the FWHM of the diffraction peak and the apparent grain size are unchanged, and comparison with the FWHM of the unmodified nanoparticles may be used to determine whether a coating heterolayer has formed on the primary particles; that is, when β becomes smaller, a coating heterolayer has formed on the particles.

REFERENCES

[1] S. H. Sun, "Recent Advances in Chemical Synthesis, Self-Assembly, and Applications of FePt Nanoparticles," *Advanced Materials*, Vol. 18, No. 4, 2006, pp. 393-403.

[2] J. Jiang and Y. M. Yang, "Facile Synthesis of Nanocrystalline Spinel $NiFe_2O_4$ via a Novel Soft Chemistry Route," *Materials Letters*, Vol. 61, No. 21, 2007, pp. 4276-4279.

[3] Q. M. Zhang, J. Li, Y. Q. Lin, X. D. Liu and H. Miao, "The Preparation and Characterization of Ni-Fe Bioxide Composite Nanoparticles," *Journal of Alloys and Compounds*, Vol. 508, No. 2, 2010, pp. 396-399.

[4] B. C. Wen, J. Li, Y. Q. Lin, X. D. Liu, J. Fu, H. Miao and Q. M. Zhang, "A Novel Preparation Method for Gamma-Fe_2O_3 Nanoparticles and Their Characterization," *Materials Chemistry and Physics*, Vol. 128, No. 1-2, 2011, pp. 35-38.

[5] H. Miao, J. Li, Y. Q. Lin, X. D. Liu, Q. M. Zhang and J. Fu, "Characterization of Gamma-Fe_2O_3 Nanoparticles Prepared by Transformation of Alpha-FeOOH," *Chinese Science Bulletin*, Vol. 56, No. 22, 2011, pp. 2383-2388.

[6] D. V. Szabó and D. Vollath, "Nanocomposites from Coated Nanoparticles," *Advanced Materials*, Vol. 11, No. 15, 1999, pp. 1313-1316.

[7] J. Liu, S. Z. Qiao, Q. H. Hu and G. Q. Lu, "Magnetic Nanocomposites with Mesoporous Structures: Synthesis and Applications," *Small*, Vol. 7, No. 4, 2011, pp. 425-443.

[8] Q. X. Liu, Z. H. Xu, J. A. Finch and R. Egerton, "A Novel Two-Step Silica-Coating Process for Engineering Magnetic Nanocomposites," *Chemistry of Materials*, Vol. 10, No. 12, 1998, pp. 3936-3940.

[9] M. A. Willard, L. K. Kurihara, E. E. Carpenter, S. Calvin and V. G. Harris, "Chemically Prepared Magnetic Nanoparticles," *International Materials Reviews*, Vol. 49, No. 3, 2004, pp. 125-170.

[10] M. Knobel, W. C. Nunes, L. M. Socolovsky, E. De Biasi, J. M. Vargas and J. C. Denardin, "Superparamagnetism and Other Magnetic Features in Granular Materials: A Review on Ideal and Real Systems," *Journal of Nanoscience and Nanotechnology*, Vol. 8, No. 6, 2008, pp. 2836-2857.

[11] T. Sato, T. Iijima, M. Seki and N. Inagaki, "Magnetic Properties of Ultrafine Ferrite Particles," *Journal of Magnetism and Magnetic Materials*, Vol. 65, No. 2-3, 1987, pp. 252-256.

[12] S. Chakrabarti, D. Ganguli and S. Chaudhuri, "Optical Properties of Gamma-Fe_2O_3 Nanoparticles Dispersed on Sol-Gel Silica Spheres," *Physica E-Low-Dimensional Systems & Nanostructures*, Vol. 24, No. 3-4, 2004, pp. 333-342.

[13] A. Naravanaswamy, H. F. Xu, N. Pradhan and X. G. Peng, "Crystalline Nanoflowers with Different Chemical Compositions and Physical Properties Grown by Limited Ligand Protection," *Angewandte Chemie-International Edition*, Vol. 45, No. 32, 2006, pp. 5361-5364.

[14] C. G. Granqvist and R. A. Buhrman, "Ultrafine Metal Particles," *Journal of Applied Physics*, Vol. 47, No. 5, 1976, pp. 2200-2219.

[15] L. L. Chen, J. Li, Y. Q. Lin, X. D. Liu and D. C. Li, "Surface Modification and Characterization of Gamma-Fe_2O_3 Nanoparticles Synthesized by Chemically-Induced Transition," *Materials Chemistry and Physics*, submitted.

[16] P. Mathur, A. Thakur and M. Singh, "Effect of Nanoparticles on the Magnetic Properties of Mn-Zn Soft Ferrite," *Journal of Magnetism and Magnetic Materials*, Vol. 320, No. 7, 2008, pp. 1364-1369.

[17] J. Li, C. Liu and Y. Yao, "Study on Preparation of $LaAlO_3$ Particles by Wet Chemical Method," *Journal of the Chinese Ceramic Society*, Vol. 22, No. 5, 1994, pp. 504-508.

[18] H. P. Klug and L. E. Alexander, "X-Ray Diffraction Procedures for Polycrystalline and Amorphous Materials," John Wiley & Sons Inc., New York, 1974, p. 299.

A New Simple Route to ZnS Quantized Particles with Tunable Size and Shape, and Size/Shape-Dependent Optical Properties

Pengfei Hu[1*], Yali Cao[2], Yanyan Lou[1], Bo Lu[1], Min Shao[1], Jiansen Ni[1], Meng Cao[3]
[1]Laboratory for Microstructures, Shanghai University, Shanghai, China
[2]Institute of Applied Chemistry, Xinjiang University, Urumqi, China
[3]School of Materials Science and Engineering, Shanghai University, Shanghai, China

ABSTRACT

With the features of convenience and eco-friendly, the low-temperature solid-state reaction synthesis was successfully developed as a new approach to prepare quantum-sized ZnS nanocrystals. One major achievement is that the size and shape of ZnS nanocrystals can be tuned by adjusting the surfactant and its feed. The UV-Vis absorption spectra of quasi-spherical and one-dimensional quantum-sized ZnS nanocrystals all showed a blue-shift from the bulk counterpart, indicating large quantum confinement effects of ZnS nanocrystals. These ZnS nanocrystals all showed well-defined excitonic emission features. Contrastive studies on photoluminescence performances indicated that the band-edge emission experienced only the size-dependent quantum confinement effect, while the trap-state emission experienced the size- and shape-dependences. So we can design a purposeful synthesis route to ZnS nanocrystals with target luminescence emission performances.

Keywords: Solid-State Nanosynthesis; Tunable; Size- and Shape-Dependence; Optical Properties

1. Introduction

Owing to fascinating size-, and shape-dependent electronic and optical properties, quantized semiconductor structures have attracted increasing attention [1-3]. It is well known, for instance, that the band gap of semiconductor nanostructure increases as the particle size decreases when the dimension of nanocrystals approach to exciton Bohr radius, due to quantum-size confinement effect [4]. As one member of the important II-VI group semiconductors, zinc sulfide (ZnS, direct band gap of 3.7 eV) motivated worldwide enthusiasms of scientists. Quantized ZnS (Q-ZnS) particles and Q-ZnS-based heterostructures have become essential light-harvesting and light-emitting components in modern technology such as imaging, diagnostics, biosensing, and display technologies [5-7]. Up to now, some methodologies for synthesizing Q-ZnS have been demonstrated [8-10]. Among most nanosynthsis techniques, surface-passivated technology in solution-process received relatively more interest [3,4,8-11]. It used surface-bonding ligands to repress the agglomeration of primary colloidal nanocrystals and obtain free-standing nanoparticles. However, this methodology generally requires complicated manipulations including delicate control of solution composition, reaction atmosphere, pH value, reaction temperature and time, even of temperature ramp. Therefore, it is necessary to explore simple and convenient strategies to produce quantized structures with tunable size/shape and performance.

The low-temperature solid-state reaction synthetic approach provided a relatively simple and powerful method for controlling the size and shape of nanoparticles [12, 13]. In view of the advantages of low cost, convenience, and lack of pollution, we introduced it to synthesize the Q-ZnS firstly. We fulfilled the size-, and shape-control of Q-ZnS through surfactant-assistanted technique, and the tunable optical behavior.

2. Experimental

The manipulation for the synthesis of Q-ZnS through one-step solid-state reaction at room temperature was recorded and illustrated in **Figure 1**. The surfactants sodium dodecyl benzene sulfonate (SDBS) and cetyltrimethyl ammonium bromide (CTAB) were introduced to assist the synthesis, respectively. Zinc acetate dihydrate ($Zn(OAc)_2 \cdot 2H_2O$) powders were firstly blended

*Corresponding author.

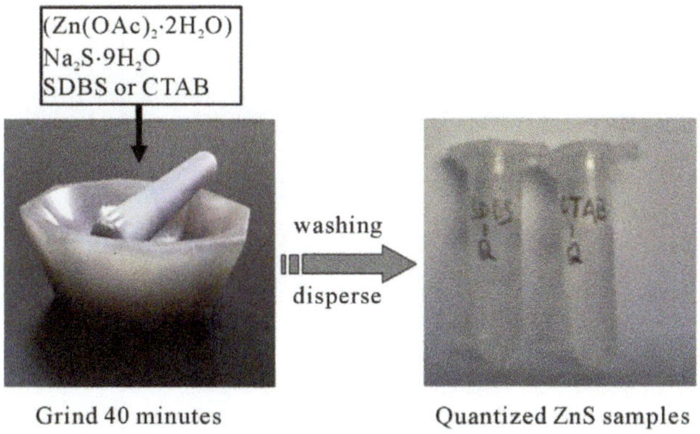

Figure 1. A convenient and effective recipe for quantized structures.

with an appreciable proportion of SDBS (Q-ZnS-i) or CTAB (Q-ZnS-ii) (MR_{SDBS} or MR_{CTAB} = 1:5, here, the molar ratio of additive/$Zn(OAc)_2 \cdot 2H_2O$ was designated as MR) and ground together for 10 min at room temperature, then sodium sulfide hydrate ($Na_2S \cdot 9H_2O$) was added with 1:1 molar ratio between $Zn(OAc)_2 \cdot 2H_2O$ and $Na_2S \cdot 9H_2O$. The mixtures were ground for about 30 min, and centrifugation washed, respectively. Finally, the Q-ZnS samples were dispersed in deionized water, respectively, and clear solutions were obtained for following characterization.

3. Results and Discussion

3.1. Crystal Structures of ZnS Nanocrystals

The X-ray diffraction (XRD) patterns for as-prepared Q-ZnS-i and -ii clearly indicates that they have cubic phase with lattice constant as cross-referenced to JCPDS 65 - 9585 card (**Figure 2(a)**). Three characteristic diffraction peaks at 2θ values of 28.7/28.9, 47.4/47.4, and 56.5/56.6 to Q-ZnS-i and -ii can be indexed to the crystal planes of (1 1 1), (2 2 0) and (3 1 1). The cell constants were calculated to be a = b = c = 0.5382/0.5375 nm from (111) peaks of Q-ZnS-i and -ii after refinement. They revealed that the lattice contractions of Δa = 0.50% and 0.63% occurred against to the reported data (a = b = c = 0.5409 nm), which implied the smaller sizes of ZnS particles and the high surface to volume ratio of nanocrystals. X-ray energy-dispersive spectroscopy (EDS) shown that the atom ratios of Zn to S are about 1:1.03 to Q-ZnS-i and 1:1.05 to Q-ZnS-ii, respectively (**Figure 2(b)**).

Additionally, the behaviors of SDBS and CTAB capping on the ZnS nanoparticles were identified by FT-IR spectra. The FTIR spectra of Q-ZnS-i and SDBS are compared in **Figure 3(a)**. It is obvious that the Q-ZnS-i hold the characteristic peaks of sulfonate around 1210 - 1025 cm^{-1} (denoted by the olive dotted ellipse in **Figure 3(a)**). In the FT-IR comparison of CTAB and Q-ZnS-ii, the transmission bands at 2920 and 2850 cm^{-1} in both are similar and respectively assigned to the asymmetric and symmetric stretching vibrations of methylene (CH_2) in the long alkyl chain, as well as that the transmission bands of Q-ZnS-ii in the ranges of 965 - 720 cm^{-1} and 1390 - 1490 cm^{-1} are close to that of CTAB (the olive dotted ellipses in **Figure 3(b)**).

As shown in **Figure 4A** and insets a-c, the Q-ZnS-i nanocrystals are the quasi-spherical nanoparticles. The average size of these nanoparticles is 2.5 - 3.5 nm. The inset c of **Figure 4A** shows clear twin and stacking fault (SF) structures in a big nanoparticle. When CTAB was used as additive, the ZnS nanocrystals put up asymmetrical growth. As shown in **Figures 4B**, insets, and S1, near one-dimensional nanocrystals were recognized clearly, such as acicula-shaped nanocrystal (inset b in

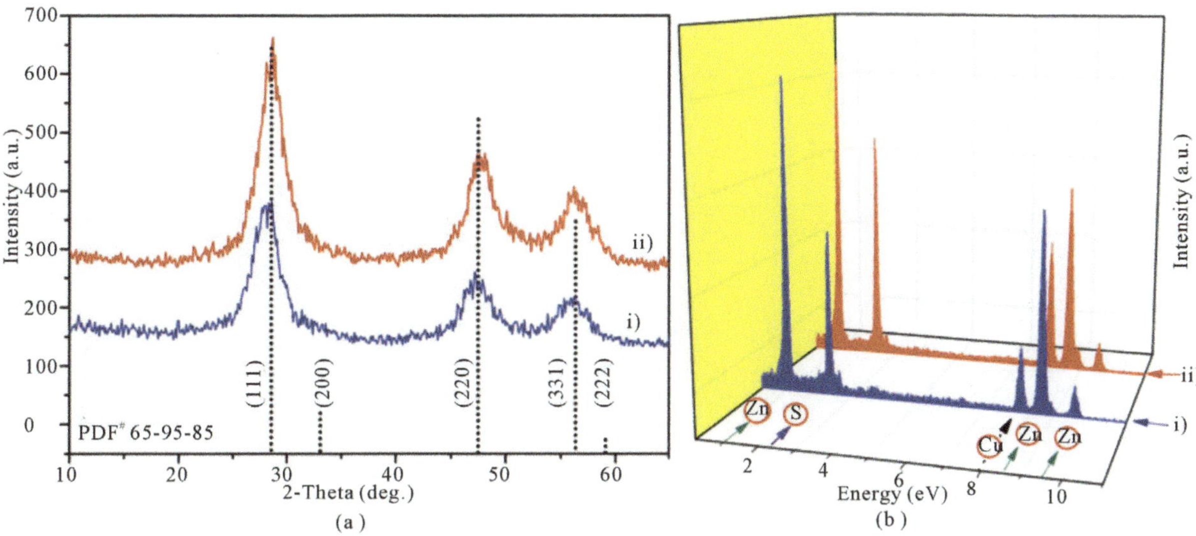

Figure 2. XRD patterns (a), and EDS spectra (b) of ZnS nanocrystals: Q-ZnS-i (blue curve-i), and Q-ZnS-ii (red curve-ii).

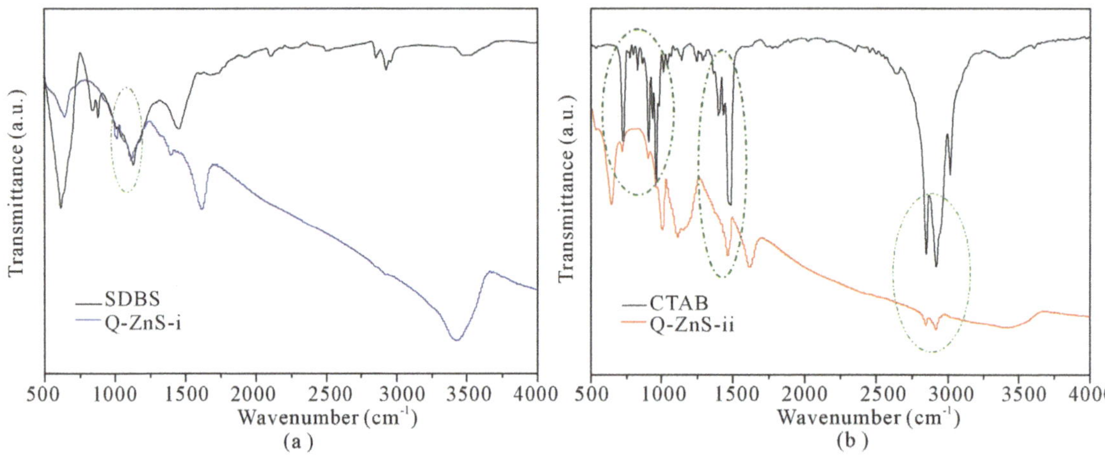

Figure 3. FT-IR spectra of SDBS and Q-ZnS-i (a), and CTAB and Q-ZnS-ii (b).

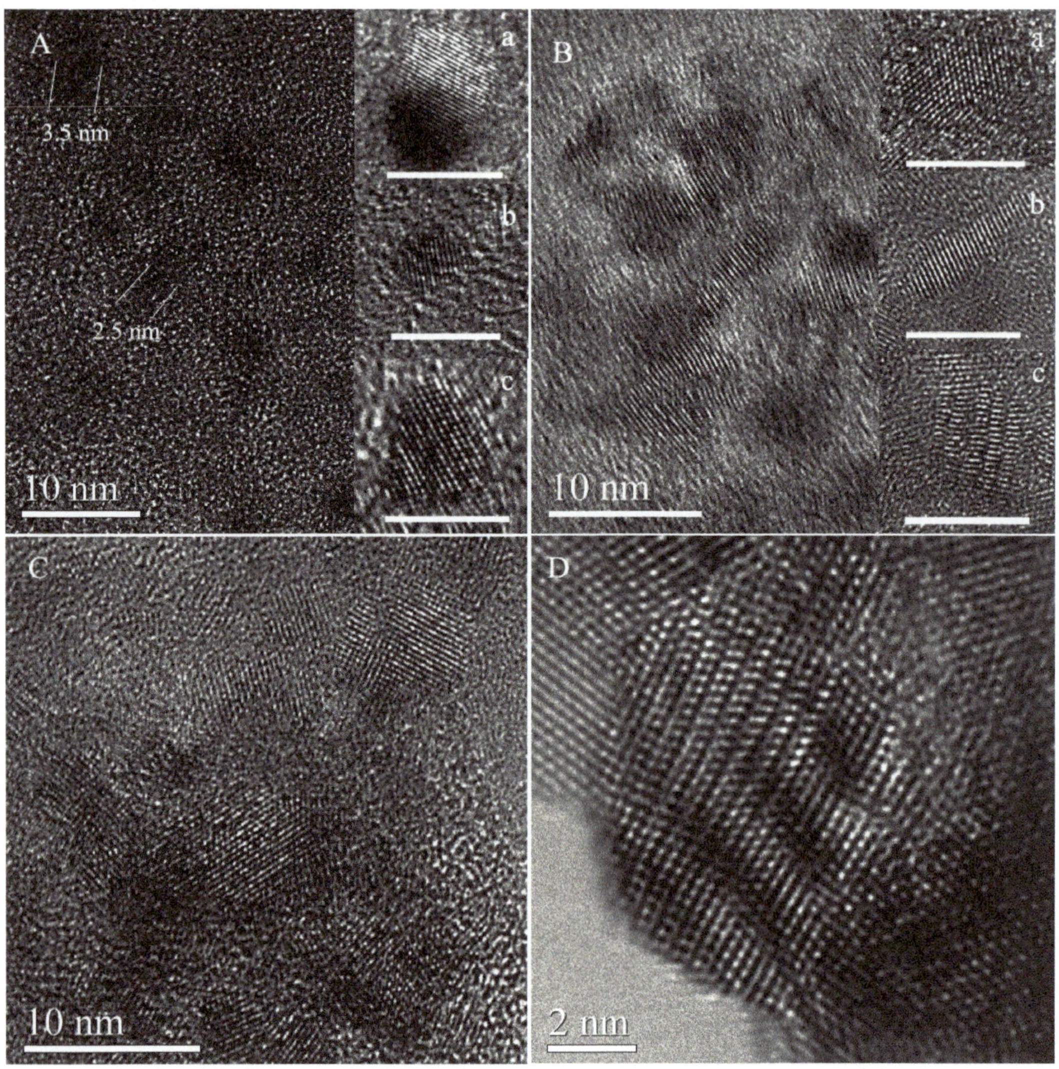

Figure 4. (A) TEM images of 3 nm quasi-spherical Q-ZnS-i nanocrystals synthesized with MRSDBS = 1:5; (B) TEM images of 3 nm near one-dimensional Q-ZnS-ii nanocrystals obtained with MRCTAB = 1:5; The insets a-c in panels A and B are HRTEM images of representative nanoparticles of Q-ZnS-i and -ii respectively (bar scale: 5 nm); (C) TEM images of 7 nm quasi-spherical Q-ZnS-iii nanocrystals synthesized with MRSDBS = 1:10; (D) Higher resolution details of the Q-ZnS nanoparticles with wave-like structure.

Figure 4B). It is worthwhile to note that some nanocrystals show the wave-like structure containing a large number of SF and dislocation at troughs (inset c in **Figures 4B**, **D** and **S1**). We shall see in a forthcoming section that these novel structures can affect the photoluminescence properties of nanocrystals. In addition, to investigate the size control of particles in the solid-state reaction synthesis, MR_{SDBS} was change to 1:10 to synthesize the Q-ZnS-iii. The Q-ZnS-iii nanocrystals are dominant with about 7 nm-sized quasi-spherical nanoparticles (**Figure 4C**).

3.2. Optical Properties of ZnS Nanocrystals

3.2.1. UV-Vis Spectral Characteristics

Figure 5 shows the UV-Vis absorption spectera of the Q-ZnS-i and -ii. They almost remain the same shape which ends near 350 nm for about 3 nm-sized nanocrystals. Each of spectra is the superimposition of a relatively well-defined band with peaks around 264 (Q-ZnS-i) or 259 nm (Q-ZnS-ii), an ill-defined broad absorption between 340 and 268 nm approximately, and a high energy intense absorption below 250 nm only displaying a tail of absorption pattern. These resolved superimpositions of experimentally obtained absorption curves imply the relatively wide size distribution of products. The band gap values were obtained from the optical absorption spectra by plotting $(\alpha h\nu)^2$ versus ($h\nu$), and then extrapolating the linear region of $(\alpha h\nu)^2$ versus ($h\nu$) on the energy axis, as shown in inset of **Figure 5**. The band gap values were estimated to be 4.73 eV for Q-ZnS-i, whereas it is 4.78 eV for Q-ZnS-ii. The blue shift of the absorption band compared to bulk counterpart causes the band gap widening which indicates that the as-prepared Q-ZnS-i and -ii offer strong quantum confinement effect.

3.2.2. Photoluminescence Studies

Photoluminescence is a luminescence phenomena that occurs in luminescent materials which can be excited to a higher energy state termed "excited state" after absorbing light at a specific wavelength, and then re-emit light with a lower energy through recombination of electron and hole [14,15]. Due to high sensitivity and non-destructive character, the photoluminescence (PL) technique has been widely used to investigate electronic structure of the surface of semiconductor nanoparticles. Many reports illustrated the luminescence mechanism of ZnS nanoparticles using the well-known energy-level diagram (**Figure S2**). The valence band is considered as made up of hybridized s-p orbitals from sulfur and zinc atoms, while the conduction band is mainly consisted of the s-p antibond states of ZnS. Theoretically, luminescence emission includes band-edge emission and trap-state emission which is divided into shallow-trap and deep-trap emission according to the difference of electron levels [16]. Trap-stated PL signals mainly originate in defects containing vacancies and interstitials of semiconductors. The defects can be categorized either shallow or deep level

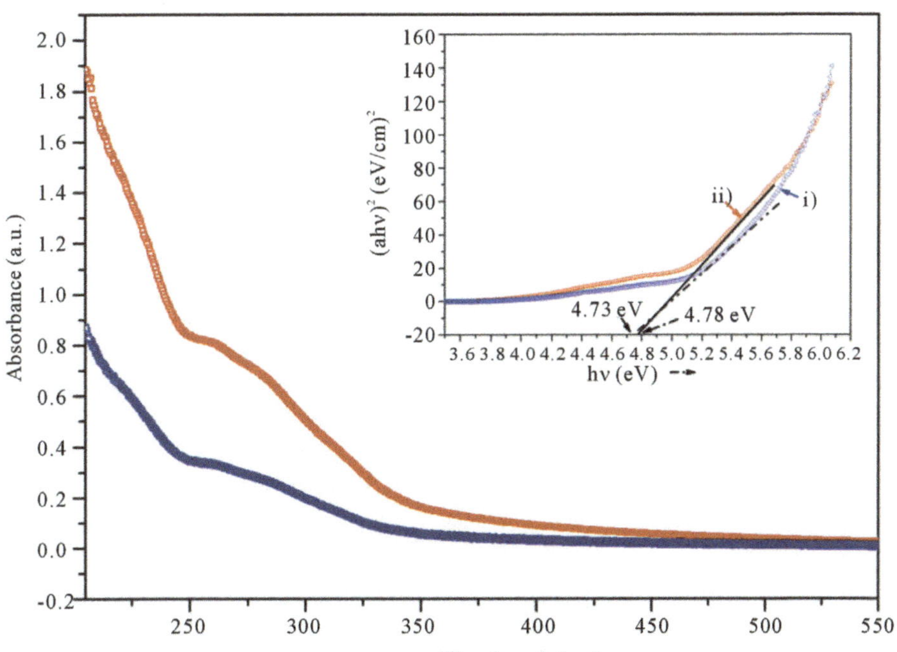

Figure 5. Comparison of UV-Vis spectra of 3 nm quasi-spherical Q-ZnS-i (blue symbol scatter-i) with one-dimensional Q-ZnS-ii (red symbol scatter-ii). The inset is the plots of $(\alpha h\nu)^2$ vs. ($h\nu$) for Q-ZnS-i (blue symbol scatter-i) and Q-ZnS-ii (red symbol scatter-ii).

defect states. Shallow level traps, which are more spatially delocalized and lie near the conduction band or valance band edge, are more likely to participate in the radiative recombination. The emission occurs when a trapped electron recombines with a hole in the valence band or in some acceptor level such as zinc vacancies [17]. On the other hand, deep level traps tend to have a tendency to undergo non-radiative recombination by emitting phonons [16].

As far as we known, although many groups have put large efforts in studying the optical properties of ZnS nanocrystals, there have been few reports on the excitonic emission feature of ZnS nanocrystals, as well as the size- and shape-dependent emission characteristics. In most cases, only trap-state emission at around 450 nm was observed and discussed which is termed "tunneling luminescence" and controlled by defect and impurity, the size- and shape-dependent quantum confinement effects were neglected. **Figure 6** shows the emission luminescence spectra of quasi-spherical (Q-ZnS-i) and one-dimensional (Q-ZnS-ii) ZnS nanocrystals, respectively, they were measured at 250 nm of excitation wavelength. The spectra all reveal the well-defined excitonic and trap-stated emission features. Many shallow trapped emissions indicated that the present nanocrystals occupied a great deal of effects.

Photoluminescence spectra of the quasi-spherical nanocrystals (Q-ZnS-i) showed broad violet emission peak centered at 342 nm accompanied by the broad asymmetry shoulder at 378 nm, and smaller hump-like blue emission peaks at 422 nm and 439 nm, and two weak shoulders at 456 and 500 nm, respectively. We attribute the 342 nm peak to a band-edge emission. The shoulder centered at 378 nm is assigned to the overlap-ping peaks involving electronic transitions from conduction band to

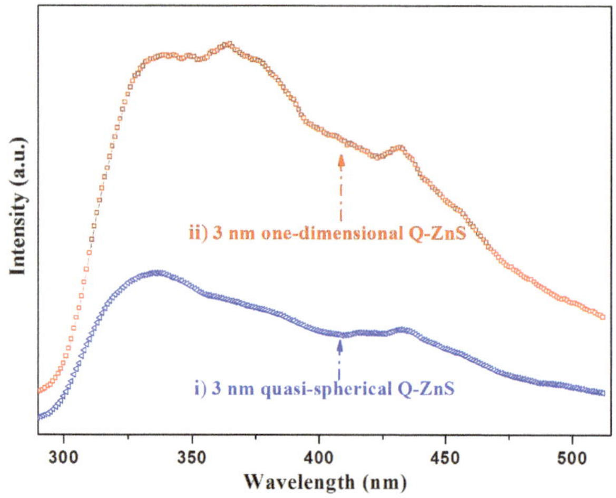

Figure 6. Comparison of photoluminescence spectra of the 3 nm quasi-spherical Q-ZnS-i (blue symbol scatter-i) with near one-dimensional Q-ZnS-ii (red symbol scatter-ii).

interstitial sulfur and interstitial zinc to valance band. In addition, blue emission peaks at 422 nm and 439 nm are assigned to the energy level of sulfur vacancies with the holes from the valance band, and interstitial zinc from interstitial sulfur, respectively, while the last peak at 500 nm which can be attributed to recombination of electrons at sulfur vacancies to the holes at zinc vacancies. It is particularly worth noting that the emission band at 456 nm was attributed to dangling sulfur bonds at the interface of ZnS in this paper. Although the deep-trap emission at around 450 nm is ascribed to surface sulfur vacant sites in some literatures [18], we can exclude the peak at 456 nm from the trap-state emission because there is excess sulfur in this report as demonstrated by the EDS data.

The emission spectra of near one-dimensional ZnS (Q-ZnS-ii) showed a band-edge emission feature at 340 nm, and high-intensity trap-state emissions about in range of 345 - 475 nm with well-defined peaks at 349, 365, and 432 nm, and broad shoulder at 455 nm. Compared to PL emission of quasi-spherical nanocrystals, the increase of intensity and shift of shallow-trapped emission implied that the near one-dimensional ZnS possess more defects, and the trap-state PL experimenced shape-dependent quantum confinements. And the intensity of emission band at 455 nm, which is ascribed to arise from the dangling sulfur bonds at the interface of ZnS, slightly increased. On the other hand, the band-edge emission of 341 nm showed almost the same wavelength as that of quasi-spherical nanocrystals (342 nm). It is concluded in this work that the band-edge emission may be mostly dependent on size rather than shape. To further demonstrate the size-dependence of quantum confinement effect to band-edge and trap-state emission, PL spectrum of 7 nm-sized quasi-spherical nanocrystals were got (**Figure 7**). The PL spectrum showed band-edge emission at 349 nm, which was red-shifted 7 nm as that of 3 nm-sized quasi-spherical nanocrystals. And the trap-state emissions all show varying degrees of red-shifts.

To estimate the applicability of approach adopted in this work, we introduce it to synthesize the CdS nanocrystals with SDBS. The 3 nm-sized quasi-spherical CdS nanocrystals were successfully obtained (**Figure S3**).

4. Conclusion

In summary, quantum-sized ZnS nanocrystals with quasi-spherical and near one-dimensional shapes were synthesized through room-temperature solid-state reaction methodology. The synthetic procedure has the advantages of convenient operation, and low cost, and lack of pollution, and for mass-production and good applicability. The size- and shape-controls can be achieved through using different additives and adjusting the MR. The quan tum-

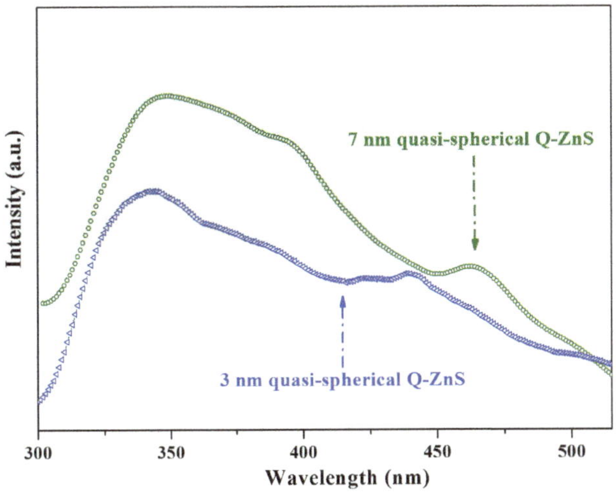

Figure 7. Comparison of photoluminescence spectra of the 3 nm quasi-spherical Q-ZnS-i (blue symbol scatter-i), and 7 nm quasi-spherical Q-ZnS-iii nanocrystals (olive symbol scatter-ii).

sized ZnS nanocrystals exhibited the quantum confinement effect and unique well-defined excitonic emission features. The band-edge PL emission showed the mostly size-dependent quantum confinement effect, but the trap-state PL experienced the size- and shape-dependences.

5. Acknowledgements

This work was funded by Shanghai University Innovation Foundation, National Nature Science Foundation of China (Grant No. 21101132), and National Nature Science Foundation of China (Grant No. 61006089).

6. Supporting Information Available

The characterization, the images of Q-ZnS-ii nanoparticles with one-dimensional and wave-like structures and CdS QDs, and Schematic energy level diagram.

REFERENCES

[1] L. E. Brus, "A Simple Model for the Ionization Potential, Electron Affinity, and Aqueous Redox Potentials of Small Semiconductor Crystallites," *Journal of Chemical Physics*, Vol. 79, No. 11, 1983, pp. 5566-5571.

[2] M. Nirmal and L. E. Brus, "Luminescence Photophysics in Semiconductor Nanocrystals," *Accounts of Chemical Research*, Vol. 32, No. 5, 1999, pp. 407-414.

[3] X. G. Peng, L. Manna, W. D. Yang, J. Wickham, E. Scher, A. Kadavanich and A. P. Alivisatos, "Shape Control of CdSe Nanocrystals," *Nature*, Vol. 404, No. 6773, 2000, pp. 59-61.

[4] V. I. Klimov, A. A. Mikhailovsky, S. Xu, A. Malko, J. A. Hollingsworth, C. A. Leatherdale, H.-J. Eisler and M. G. Bawendi, "Optical Gain and Stimulated Emission in Nanocrystal Quantum Dots," *Science*, Vol. 290, No. 5490, 2000, pp. 314-317.

[5] X. Michalet, F. F. Pinaud, L. A. Bentolila, J. M. Tsay, S. Doose, J. J. Li, G. Sundaresan, A. M. Wu, S. S. Gambhir and S. Weiss, "Quantum Dots for Live Cells, *in Vivo* Imaging, and Diagnostics," *Science*, Vol. 307, No. 5709, 2005, pp. 538-544.

[6] T. H. Kim, K. S. Cho, E. K. Lee, S. J. Lee, J. Chae, J. W. Kim, D. H. Kim, J. Y. Kwon, G. Amaratunga, S. Y. Lee, B. L. Choi, Y. Kuk, J. M. Kim and K. Kim, "Full-Colour Quantum Dot Displays Fabricated by Transfer Printing," *Nature Photonics*, Vol. 5, No. 3, 2011, pp. 176-182.

[7] C. L. Cowles and X. S. Zhu, "Sensitive Detection of Cardiac Biomarker Using ZnS Nanoparticles as Novel Signal Transducers," *Biosensors and Bioelectronics*, Vol. 30, No. 1, 2011, pp. 342-346.

[8] K. E. Sapsford, I. L. Medintz, J. P. Golden, J. R. Deschamps, H. T. Uyeda and H. Mattoussi, "Surface-Immobilized Self-Assembled Protein-Based Quantum Dot Nanoassemblies," *Langmuir*, Vol. 20, No. 18, 2004, pp. 7720-7728.

[9] Y. W. Zhao, Y. Zhang, H. Zhu, G. C. Hadjipanayis and J. Q. Xiao, "Low-Temperature Synthesis of Hexagonal (Wurtzite) ZnS Nanocrystals," *Journal of the American Chemical Society*, Vol. 126, No. 22, 2004, pp. 6874-6875.

[10] Y. J. Zhang, H. R. Xu and Q. B. Wang, "Ultrathin Single Crystal ZnS Nanowires," *Chemical Communications*, Vol. 46, No. 47, 2010, pp. 8941-8943.

[11] C. B. Murray, C. R. Kagan and M. G. Bawendi, "Synthesis and Characterization of Monodisperse Nanocrystals and Close-Packed Nanocrystal Assemblies," *Annual Review of Materials Science*, Vol. 30, No. 1, 2000, pp. 545-610.

[12] R. F. Service, "Don't Sweat the Small Stuff," *Science*, Vol. 320, 2008, pp. 1584-1585.

[13] P. F. Hu and Y. L. Cao, "A New Chemical Route to a Hybrid Nanostructure: Room-Temperature Solid-State Reaction Synthesis of Ag@AgCl with Efficient Photocatalysis," *Dalton Trans*actions, Vol. 41, No. 29, 2012, pp. 8908-8912.

[14] S. Kulmala and J. Suomi, "Current Status of Modern Analytical Luminescence Methods," *Analytica Chimica Acta*, Vol. 500, No. 1-2, 2003, pp. 21-69.

[15] W. G. Becker and A. J. Bard, "Photoluminescence and Photoinduced Oxygen Adsorption of Colloidal Zinc Sulfide Dispersions," *The Journal of Physical Chemistry*, Vol. 87, No. 24, 1983, pp. 4888-4893.

[16] N. Chestnoy, T. D. Harris, R. Hull and L. E. Brus, "Luminescence and Photophysics of CdS Semiconductor Clusters: The Nature of the Emitting Electronic State," *The Journal of Physical Chemistry*, Vol. 90, No. 15, 1986, pp. 3393-3399.

[17] S. Wageh, Z. S. Ling and X. X. Rong, "Growth and Optical Properties of Colloidal ZnS Nanoparticles," *Journal of Crystal Growth*, Vol. 255, No. 3-4, 2003, pp. 332-337.

[18] J. H. Yu, J. Joo, H. M. Park, S. Baik, Y. W. Kim, S. C. Kim and T. Hyeon, "Synthesis of Quantum-Sized Cubic ZnS Nanorods by the Oriented Attachment Mechanism," *Journal of the American Chemical Society*, Vol. 127, No. 15, 2005, pp. 5662-5670.

Characterization

The crystalline structures of the products were analyzed by a powder X-ray diffractometer (XRD, MXP18AHF, MAC) with Cu-Kα radiation (λ = 0.154056 nm). The morphologies, microstructures, and crystal lattice of the obtained samples were characterized by transmission electron microscopy (TEM, JEM-2010F at 200 KV). The UV spectra were recorded on a spectrophotometer (HITACHI U-3310) at room temperature. The photoluminescence spectra were obtained by using a HITACHI F-4500 fluorescence spectrophotometer at room temperature.

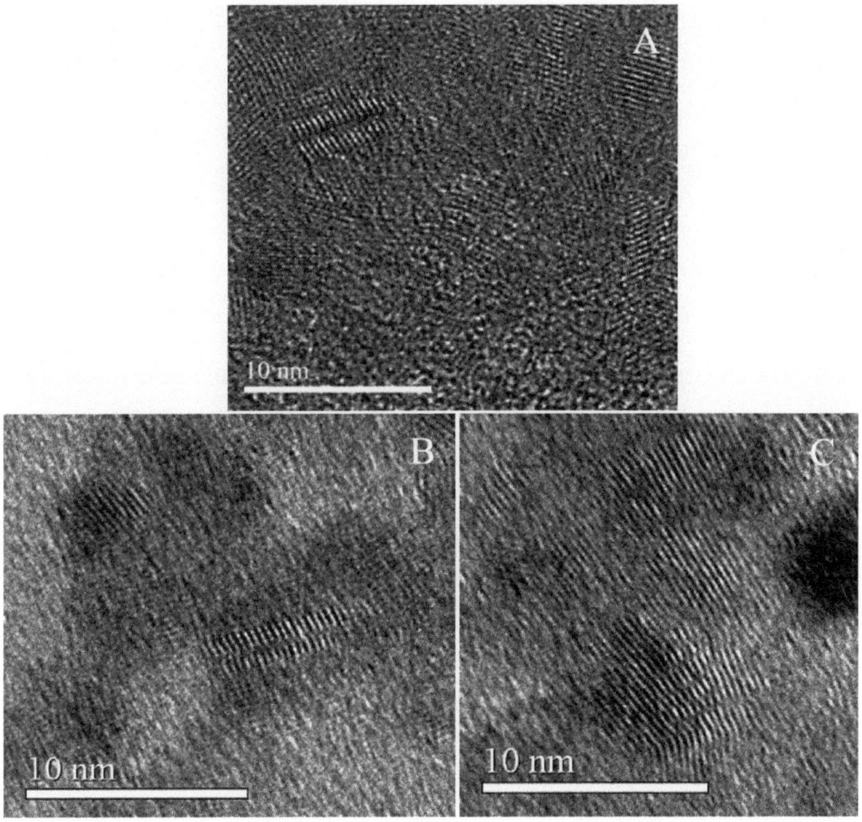

Figure S1. HRTEM images of Q-ZnS-ii nanoparticles with one-dimensional and wave-like structures.

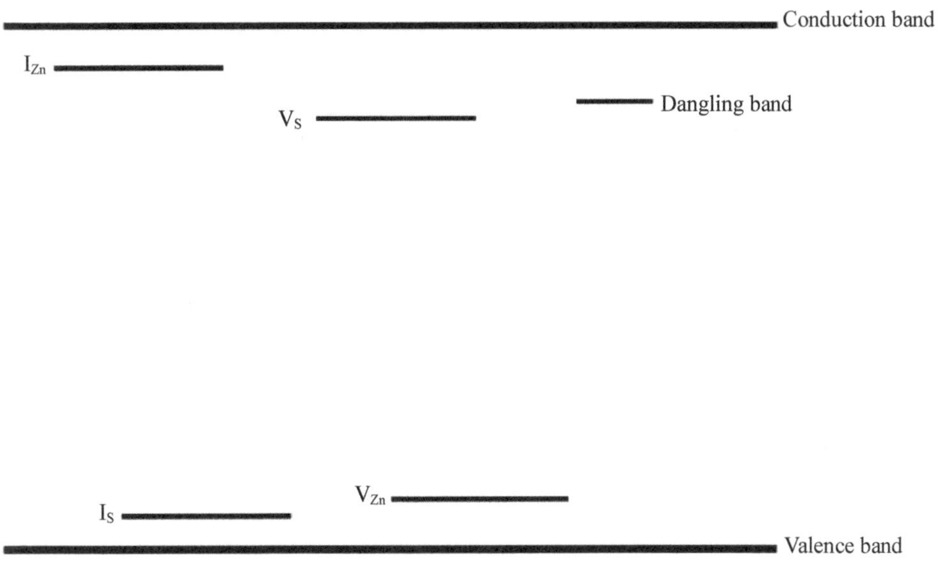

Figure S2. Schematic energy level diagram.

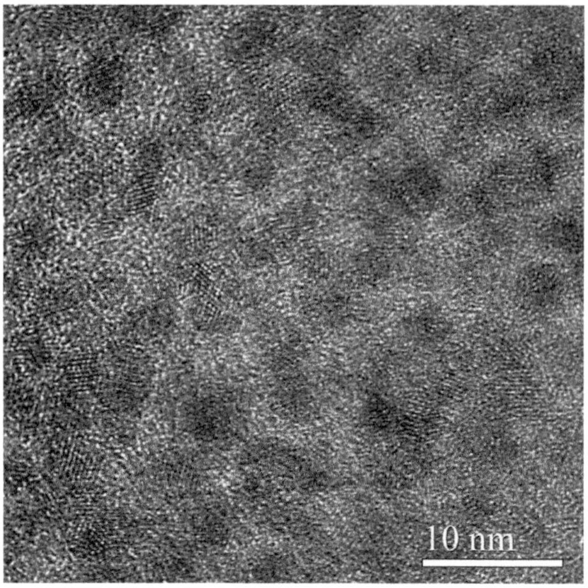

Figure S3. HRTEM image of CdS QDs synthesized through room-temperature solid-state route with SDBS (molar ratio of 1:10 for SDBS to zinc salt).

18

Structural and Surface Morphology Analysis of Nickel Phthalocyanine Thin Films

Mohammed T. Hussein, Eman M. Nasir, Addnan H. Al-Aarajiy
Department of Physics, University of Baghdad, Baghdad, Iraq

ABSTRACT

The thin films of Nickel Phthalocyanine (NiPc) on glass substrates were prepared by vacuum evaporation at different substrates temperatures (300, 325, 350, 400, 450) K. The structure and surface morphology of NiPc in powder and thin film forms (265 nm) were studied using X-ray diffraction and atomic force microscope (AFM), and showed that there was a change and enhance in the crystallinity and surface morphology due to change in the substrates temperatures. Analysis of X-rays diffraction patterns of NiPc in powder form showed that it had an α-polycrystalline phase with monoclinic system with lattice constants a = 1.513 nm, b = 0.462 nm, c = 2.03 nm and β = 123.46°. Thermal evaporation of NiPc at different substrates temperatures led to β-crystalline films oriented preferentially to the (100) plane with different substrate temperatures. The mean crystallite size increased with substrates temperatures from 300 K to 450 K. This result was supported by AFM measurements, which exhibited a relatively larger grain size.

Keywords: Nickel Phthalocyanine; Organic; XRD; AFM; Morphology; Thin Film; Structure

1. Introduction

The materials considered of most importance are the phthalocyanines in the organic dyes and pigments. Phthalocyanines have found applications as semiconducting devices, photovoltoic cells, rectifying devices, liquid crystal displays and gas sensors. The electrical, optical and structural properties of phthalocyanine thin films are dependent on various parameters such as evaporation rate, substrate temperature and post-deposition annealing [1]. Phthalocyanines are found in different phases *i.e.* α-, β- and γ-phases. Metal-substituted and metal free phthalocyanines are readily used as active layer in several applications such as gas sensors, photocapacitive and photoresistive detectors, Organic Thin Film Transistors (OTFTs), colour filters and organic laser materials. Among the various metal-substituted phthalocyanines, nickel phthalocyanine has received considerably less attention. In the present study, we report the structural properties of nickel phthalocyanine thin films prepared at different substrates temperatures. To explore the potentially interesting properties of the material, structural and morphological characterization is thus considered as a prerequisite for deep insight knowledge of the material. These structural properties mostly depend on the deposition technique, the heat-treatment temperature and the conditions during film deposition. Annealing process is a broadly employed method to enhance the quality of the crystal and exploit structural defects in the material. During thermal annealing, the morphology and structural properties of the material change [2-4]. For semiconductor devices like light emitting diodes the structural properties are very important, therefore, it is essential to investigate how thermal annealing and substrate temperatures process affects these properties [5,6]. In the present work, thin films of NiPc have been fabricated on glass substrates by vacuum evaporation techniques. A detailed study on the effects of heating substrate temperatures on the structural properties and surface morphology has been performed by X-ray diffraction and atomic force microscopy (AFM). Hence, the purpose of our work is to find the optimal temperature for which the film produces best structural properties.

2. Experimental

The NiPc powder used in this study was obtained from Aldrich chemical company and was used as the source material for thermal evaporation. Its molecular structure is shown in **Figure 1**. Before starting the deposition, the glass substrate was cleaned in an ultrasonic bath for 10 min using acetone, followed by rinsing in distilled water.

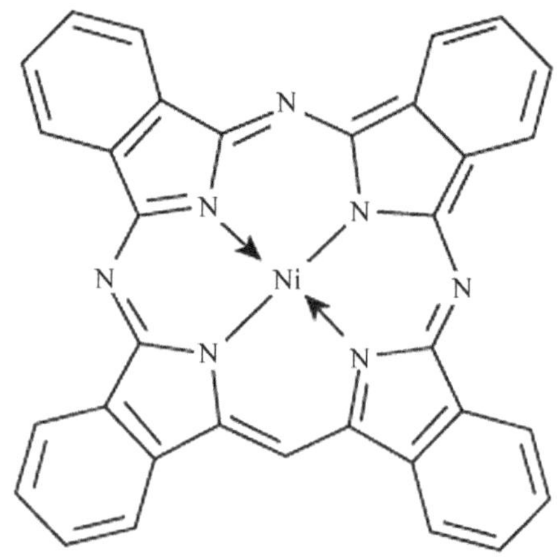

Figure 1. Molecular structure of Ni-phthalocyanine (NiPc).

The substrate was dried in open air in a cleaned room. A molybdenum boat was used as a heating source. Thin films of NiPc, were deposited by vacuum evaporation technique on thoroughly cleaned glass substrate at different substrate temperatures (300, 325, 350, 400, 450) K using Edwards 306 system. During deposition, the pressure in the vacuum chamber was kept constant at about 10^{-5} mbar and the deposition rates of all the films were kept at about 0.2 nm/s, Also the substrate are placed at a distance of 15 cm from the source. The thickness of the films was measured using Michelson interferometer technique. The thickness of the films was 265 nm. The adhesion of the films to the substrate seems to be extremely good. The structural analysis of films is performed by X-ray diffractometer with Cu Kα radiation (λ = 1.54 Å). The surface morphology of the NiPc films at different substrate temperatures (T_s) is investigated by means of atomic force microscopy (AFM).

The lattice parameters of the films were calculated using the Bragg's formula:

$$2d\sin\theta = n\lambda \quad (1)$$

The grain size of the crystallites were calculated from the XRD using Scherer's relation,

$$D = K\lambda/\beta\cos\theta \quad (2)$$

Where K = 0.94 is a constant, λ the wavelength of X-ray, β the full width half maximum and θ the diffraction angle. The micro strain (ε) and the dislocation density (δ) of films were estimated using the equations

$$\varepsilon = \beta\cos\theta/4 \quad (3)$$

$$\delta = 1/D^2 \quad (4)$$

3. Results and Discussions

The most common polymorphic crystal phases occur in Ni-phthalocyanines are α, and β, that have two crystallite structures. The α phase is metastable and obtained either as a polycrystalline powder or as a thin film deposited on a cold substrate in vacuum [7]. The phase β can be obtained in single crystal form or as a thin film formed by deposition of phthalocyanine on heated substrates, which are thermodynamically stable. Phase separation could be responsible for the de gradation of the optical properties. The α form is metastable and can be converted to β form by temperature treatments with structures tetragonal and monoclinic respectively. The differences in the d values are attributed to higher X-ray absorption, sample purity, particle size, preferred orientation and crystal texture [4, 7].

The X-ray diffraction of the NiPc powder used as the source material for sublimation is given in **Figure 2**. The peaks are identified using the standard JCPDS File No. 11-0744 data. The preferential orientation is found to be in the (100) direction at 2θ = 6.938° and that it has an α-crystalline phase with monoclinic system with lattice constants a = 1.513 nm, b = 0.462 nm, c = 2.03 nm and β = 123.46°. The close agreement showed that the powder is in the monoclinic structure as reported for other phthalocyanines [1-7]. **Table 1** gives the lattice constant of present observed and the standard data of NiPc.

X-ray diffraction pattern of NiPc thin film of thickness (265 nm) prepared at different substrate temperatures (300, 325, 350, 400, 450) K are shown in the **Figure 3** respectively. XRD pattern indicates that all the samples are single phase and that it has a β-crystalline phase [4-8]. The difference between β- and α-phases is attributed to the tilt angle of b-axis of the unit cell. The structure of the NiPc thin film is determined as tetragonal with preferential orientation along the (100) direction. Well-defined diffraction peaks by (100) gives the direction of the preferential orientation as deposited film (JCPDS, File No. 11-0744). The preferential orientation is at 2θ = 6.859°. The spacing between the planes, d = 12.876 Å. The d values obtained from diffractogram are in perfect match with JCPDS record. The diffraction peaks are well matched with the previous observations [1-8]. The increased crystallinity is attributed to the destruction of pseudomorphic layers present in the film upon high substrate temperature. **Table 2** shows the structure parameters of NiPc films, The full width at half maximum (FWHM) intensity of preferential orientation, shows that the FWHM are reduced with increasing substrate temperatures, which represents better lattice quality [1,8-10].

From the **Table 2** it is observed that the grain size increases which intern decreases strain and dislocation density. The grain size D of the film is calculated using the Scherrer relation, the grain size increases from (22.2 - 71.8 nm) for substrate temperatures (300 - 450 K). This is because the pseudomorphic layers formed at room temperature, which are in the metastable state, are destroyed by increasing the substrate temperatures and lar-

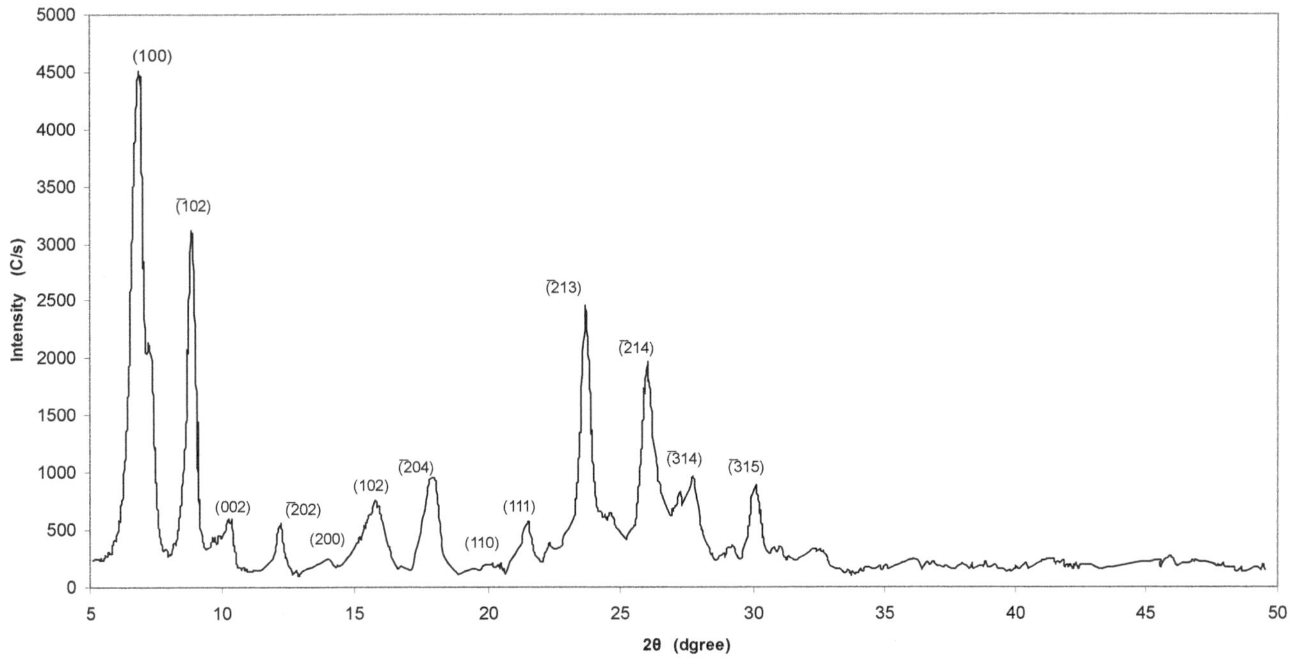

Figure 2. X-ray diffraction of the NiPc powder.

Table 1. The present observed and the standard data NiPc.

Lattice constant	Present work	Pcpdf	Error ratio %
a (Å)	15.128	14.9	1.53
b (Å)	4.621	4.7	1.6
c (Å)	20.355	19.9	2.28
β (deg)	123.46	121.15 (deg)	−1.9
d (Å)	7.370124	7.266181	−1.43

ger crystallites are formed or this can be attributed to the onset of phase transition occurring in NiPc. Also this may be attributed to a phase transition or to the regrowth of phthalocyanine. This mean that the crystallization of the films is a function of substrate temperatures and the optimal temperature for which the film produces best structural properties is at 400 - 450 K. Also from **Figure 3** two distinct peaks (200) and (314) are visible at high diffraction angles for heating substrate 400 and 450 K respectively, and this is agreement with other literatures [6-9], and this is may be attributed to enhancement and recrystallization these films by increasing the substrate temperatures.

Surface morphology of NiPc thin film is shown in **Figures 4** and **5** which prepared at different substrate temperatures (300, 325, 350, 400, 450) K. It shows that the morphology of the NiPc thin film has larger number of grain size and are homogeneously distributed, which indicates the crystalline nature of the film. It is obvious because the films were prepared at different substrate temperatures, due to the heating grain growth will takes place which gives the crystallinity and good surface morphology [2]. These results are in good agreement with XRD characteristics. The crystal morphology and molecular orientation change with substrate temperature. The average grain size increases as temperature increases. Gas sensitivity depends both up on material parameters, such as the crystal phase of the material and film parameters, such as the surface topography of the film. These structural characterizations confirm that the substrate temperature allows us to control the structural organization of nickel phthalocyanine molecules in their solid state, from which a large improvement of the properties can be expected. **Table 3** shows that the grain size increased by increasing the substrate temperature.

The morphology of these films, has also confirmed the appearance of polycrystallites and increasing grain size when substrate temperature is raised. This is caused due to the nucleating behavior of the substance [7]. The root mean square (rms) roughness of the film at different substrate temperatures are shown in **Table 3** which increases with increasing T_s. The best results can be seen for the film prepared at heating substrate of 400 - 450 K. The NiPc thin films, after increasing substrate temperatures to 400 - 450, show relatively rough and non-uniform surface, suggesting the formation of interpenetrating molecular network of NiPc. 400 - 450 K and above, deterioration feature is observed due to degradation of the film. From the application point of view, the higher efficiency devices are obtained from the film of greater roughness.

4. Conclusions

Thin films of NiPc were fabricated by the thermal

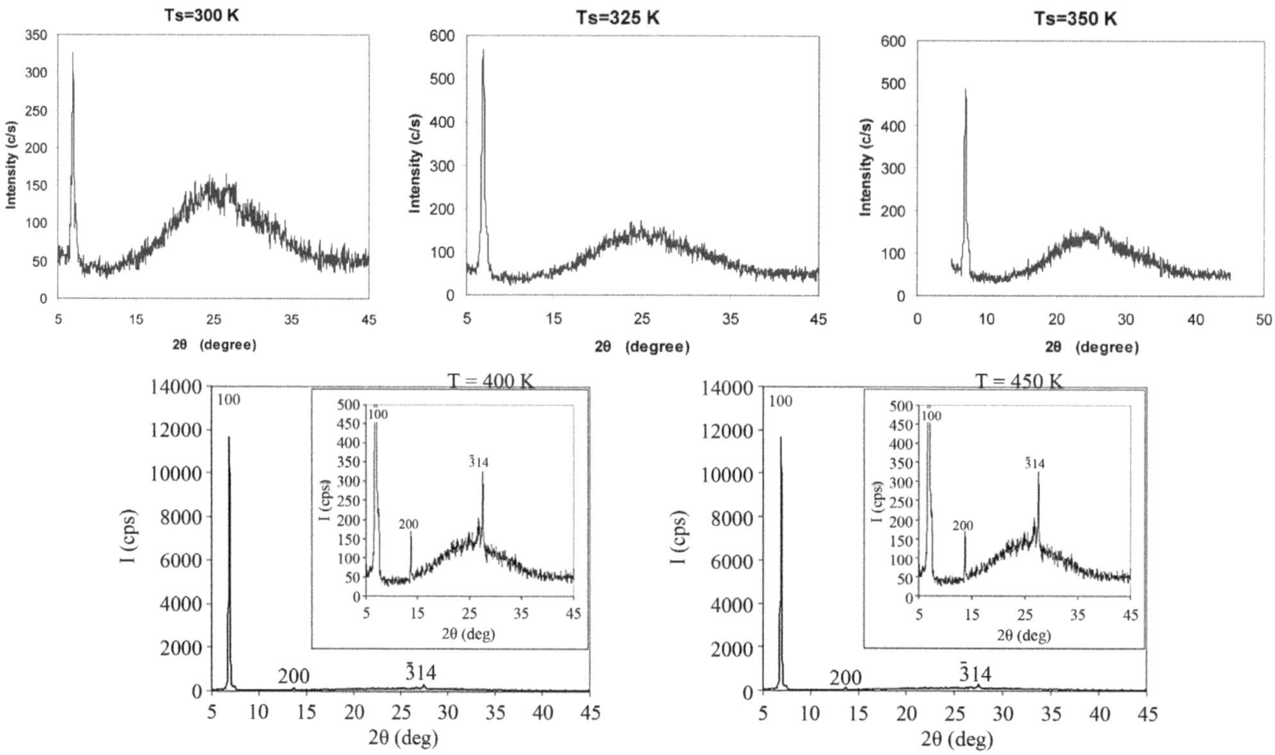

Figure 3. X-ray diffraction pattern of NiPc thin film at different substrate temperatures (300, 325, 350, 400, 450) K.

Table 2. Shows the structure parameters of NiPc films.

T_s (k)	(hkl)	2θ (degree)	d (Å)	FWHM (degree)	D (nm)	$\varepsilon \times 10^{-3}$	$\delta \times 10^{-14}$ lines/m^2
300	100	7.060	12.51	0.3741	22.2	1.62	20.2
325	100	7.028	12.56	0.3746	22.19	1.63	20.3
350	100	7.030	12.56	0.3583	23.2	1.55	18.5
400	100	6.920	12.76	0.1115	45.0	0.485	1.79
450	100	6.990	12.63	0.1157	71.8	0.503	1.93

evaporation technique on glass substrates with thickness (265 nm) prepared at different substrate temperatures successfully. The outcome of this investigation can be summarized as follows:

- X-ray diffraction patterns obtained for NiPc powder showed α-polycrystalline phase with monoclinic system, while the thin film of thermally evaporated NiPc showed it has a single crystalline β-phase film oriented preferentially (100) plane as tetragonal. It can be seen that the intensity, d value and sharpening of the main peak increase with substrate temperature which shows crystallization of the films when substrate temperature increases.
- Diffraction peaks obtained for the films prepared at higher substrate temperatures are sharp and that is due to the increased crystallinity.
- The full width at half maximum (FWHM) intensity of preferential orientation shows that the FWHM are reduced with increasing substrate temperatures and the grain size increases which intern decreases strain and dislocation density, which represents better lattice quality.
- The grain size increases from (22.2 - 71.8 nm) for substrate temperatures (300 - 450 K), this can be attributed to the onset of phase transition occurring in NiPc. Also this may be attributed to a phase transition or to the regrowth of phthalocyanine. This mean that the crystallization of the films is a function of substrate temperatures and the optimal temperature for which the film produces best structural properties is at 400 - 450 K.
- Rearrangements of molecules are taking place with different substrate temperatures. The change in grain size shows that the properties of NiPc are controlled by its structure and morphology depending on increases the substrate temperatures.
- By analyzing the AFM topographic images, we found that there is a significant influence of substrate temperatures on the surface morphology of NiPc films and the crystal morphology and molecular orientation

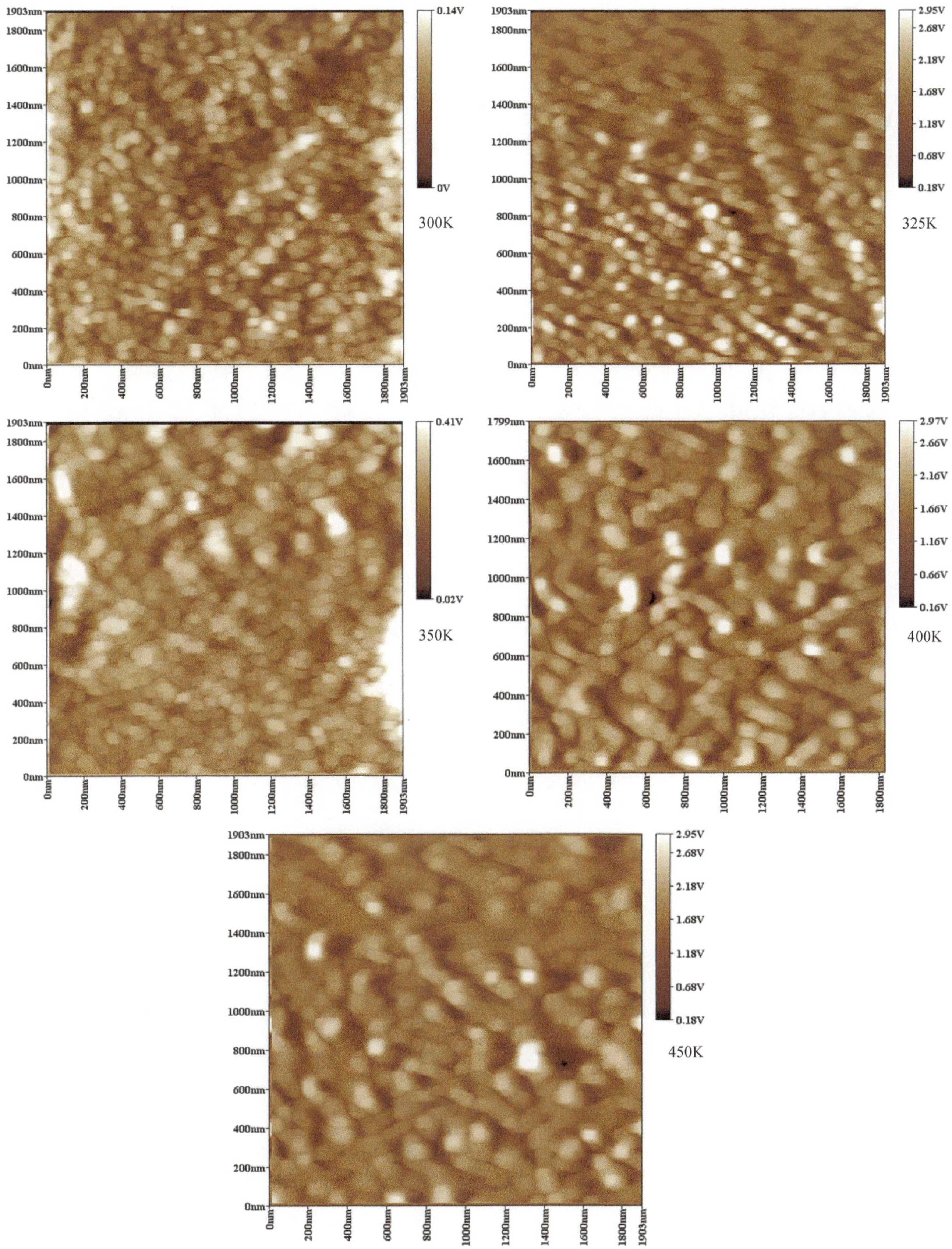

Figure 4. 2-D AFM images of NiPc thin films at different substrate temperatures (300, 325, 350, 400, 450) K.

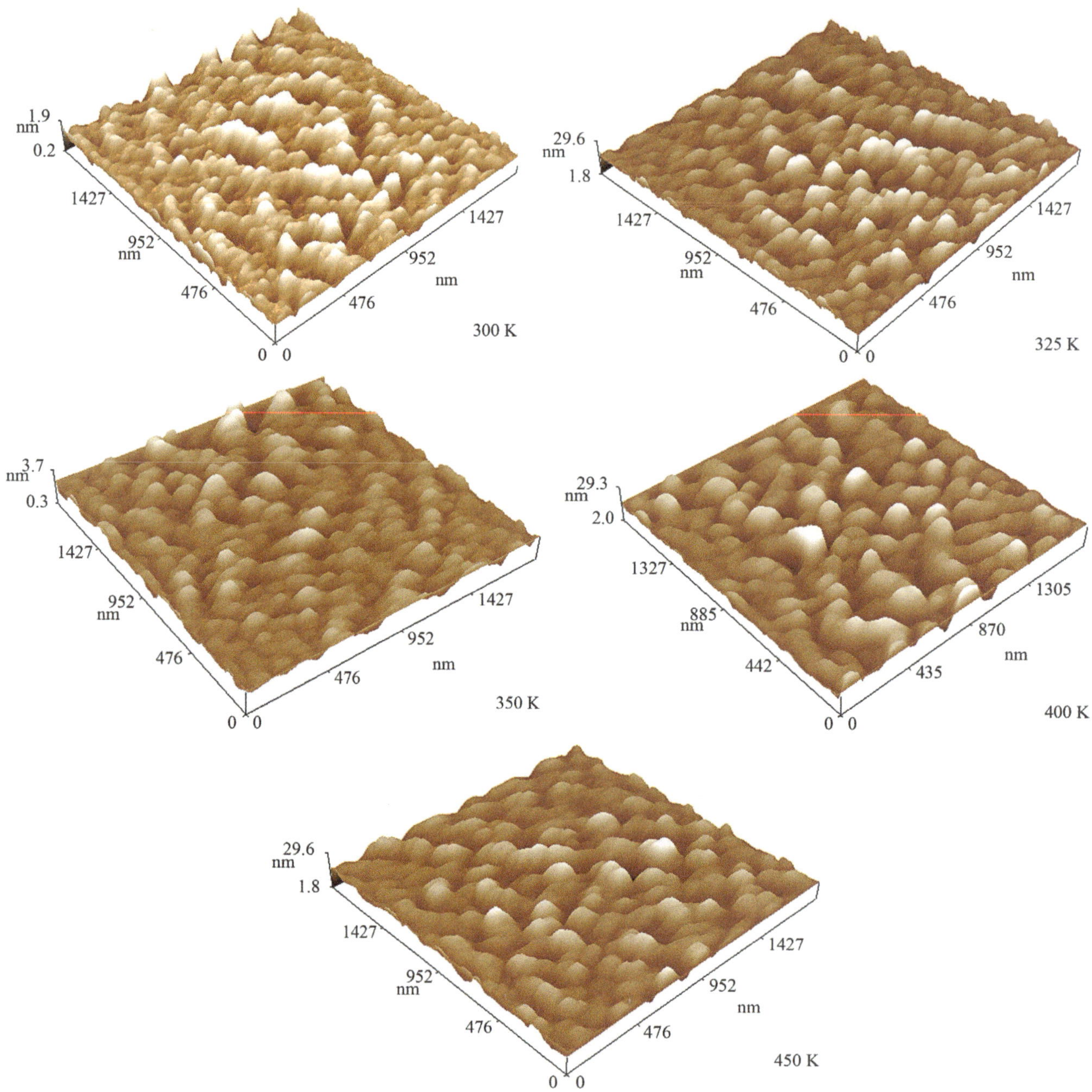

Figure 5. 3-D AFM images of NiPc thin films at different substrate temperatures (300, 325, 350, 400, 450) K.

Table 3. Shows the grain size and roughness of NiPc.

Ts (k)	Roughness Average (nm)	D (nm)
300	0.216	92.42
325	0.275	108.18
350	2.9	94.04
400	2.86	116.26
450	2.54	116.59

change with substrate temperature. These films showed a larger number of grain size, homogeneously distributed, rough which indicates the crystalline nature of the film. The surface roughness of Ni-Pc thin films is increased by increasing substrate temperatures, which makes it useful for higher efficiency devices. The increase in the temperature beyond 400 K has higher roughness and grain size.

- These structural characterizations confirm that the substrate temperature allows us to control the structural organization of nickel phthalocyanine molecules in their solid state, from which a large improvement of the properties by increasing the substrate temperatures, therefore, the best results can be seen for the film prepared at heating substrate of 400 - 450 K.

These results are in good agreement with XRD characteristics.

5. Acknowledgements

The work was supported by Department of Physics college of Science, thin films and opto-electronic laboratories, Baghdad University.

REFERENCES

[1] K. N. N. Unni and C. S. Menon, "Electrical, Optical and Structural Studies on Nickel Phthalocyanine Thin Films," *Materials Letters*, Vol. 45, No. 6, 2000, pp. 326-330.

[2] F. Aziza, K. Sulaimana, M. R. Muhammad, M. H. Sayyad and Kh. Karimov, "Influence of Thermal Annealing on the Structural Properties of Vanadyl Phthalocyanine Thin Films: A Comparative Study," *World Academy of Science, Engineering and Technology*, Vol. 56, 2011, pp. 852-854.

[3] D. Gu, Q. Y. Chen, J. P. Shu, X. D. Tang, F. X. Gan, S. Y. Shenb, K. Liu and H. J. Xu, "Optical Recording Performance of Thin Films of Phthalocyanine Compounds," *Thin Solid Films*, Vol. 257, No. 1, 1995, pp. 88-93.

[4] S. Shihub and R. Gould, "Structure and Phase-Change Phenomena in Evaporated Thin Films of Cobalt Phthalocyanine," *Physica Status Solidi A*, Vol. 139, No. 1, 1993, pp. 129-138.

[5] A. El-Bosaty, *et al.*, "Surface Plasmon-Cobalt Phthalocyanine Sensor for NO_2 Gas," *Egypt. J. Solids*, Vol. 29, No. 1, 2006, pp. 121-129.

[6] M. Shah, Kh. S. Karimov and M. H. Sayyad, "Organic Semiconductor Nickel Phthalocyanine-Based Photocapacitive and Photoresistive Detector," *Semiconductor Science and Technology*, Vol. 25, No. 7, 2010, pp. 75014-75018.

[7] B. Joseph and C. S. Menon, "Studies on the Optical Properties and Surface Morphology of Nickel Phthalocyanine Thin Films," *E-Journal of Chemistry*, Vol. 4, No. 2, 2007, pp. 255-264.

[8] P. Kalugasalama and S. Ganesanb, "Surface Morphology of Annealed Lead Phthalocyanine Thin Films," *International Journal of Engineering Science and Technology*, Vol. 2, No. 6, 2010, pp. 1773-1779.

[9] Z. G. Ji, K. W. Wongb, P. K. Tseb, R. W. M. Kwoka and W. M. Laub, "Copper Phthalocyanine Film Grown by Vacuum Deposition under Magnetic Field," *Thin Solid Films*, Vol. 402, No. 1-2, 2002, pp. 79-82.

[10] P. Kalugasalam and S. Ganesanb, "Structural Analysis of Annealed Lead Phthalocyanine Thin Films," *International Journal of Engineering Science and Technology*, Vol. 2, No. 7, 2010, pp. 2567-2573.

Synthesis and Analysis of PZT Using Impedance Method of Reactance Estimation

Witsarut Sriratana[1], Riichi Murayama[1], Lerdlekha Tanachaikhan[2]
[1]Department of Intelligent Mechanical Engineering, Faculty of Engineering, Fukuoka Institute of Technology, Fukuoka, Japan
[2]Faculty of Engineering, Ramkhamhaeng University, Bangkok, Thailand

ABSTRACT

This study presents an analysis of equivalent circuit namely Butterworth Van Dyke (BVD) [1,2] by using impedance method to stimulate Zirconate Titanate (Piezoelectric ceramic) which is initially synthesized from Lead Oxide (PbO), Zirconium Dioxide (ZrO_2) and Titanium Dioxide (TiO_2) and vibrated in thickness mode. The reactance was estimated in the frequency range lower than the resonance frequency and then compared to the impedance obtained from measurement using impedance analysis machine model HP4192A and HP4194 [3]. The results from HP4194 were analyzed for BVD parameters: Motional resistance (R_1), Inductor (L_1), Capacitor (C_1), and Capacitor corresponds to the electrostatic capacitance (C_o). Another accuracy analysis was compared by the calculation results using the method of IEEE 176-1987 [4] to the impedance values measured by HP4192A. In this study, there were two conditions for experiment and consideration of parameter variation in BVD equivalent circuit: variation of temperature and mechanical force. These parameters are evaluated to design the efficient circuit for PZT utilization to obtain the optimal efficiency.

Keywords: BVD; Capacitance; Force; Frequency; Impedance; Inductance; PZT; Temperature

1. Introduction

Electronic circuit using BVD analysis for effectively stimulating PZT can be designed by adopting, for example, IEEE Standard [4,5], Smits [6], Sherrit [7], and Xu [8]. Smith's method and Sherrit's method would consider those parameters as complex numbers. Xu's method would be similar to Smith's method which considers the impedance of piezoelectric measured at the frequency values ranging from resonance frequency (f_r) and antiresonance frequency (f_a) as well as under mechanical force *i.e.* elastic, piezoelectric constant and coupling coefficient. In this study, five pieces of PZT, namely O01, N01, N02, N03, and N04, were synthesized from ratio of PbO:ZrO_2:TiO_2 (100:52:48) and burned at 700˚C for 1 hour [9,10] in order to experiment the density (ρ), PZT constant (d_{33}), dielectric constant (K) at 1 kHz, electromechanical coupling coefficient (k_t) and mechanical quality factor (Q_m). In addition, the PZT were utilized in an ultrasonic cleaner.

2. Theory and Fundamental Method

2.1. Equivalent Circuit of BVD

Electrical characteristics in thickness mode of PZT [3] tested in this study were comparable to the BVD equivalent circuit, which consists of R_1 connected serially to C_1 and L_1 then all parallelly connected to C_o [1,2,11] as shown in **Figure 1(a)**. The resistance, the capacitance, and the inductance were comparable to the force friction (η), spring constant (k), and mass, respectively as shown in **Figure 1(b)**.

Assuming R_1, C_1, L_1, and C_o were constants and independent on frequency and impedance [5] as given in Equation (1), the impedance equation at PZT resonance frequency (Z_r), f_r and f_a, in terms of BVD relating parameters was obtained. In **Figure 2**, the impedance responding to f_r and f_a of PZT is shown.

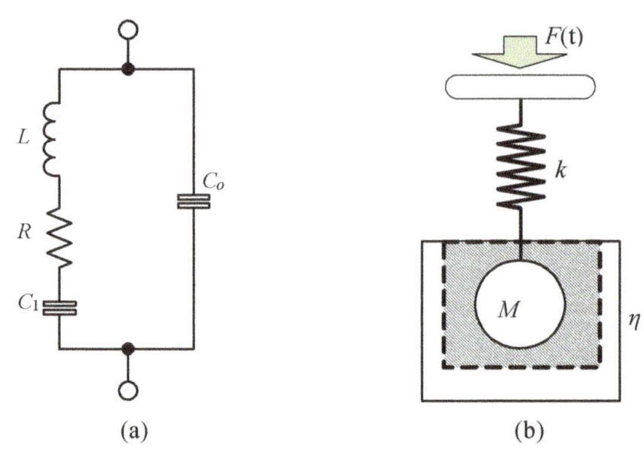

Figure 1. Equivalent circuit of PZT compared to mechanical characteristics: (a) The BVD model; (b) Mechanical model.

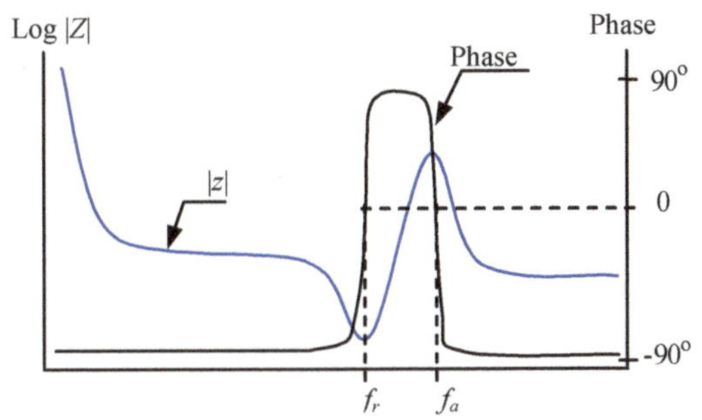

Figure 2. Frequency response characteristic of piezoelectric ceramic.

$$Z = \frac{R_1 + j(\omega L_1 - 1/\omega C_1)}{1 + j\omega\omega_0\left[R_1 + j(\omega L_1 - 1/\omega C_1)\right]} \quad (1)$$

The calculation of BVD circuit using impedance method by the capacitor ratio (r) is as shown in Equation (2), where ω_a is the angular velocity of the anti-resonance frequency ($2\pi f_a$) and ω_r is the angular velocity of the resonance frequency ($2\pi f_r$).

$$r = \frac{C_1}{C_o} = \frac{\omega_a^2 - \omega_r^2}{\omega_r^2} \quad (2)$$

The reactance values of C_1 and L_1 in BVD circuit at low frequency range was observed. It was noted that the reactance of PZT was performed as a capacitor as shown in Equation (3).

$$j/\omega C_1 \qquad j\omega\omega_1 \quad (3)$$

Therefore,

$$(R_1 + j\omega\omega_1 - j/\omega C_1) \approx (R - j/\omega C_1) \quad (4)$$

Then the impedance equation of PZT can be revised as shown in Equation (5).

$$Z = \frac{R_1 - j/\omega C_1}{\left[\frac{\omega_r^2 - \omega^2}{\omega_a^2 - \omega_r^2}\right] + jR\omega R_0} \quad (5)$$

C_1 can be calculated as given in Equation (6) by choosing the impedance measured at low frequency range where ω is the angular velocity of the selected frequency and R is $|Z|$ measured at resonance frequency.

$$C_1 = \frac{1}{\omega\sqrt{\left[\frac{\omega_r^2 - \omega^2}{\omega_a^2 - \omega_r^2}\right]^2 |Z|^2 - R^2}} \quad (6)$$

2.2. Measurement of Dielectric Constant (K)

Measurement of K or relative permittivity (ε_r) of PZT is to test for the electrical characteristics which can be calculated by using Equation (7), where C_f is the capacitance at 1 kHz, t is the thickness of PZT measured from both metal poles, A is the area of metal pole, and ε_o is the dielectric constant of vacuum. The dielectric constant can be varied depending on the temperature (T) of substance based on Curie-Weiss Law [12] where T_c is the temperature of PZT with zero polarization, and C is Curie constant (10^3 to 10^5) as given in Equation (8).

$$K = \varepsilon_r = \frac{C_f t}{\varepsilon_o A} \quad (7)$$

$$K - 1 = \frac{C}{T - T_c} \quad (8)$$

Equations (7) and (8) can also be revised as shown in Equation (9).

$$\varepsilon_r = \left[\frac{C}{T - T_c}\right] + 1 \quad (9)$$

2.3. Measurement of PZT Constant (d_{33})

The value of d_{33}, measured by Piezo-D meter model CADT (Channel Products Inc.), implies the ability to emit the electric charge of PZT when it is stimulated by mechanical force. Number 33 represents the three perpendicular directions of compression force on PZT.

2.4. Measurement of Electromechanical Coupling Coefficient (k_t)

Electromechanical coupling coefficient (k_t) represents the efficiency of transformation from electrical energy to mechanical force. In Equation (10), Δf is the difference between f_r and f_a.

$$K_t^2 = \frac{\pi}{2}\frac{f_r}{f_a}\tan\left(\frac{\pi}{2}\frac{\Delta f}{f_a}\right) \quad (10)$$

2.5. Measurement of Mechanical Quality Factor (Q_m)

Q_m represents the maximum ability of mechanical vibration in f_r range where R_1 is the impedance at resonance frequency as given in Equation (11).

$$Q_m = \frac{f_a^2}{2\pi f_r R_1 C_f \left(f_a^2 - f_r^2\right)} \quad (11)$$

3. Analysis of BVD Equivalent Circuit

The PZT with 40 mm of diameter and 3 mm of thickness is electrode by silver paste at the top and the bottom as shown in **Figure 3**, and then connected to the system as

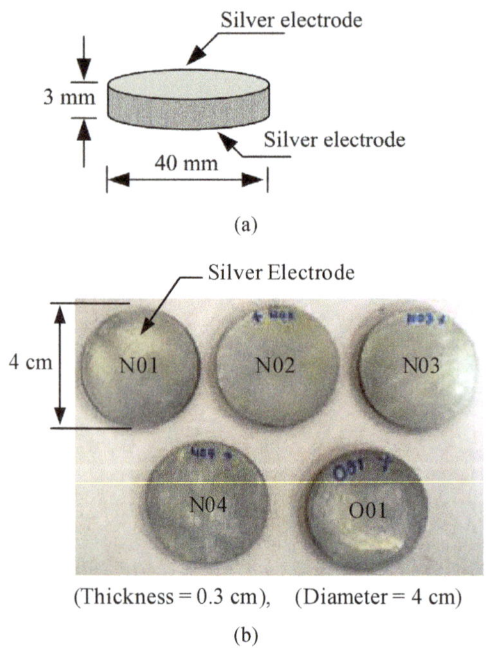

Figure 3. Physical of PZT sample: (a) Dimension of PZT; (b) Top view.

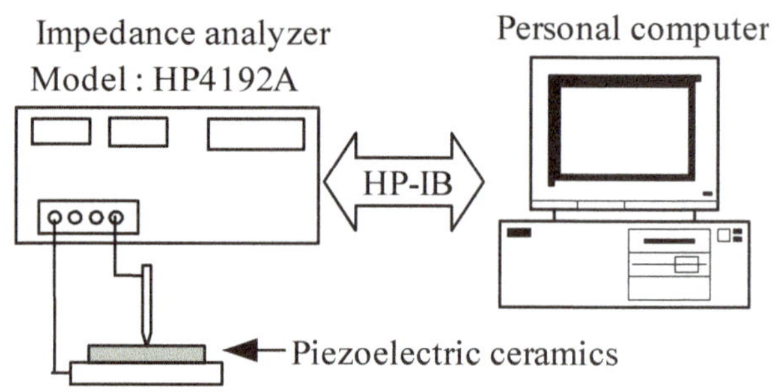

Figure 4. Impedance of PZT measurement diagram.

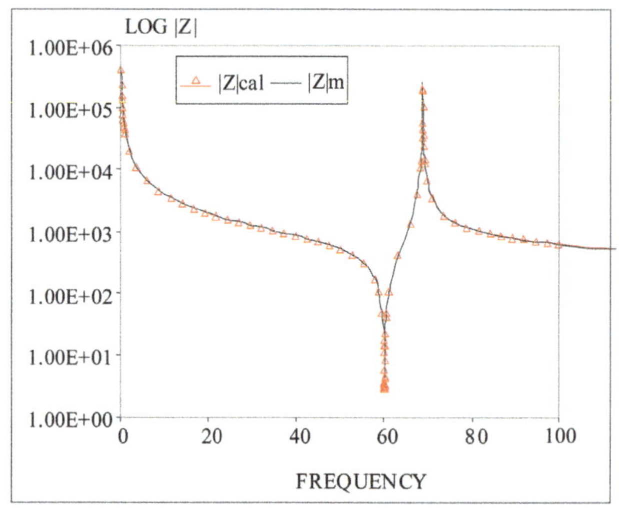

Figure 5. Gain impedance of PZT measurement at various frequency.

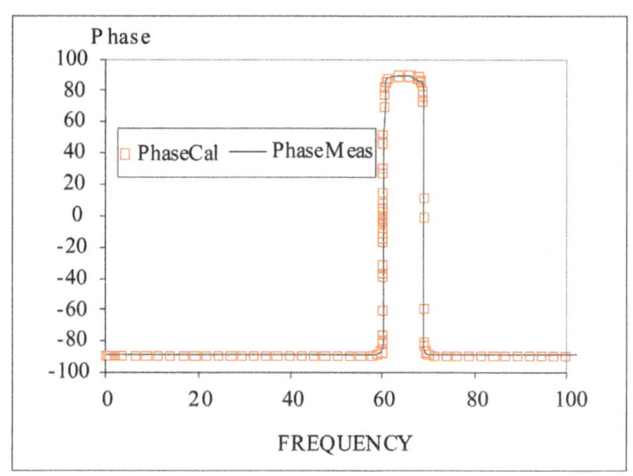

Figure 6. Phase impedance of PZT at various frequency.

shown in **Figure 4**. The HP Vee program is applied to control the impedance analyzer model HP4192A via HP-IB interface and collect the impedance at 100 Hz to 1.3 MHz [4]. Then, the resonance frequency and the impedance at resonance frequency and lower resonance frequency will be applied to calculate R, L, and C_o. The calculated value will be evaluated and compared with the value calculated by using Equation (1). In this paper, only the fundamental resonance is considered regardless the resonance frequency at harmonic period.

From experiment and measurement of PZT impedance, the calculated parameters from BVD model are as shown in **Table 1**. **Figures 5** and **6** are illustrated the comparison of size ($|Z|m$) and phase of the impedance at 100 Hz to 100 kHz measured by impedance analyzer model HP4192A and the calculated impedance ($|Z|cal$ and Phase Cal). The impedance at 1 kHz is selected to calculate in Equation (6). To obtain the appropriate value of C, the impedance at lower resonance frequency is suggested such as 100 Hz, 1 kHz, 2 kHz, 3 kHz, 6 kHz, and 14 kHz. The BVD parameters are as shown in **Table 2** and the percentage of error is shown in **Figure 7**.

4. Experiment of PZT under Temperature Control

Physical characteristics of PZT were experimented by considering ρ, PZT constant (d_{33}), k_t, and Q_m of electrical charge at 1 kHz. In this study, only the results from N01 and N04 (details are given in **Table 3**) were compared under room and controlled temperatures. **Figure 8** shows the PZT experimental system connected with impedance analysis model HP4192A under temperature conditions.

Measuring the PZT impedance under room temperature was for assessing the influences of temperature on PZT impedance and comparing the results under room temperature and controlled temperature at various frequency. **Figures 9** and **10** illustrate the PZT impedance

Table 1. The PZT properties.

PZT properties	
Fundamental resonance frequency (f_r)	60.329 kHz
Fundamental anti-resonance frequency (f_a)	68.971 kHz
Impedance at f_r	2.847 Ω
Impedance at 1 kHz	41.040 kΩ
Impedance at 2 kHz	20.540 kΩ
Impedance at 3 kHz	11.245 kΩ
Impedance at 6 kHz	6.526 kΩ
Impedance at 14 kHz	2.863 kΩ

Table 2. Calculated parameters of BVD circuit at 100 Hz, 1 kHz, 2 kHz, 3 kHz, 6 kHz, and 14 kHz.

Parameters	Frequency					
	100 Hz	1 kHz	2 kHz	3 kHz	6 kHz	14 kHz
R_1 (Ω)	2.847	2.847	2.847	2.847	2.847	2.847
C_o (nF)	2.9715	2.9693	2.9697	3.019	3.037	3.172
C_1 (pF)	91.22	91.16	91.17	92.69	91.33	97.39
L_1 (mH)	7.63	7.64	7.64	7.51	7.47	7.15

Table 3. Physical and electrical characteristics of PZT (N01 and N04).

PZT	Density	d_{33} (10^{-12} C/N)	C_f (nF)	K (F/m)	k_t	Q_m
N01	7.38	275	4.7888	1040	0.48	846
N04	7.40	271	3.6474	1016	0.51	844

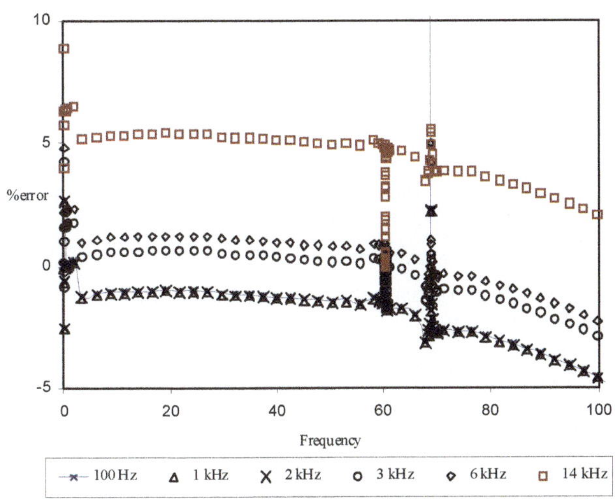

Figure 7. Error percentage of calculated impedance at various frequency.

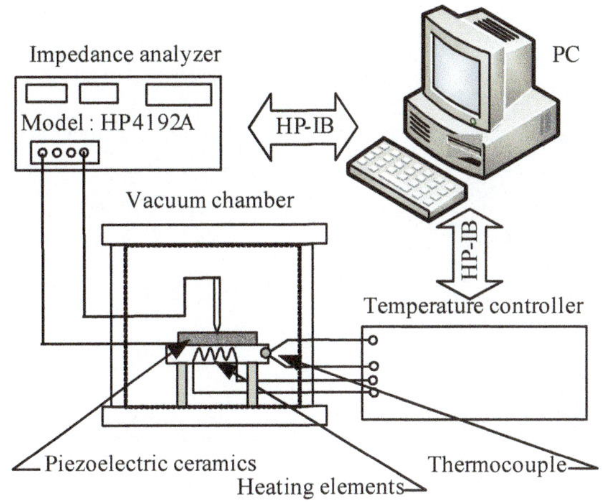

Figure 8. PZT experimental system and impedance analysis model HP4192A.

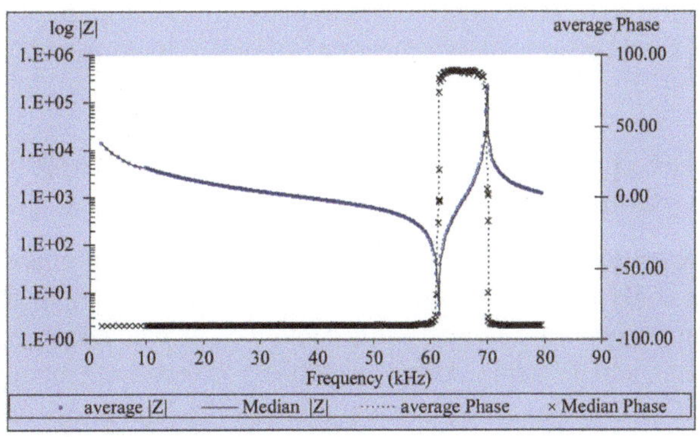

Figure 9. Impedance and phase at the frequency ranging from 100 Hz to 80 kHz of N01.

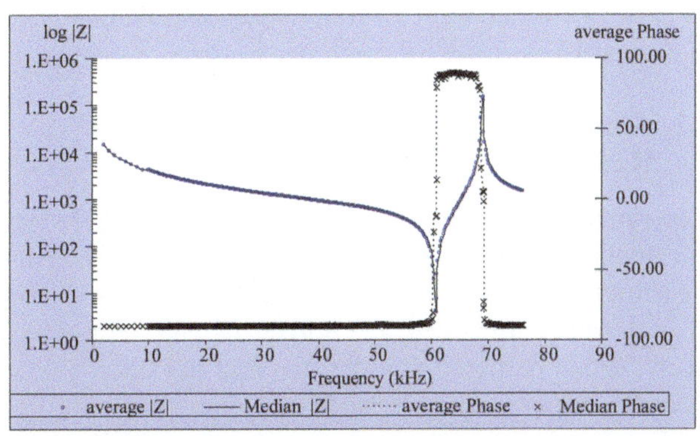

Figure 10. Impedance and phase at the frequency ranging from 100 Hz to 80 kHz of N04.

of N01 and N04 repeatedly measured 5 times by impedance analysis model HP4192A under room temperature at the frequency of 1 kHz to 80 kHz.

PZT was experimented by placing on heater plate in vacuum chamber under controlled temperature. The temperature can be controlled by a program developed on HP Vee via HP-IB port as shown in **Figure 11**. The temperature was varied from 25˚C to 230˚C with 20˚C

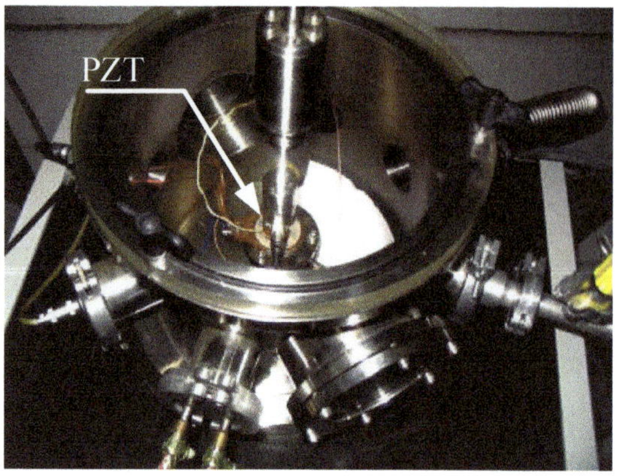

Figure 11. Position of PZT in temperature controlling chamber.

interval of measurement and the frequency was varied from 2 kHz to 80 kHz. Uncertainty of measurement and PZT was estimated by repeating 3 times of measurement.

4.1. Experimental Results under Temperature Conditions

The impedance of PZT in temperature controlled chamber (only N01) under specific temperature was similar to that under room temperature. However, it was observed that f_r and f_a were drifted. At 230°C, f_r increased about ≈1.6 kHz while f_a decreased about ≈1 kHz. Therefore, the difference between f_r and f_a decreased about 2.6 kHz as shown in **Figure 12**. Moreover, when the temperature was increasing, the impedance at f_r was also increasing up until 190°C and then starting to decrease as illustrated in **Figure 13**.

4.2. Analysis of Impedance Variation in the Range of 100 Hz to 10 kHz

Experiment of PZT in the range of 100 Hz to 10 kHz was to calculate the impedance of BVD circuit by repeatedly measuring 30 times. An average and standard deviation of impedance are given in **Figures 14** and **15**, respectively. It can be noted that the standard deviation would increase at low frequency range and decrease as the frequency is rising.

4.3. Calculation of BVD Circuit Using Impedance Method

To calculate BVD circuit, average experimental values of f_r, f_a, and Z_r at selected frequency ranging from 0.1 kHz to 10 kHz were used and then BVD parameters were obtained. The results were ranged as follows: $L_1 \approx 0.16$ mH to 0.2 mH, $C_1 \approx 0.018$ nF to 0.040 nF, $C_o \approx 0.06$ nF to 0.18 nF. However, R_1 was uniform due to the identical impedance at f_r as shown in **Table 4**.

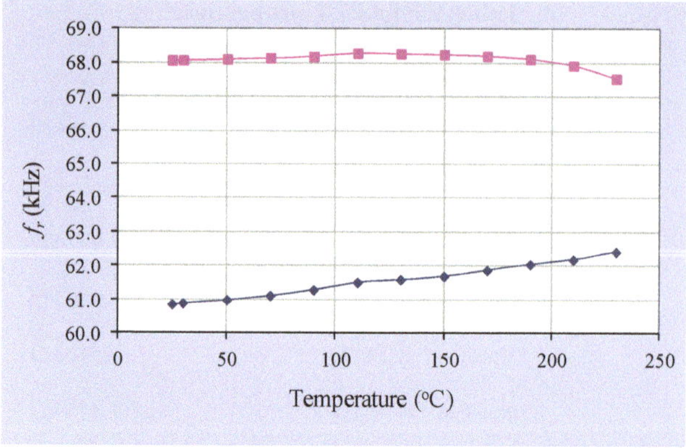

Figure 12. f_r of PZT at various temperatures.

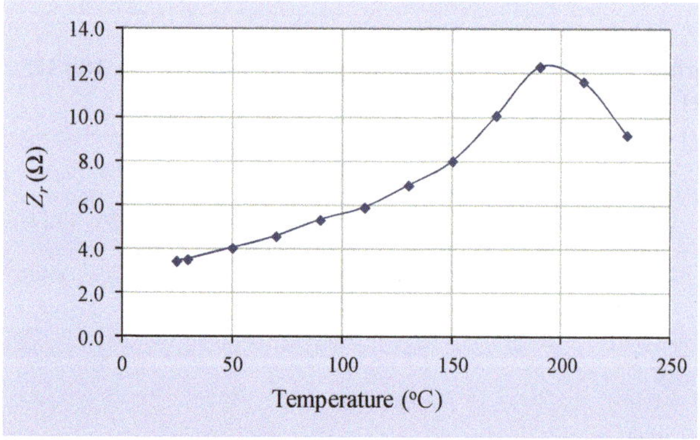

Figure 13. Impedance of PZT at f_r at various temperatures.

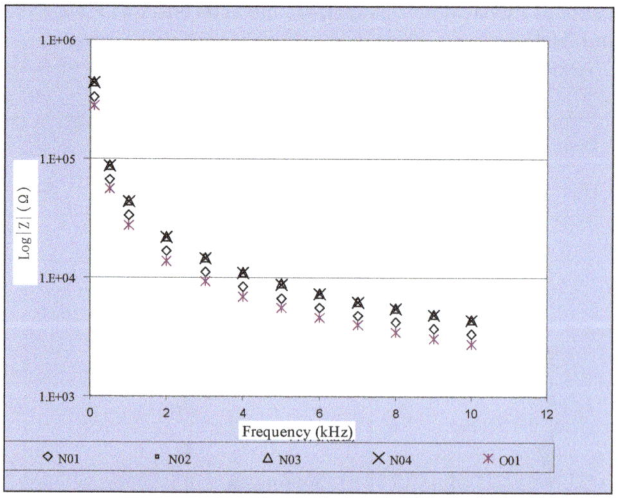

Figure 14. Average values at 100 Hz to 10 kHz of N01, N02, N03, N04, and O01.

4.4. Influences of Temperature Variation on BVD Parameters

When PZT was placed in the temperature controlling chamber at 23°C, C_1 would be decrease by 72% or 0.726%/°C that leads to 255% of increase in L_1. Therefore, the temperature would directly influence the char-

Table 4. BVD calculating parameters of N01 and N04 using impedance method.

PZT	BVD	Frequency (kHz)												
		0.1	0.5	1	2	3	4	5	6	7	8	9	10	σ
N01	C_o(nF)	3.870	3.850	3.850	3.840	3.840	3.850	3.860	3.870	3.880	3.890	3.910	3.930	0.090
	C_1(nF)	0.947	0.941	0.940	0.940	0.940	0.941	0.943	0.946	0.948	0.952	0.956	0.961	0.021
	L_1(mH)	7.146	7.190	7.199	7.203	7.202	7.192	7.179	7.159	7.142	7.112	7.079	7.044	0.159
	R_1(Ω)	3.331	3.331	3.331	3.331	3.331	3.331	3.331	3.331	3.331	3.331	3.331	3.331	0.000
N04	C_o(nF)	2.840	2.840	2.840	2.840	2.840	2.840	2.850	2.850	2.860	2.870	2.890	2.900	0.060
	C_1(nF)	0.804	0.806	0.805	0.804	0.805	0.806	0.808	0.809	0.811	0.814	0.818	0.822	0.018
	L_1(mH)	8.449	8.429	8.439	8.441	8.439	8.429	8.403	8.396	8.373	8.337	8.298	8.258	0.191
	R_1(Ω)	3.828	3.828	3.828	3.828	3.828	3.828	3.828	3.828	3.828	3.828	3.828	3.828	0.000

where σ is the standard deviation.

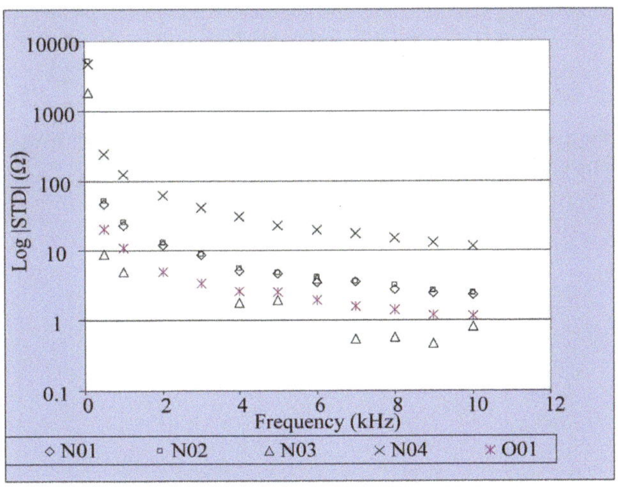

Figure 15. Standard deviation values at 100 Hz to 10 kHz of N01, N02, N03, N04, and O01.

acteristics of PZT as provided in **Table 5**.

5. Experiment of PZT under Mechanical Force

The parameters which was use in consideration of physical and electrical characteristics of PZT consists of ρ, d_{33}, k_t, and Q_m at 1 kHz. In this case, comparison of N02 and N03 is as shown in **Table 6**.

Measurement of Z_r under mechanical force was performed by using 0.5 lbs, 1.0 lbs, and 1.5 lbs of brass standard weights at room temperature. Each measurement was varied the frequency starting from 2 kHz to 80 kHz and repeated 30 times for calculating an average value. **Figures 16(a)** and **(b)** illustrate the experimental system and the brass standard weights, respectively.

Table 5. Percentage of BVD parameter changes as PZT temperature increasing by 230˚C.

Parameter of BVD	% Difference	% Difference/°C
R_1	167.0	0.726
C_1	−72.0	−0.313
L_1	255.0	1.109
C_o	18.1	0.079

Table 6. Physical and electrical characteristics of PZT (N02 and N03).

PZT	Density	d_{33} (10^{-12} C/N)	C_f(nF)	K (F/m)	k_t	Q_m
N02	7.42	256	3.5619	996	0.51	822
N03	7.40	262	3.5812	1002	0.51	919

5.1. Experiment of PZT at f_r

Standard weights (0.5 lbs, 1.0 lbs, and 1.5 lbs) were pressed to estimate f_r, f_a, and Z_r as shown in **Table 7** compared to weightless condition. The comparison was shown in **Table 8**. It can be noted that PZT pressed by 1.5 lbs standard weight would express higher f_r and f_a by ≈0.24 kHz and 2 kHz, respectively. However, Z_r would be varied by the weight pressing and become stable at ≈6.5 Ω.

5.2. Experiment of Z_r of PZT Connected Serially with Specific Capacitors

A capacitor (1.097 nF, 1.46 nF, 1.92 nF, 4.48 nF, 5.92 nF, 8.97 nF, and 33.016 nF) was connected serially with a given PZT (N02 and N03) and then, f_r, f_a, and Z_r were

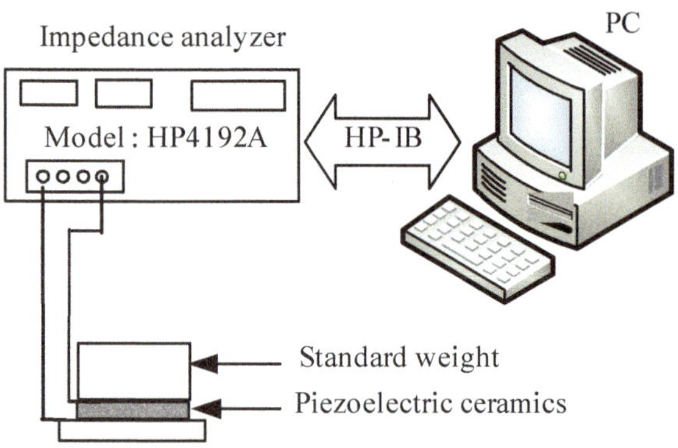

(a)

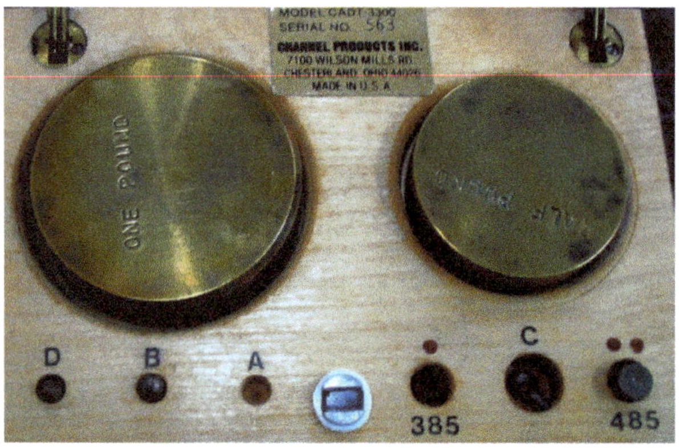

(b)

Figure 16. Impedance measurement using Impedance analysis model HP4192A and standard weight pressing: (a) System diagram; (b) Brass standard weights.

Table 7. f_r, f_a, and Z_r of PZT under mechanical force loads.

PZT	Parameter	Under force load (lb)			
		0	0.5	1	1.5
	Z_r	3.601	8.770	7.536	10.553
N02	f_r	61.550	61.850	61.700	61.850
	f_a	69.800	69.650	69.950	79.099
	Z_r	3.964	5.575	10.587	5.506
N03	f_r	61.250	61.400	61.400	61.400
	f_a	69.350	69.500	69.350	69.350

Table 8. Variation of f_r, f_a, and Z_r of PZT under mechanical force loads.

Parameter	Under force load (lb)		
	0.5	1	1.5
ΔZ_r	2.602	5.380	6.560
Δf_r	0.180	0.180	0.240
Δf_a	0.210	0.300	2.130

measured. The relationships of $f_s/2$ $(f_{sCL}\text{-}f_s)$ are as illustrated in **Figures 17-20** where f_s is the serial resonance frequency based on IEEE method [4,5]. The slope of linear relationship is $1/C_1$ as shown in **Table 9**.

5.3. Calculation of BVD Using Impedance Method

The average values of f_r, f_a, and Z_r at resonance frequency obtained from experiments of N02 and N03 were used for analyzing BVD. By selecting the frequency ranging from 0.1 kHz to 10 kHz, it can be observed that

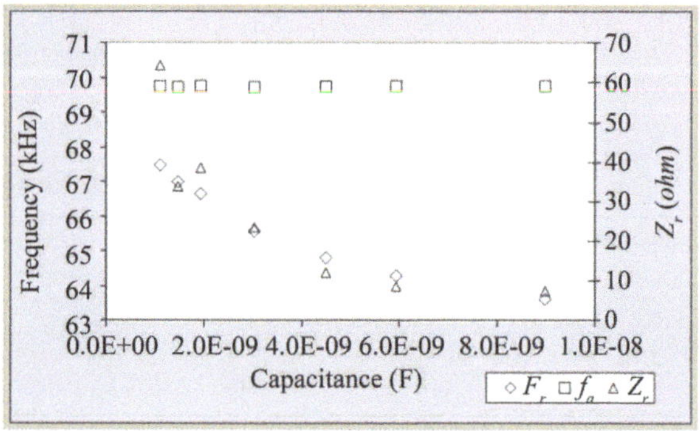

Figure 17. f_r, f_a, and Z_r of N02 connected serially with various capacitors.

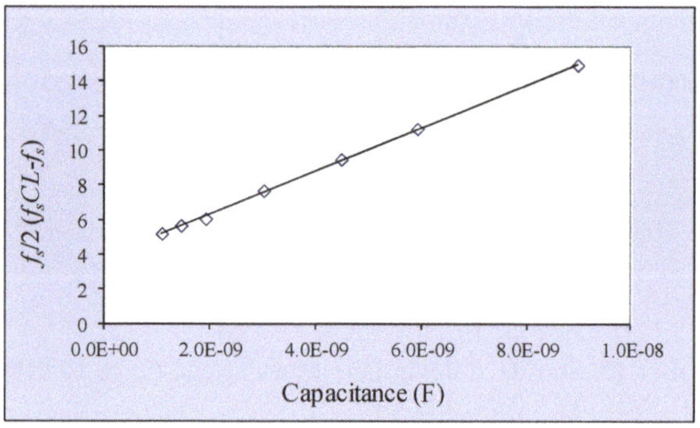

Figure 18. $f_s/2(f_{sCL}\text{-}f_s)$ of N02 connected serially with various capacitors.

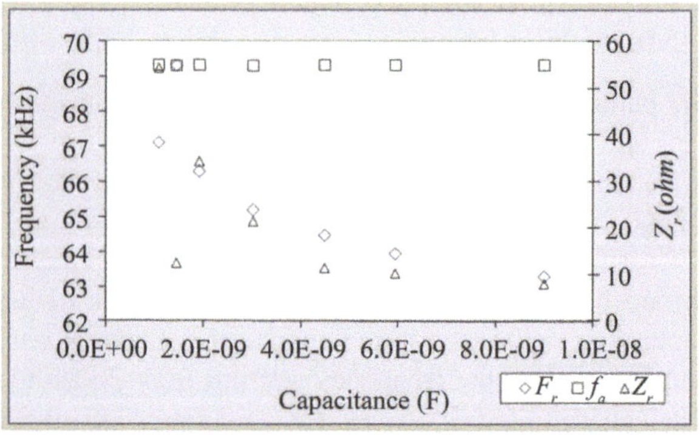

Figure 19. f_r, f_a, and Z_r of N03 connected serially with various capacitors.

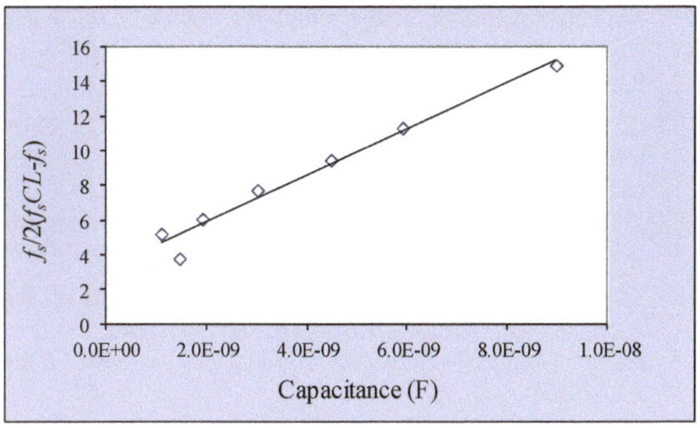

Figure 20. $f_s/2(f_{sCL}\text{-}f_s)$ **of N02 connected serially with various capacitors.**

Table 9. Slopes of $f_s/2(f_{sCL}\text{-}f_s)$**.**

PZT	Slope = $1/C_1$ (10^9)	C_1 (nF)
N02	1.237	0.808
N03	1.235	0.810

$C_o \approx 0.06$ nF, $C_1 \approx 0.018$ nF, $L_1 \approx 0.184$ mH to 0.190 mH while R_1 was identical due to the same impedance at f_r as given in **Table 10**. The **Table 11** shows the comparison of BVD parameters calculated by using impedance method, IEEE method, and HP4194 where IEEE method is the method of IEEE 176-1987, Δ_1 is the percentage of difference between values obtained from impedance and IEEE 176-1987 method, and Δ_2 is the percentage of difference between values obtained from impedance and HP4194 method.

5.4. Influences of Temperature on BVD Parameters under Mechanical Force

Variation of f_r, f_a, and Z_r when PZT was pressed by standard weights was calculated for C_o, C_1, and L_1 by impedance method. The differences are given in **Table 12**.

6. Conclusion

From calculation, it was observed that $|Z|$ and $Z(\theta)$ of the impedance ranging from low frequency to f_r and f_a tended to be equal to the impedance measured by impedance analysis model HP4192A. Selecting low frequency range (100 Hz, 1 kHz, and 2 kHz) for calculating C would cause the error of $|Z|$ by $\approx -1\%$, On the other hand, selecting the impedance and frequency at 3 kHz the error would be exhibit less than 1%. However, the impedance and frequency at 14 kHz would exhibit the greater error by $\approx 5\%$ because ωL can much influence the impedance while $1/\omega L$ decreases. Therefore, the impedance selected for calculating should have the frequency lower than f $f_a^2 - f_r^2$. Considering the percentage of impedance error, it can noted that the impedance at the frequency of 100 Hz, 1 kHz, and 2 kHz would provide the similar results with 20% - 25% of the maximum error of f_a. From experiment of the calculated impedance, the accuracy of f_r and f_a was ± 0.0033 kHz with the standard deviation of 0.00129 kHz which would be reliable for calculating BVD circuit. From experiment of accuracy which was compared between Z_r of BVD equivalent circuit and Z_r from measurement, it can be noted that Z_r at f_r of BVD circuit calculated by impedance method and IEEE 176-1987 would exhibit the error by $\approx 2.37\%$ which was lower than HP4194 method ($\approx 4\%$). It would be because the impedance analysis using impedance analysis model HP4192A is applied the program to estimate f_r more accurately. However, R_1 was less different as it was a specific resistance at resonance frequency. Considering the influence of temperature on PZT characteristic, it was noted that the increase of temperature would contribute to less difference of f_r and f_a.

Table 10. Calculation of BVD parameters based on impedance method: a case of N02 and N03.

PZT	BVD	Frequency (kHz)												
		0.1	0.5	1	2	3	4	5	6	7	8	9	10	σ
N02	C_o(nF)	2.780	2.780	2.770	2.770	2.770	2.780	2.780	2.790	2.800	2.810	2.820	2.830	0.060
	C_1(nF)	0.788	0.787	0.786	0.785	0.785	0.786	0.788	0.790	0.792	0.795	0.799	0.803	0.018
	L_1(mH)	8.469	8.482	8.494	8.497	8.497	8.489	8.470	8.448	8.425	8.391	8.354	8.313	0.184
	R_1 (Ω)	4.035	4.035	4.035	4.035	4.035	4.035	4.035	4.035	4.035	4.035	4.035	4.035	0.000
N03	C_o(nF)	2.820	2.790	2.790	2.790	2.790	2.790	2.800	2.810	2.820	2.830	2.840	2.850	0.060
	C_1(nF)	0.795	0.788	0.787	0.787	0.787	0.788	0.789	0.792	0.794	0.797	0.801	0.805	0.018
	L_1(mH)	8.488	8.559	8.570	8.571	8.572	8.561	8.549	8.518	8.496	8.463	8.425	8.382	0.190
	R_1 (Ω)	3.590	3.590	3.590	3.590	3.590	3.590	3.590	3.590	3.590	3.590	3.590	3.590	0.000

Table 11. Comparison of average BVD parameters from each method.

PZT	BVD	Impedance method (1 kHz)	IEEE	HP4194	Δ_1	Δ_2
N02	C_o(nF)	2.770	2.852	2.707	2.96%	2.27%
	C_1(nF)	0.786	0.808	0.791	2.80%	0.64%
	L_1(mH)	8.494	8.258	8.483	2.78%	0.13%
	R_1(Ω)	4.035	4.035	3.415	0.00%	15.37%
N03	C_o(nF)	2.790	2.872	2.719	2.94%	2.54%
	C_1(nF)	0.787	0.810	0.791	2.92%	0.51%
	L_1(mH)	8.570	8.328	8.572	2.82%	0.02%
	R_1(Ω)	3.590	3.590	4.056	0.00%	12.98%

Table 12. Percentage of BVD parameter variation with 1.5 lbs standard weight pressing.

Parameter of BVD	% Difference	% Difference/°C
R_1	144.0	96
C_o	−4.96	−3.30667
C_1	21.22	14.14667
L_1	−18.14	−12.0933

However, the increase of Z_r would cause the variation of BVD calculated parameters such as decrease of C_1, increase of R_1, L_1, and C_o. On the other hand, the influences of mechanical force load on PZT parameters would cause f_r, f_a, and Z_r increase that also contribute to R_1 and C_1 increase while L_1 and C_o decrease.

REFERENCES

[1] S. Sherrit, H. D. Wiederick and B. K. Mukherjee, "Accurate Equivalent Circuits for Unloaded Piezoelectric Resonators," *IEEE Ultrasonics Symposium Proceedings*, Vol. 2, 1997, pp. 931-935.

[2] S. Sherrit, H. D. Wiederick, B. K. Mukherjee and M. Sayer, "An Accurate Equivalent Circuits for the Unloaded Piezoelectric Vibrator in the Thickness Mode," *Journal of Physics D: Applied Physics*, Vol. 30, No. 16, 1997, pp. 2354-2363.

[3] C. Jeerapan, W. Sriratana, P. Julsereewong and S. Kummool, "Analysis of Appropriate Parameters for Piezoelectric Ceramic Utilization by Using BVD Model," *ICCAS* 2005, *International Conference on Control, Automation and System*, Gyeonggi-Do, 2-5 June 2005, pp. 2067-2070.

[4] "An American National Standard IEEE Standard on Piezoelectricity," *IEEE Ultrasonics, Ferroelectrics, and Frequency Control Society*, New York, 1988, ANSI/IEEE Std 176-1987.

[5] "Standard Definitions and Method of Measurement for Piezoelectric Vibrators," The Institute of Electrical and Electronics Engineers Inc., New York, 1966, IEEE No. 177.

[6] J. G. Smits, "Iterative Method for Accurate Determination of Real and Imaginary Parts of the Materials Coefficients of Piezoelectric Ceramics," *IEEE Transactions on Sonics and Ultrasonics*, Vol. 23, No. 6, 1976, pp. 393-402.

[7] S. Sherrit, H. D. Wiederick and B. K. Mukherjee, "Non-Iterative Evaluation of the Real and Imaginary Material Constants of Piezoelectric Resonators," *Ferroelectrics*, Vol. 134, No. 1-4, 1992, pp. 111-119.

[8] Q. C. Xu, A. R. Ramachandran and R. E. Newnham, "Resonance Measuring Technique for Complex Coefficients of Piezoelectric Composites," *Journal of Wave-Material Interaction*, Vol. 2, 1987, pp. 105-122.

[9] B. Jaffe, W. R. Cook Jr and H. Jaffe, "Piezoelectric Ceramics," *Journal of Sound and Vibration*, Vol. 20, No. 4, 1972, pp. 562-563.

[10] T. Ikeda, "Fundamentals of Piezoelectricity," 19th Edition, Oxford University Press, New York, 1990.

[11] S. Fujishima, K. Togawa and S. Ohta, "Analysis and Design of the Piezoelectric Ceramic Resonator Osillators," 41*st Annual Symposium on Frequency Control*, Pennsylvania, 27-29 May 1987, pp. 391-397.

[12] M. Trainer, "Ferroelectrics and the Curie-Weiss Law," *European Journal of Physics*, Vol. 21, No. 5, 2000, pp. 459-469.

Growth and Improvement of ZnO Nanostructure Using Aged Solution by Flow Coating Process

Kasimayan Uma[1], Solaiappan Ananthakumar[2], Ramalinga Viswanathan Mangalaraja[3*], Tetsuo Soga[1,4*], Takashi Jimbo[4]

[1]Centre for Social Contribution and Collaboration, Nagoya Institute of Technology, Nagoya, Japan
[2]Materials and Minerals Division, National Institute for Interdisciplinary Science and Technology (CSIR), Kerala, India
[3]Department of Materials Engineering, University of Concepcion, Concepcion, Chile
[4]Department of Frontier Materials, Nagoya Institute of Technology, Nagoya, Japan

ABSTRACT

ZnO nanostructures were prepared on corning glass substrate by flow coating process with different annealing temperature from 100˚C to 600˚C. Fresh and two days aged solutions were used to investigate the growth behavior and to evaluate the nanostructure of ZnO. The effect of preparation conditions on the deposition of ZnO nanostructure was investigated by scanning electron microscopy (SEM), X-ray diffraction (XRD), Raman, and photoluminescence spectroscopy (PL). The results indicated that the solution aging condition and annealing temperature have a strong influence on the morphology and structural properties of the ZnO nanostructure. The solution aged after two days shows the different morphologies compared with the freshly prepared solution.

Keywords: ZnO Nanostructure; Flow Coating; Photo Luminescence (PL)

1. Introduction

Synthesis of one-dimensional (1-D) nanostructures with interesting hierarchical organization has attracted much attention in recent years because of fundamental curiosity as well as practical reasons [1-4]. 1-D nanostructure has been studied for optoelectronic nanodevice applications as a promising candidate for UV light emitting diodes and laser diodes. The ability to precisely control their dimensions, chemical composition, surface properties, phase purity and crystal structure has become more and more important for the utilization of 1-D nanostructures in a broad range of areas [5]. Semiconducting oxides such as ZnO, In_2O_3 and SnO_2 have a wide range of application in transparent conductive films, optical devices, solar cells, sensors, photocatalysis, transparent field effect transistors and bulk acoustic wave devices etc., [6-9]. Among these, ZnO has attracted much attention for its potential applications in optoelectronics and microelectronics. With the development of research on fabrication and applications of ZnO particles and films, more and more valuable research results have confirmed that the morphology of ZnO particles and films plays the key role for the applications of ZnO in special fields [10,11].

A wide variety of ZnO nanostructures have been synthesized by the various techniques and are reported in the literature till date [12-14]. Recently, self-catalysis of Zn or ZnO was discovered in the catalyst-free methods for preparing ZnO nanostructures. So far, nanocrystalline ZnO with different particle morphologies and sizes has been obtained by several preparation approaches including thermal decomposition, chemical vapor deposition (CVD), pulsed laser deposition (PLD), gas phase reaction, hydrothermal synthesis and so on. However, these methods are expensive and require high vacuum and formation controlling conditions. Recently, solution phase routes including microemulsion, solvothermal, hydrothermal, self-assembly and template assisted sol-gel process have been employed to synthesize ZnO nanostructure [15]. Among the fabrication methods, solution deposition method is the simplest, cheapest and the most attractive one. Our work is focused on the growth of ZnO nanostructure using aged sol-gel solution of zinc acetate di-hydrate. This work reports on improvement in optical and crystallographic properties of ZnO nanostructure using aged solution with different annealing temperature. The ZnO nanostructures have been grown in large areas without a

*Corresponding author.

catalyst by this simple and low cost method. Here ZnO nanostructures were successfully grown on glass substrates using aged solution giving strong photoluminescence around 383 nm wavelengths when compared with visible region around 525 nm.

2. Experimental Method

A coating precursor solution was prepared by dissolving zinc acetate di-hydrate ($Zn(CH_3COO)_2{\cdot}2H_2O$, Nacalai Tesque Chemical Co. Ltd., 99.5%) and equivalent molar quantity of monoethanolamine (MEA, Nacalai Tesque Chemical Co. Ltd., 99.5%) in 2-methoxyethanol (Nacalai Tesque Chemical Co. Ltd.). The precursor solution was homogenously mixed by simple stirring for 3 hours at room temperature and then at 60˚C for 2 hours. The precursor sol was deposited onto the preheated Corning glass by spin coating at 2000 rpm for 20 seconds and then at 3000 rpm for another 20 seconds at room temperature. Immediately after the coating, the precursor films were subjected to curing at 300˚C for 10 min to remove the excess solvents and residual moisture. Finally, the coated substrates were annealed at 600˚C for 2 hr in air atmosphere.

The morphology of the as-prepared films was observed by SEM (SEM, S-3000H Hitachi). The crystallinity of each ZnO film was analyzed by an XRD (X-ray diffractometer—JEOL) with Cu Kα radiation. Raman spectra were measured with Raman spectrometer (NRS 1500 W Raman Spectrometer—JASCO) with the excitation wavelength of 523 nm and PL (JASCO) were measured at room temperature of He-Cd laser at 325 nm excitation.

3. Results and Discussion

The XRD patterns of the ZnO nanostructure are shown in **Figures 1** and **2** for different annealing temperatures using fresh and two days aged solution coated on Corning glass substrates. For fresh and two days aged solution, the diffraction peaks of ZnO nanostructure were not clearly observed for the samples annealed below 300˚C which confirms that the crystal was in the amorphous state. This is due to the thermal decomposition of the zinc acetate at 240˚C. When the ZnO nanostructure is prepared from zinc acetate-2-methoxyethanol-MEA solutions, the crystal begins to grow only at 200˚C - 300˚C. Therefore the diffraction peaks were not observed clearly at the annealing temperature below 300˚C. The diffraction peaks such as (101), (100) and (002) planes were observed for the ZnO samples annealed above 300˚C and this is due to the annealing treatment which results in the formation of oriented ZnO nanoparticles on the substrates. No characteristics peaks were observed for impurities from **Figures 1** and **2** which illustrate that the entire crystalline precursor has been decomposed and grown into ZnO crystal [16]. Increasing the annealing temperature enhances uniform crystal grow at 600˚C and the diffraction peaks were stronger by comparing the samples annealed from 100˚C to 500˚C. **Figure 3** shows the XRD patterns of the ZnO nanostructure using (a) fresh and (b) two days aged solution annealed at 600˚C. For the samples prepared from fresh solution, the diffraction peaks such as (100), (101) and (002) are broader. But after two days aging the diffraction peaks were intense and narrow which indicates that material has good crystallinity [17] and large grain size. All the diffraction peaks from the film prepared by aged solution can be well indexed to the hexagonal phase of ZnO reported in JCDDS card (No. 36-1451). The lower intensities and broader diffraction peaks from fresh solution samples show that the films may have finer crystalline size when compared to aged samples. The lowest (0.435˚) and highest (0.528˚) full width at half maximum (FWHM) were obtained for films made from aged and fresh solution as shown in **Figure 3**. However the value of FWHM is inversely proportional to the grain size implying that the grain size of the samples was improved in the aged solution when compared with

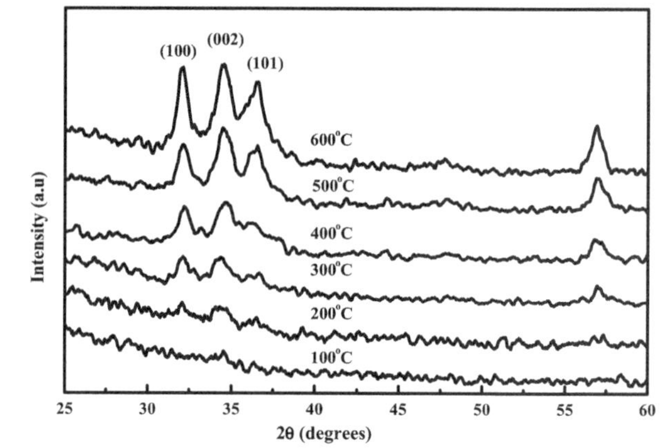

Figure 1. XRD patterns of the ZnO nanostructure prepared from fresh solution with different annealing temperatures.

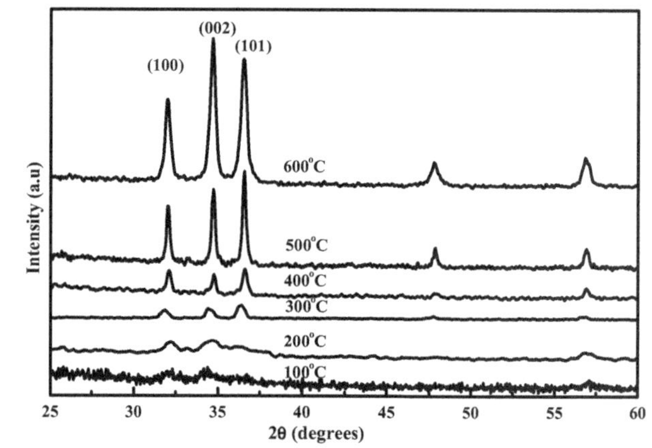

Figure 2. XRD patterns of the ZnO nanostructure prepared from two days aged solution with different annealing temperatures.

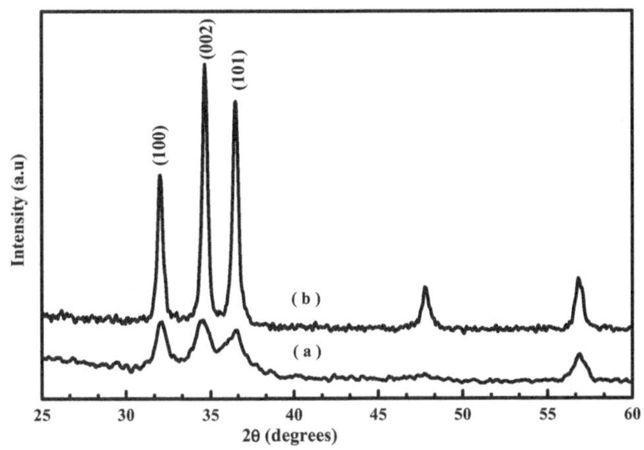

Figure 3. XRD patterns of the ZnO nanostructure prepared from (a) fresh and (b) two days aged solution annealed at 600°C.

fresh solution. The grain size of the films was calculated from the FWHM of the peak corresponding to (002) plane using Scherrer formula [18].

$$l = 0.94\lambda / B\cos\theta$$

where λ is radiation wavelength (1.54 Å), B is FWHM of the diffraction peak and θ is a diffracting angle. The grain size was 15.8 and 19 nm for fresh and aged solution respectively. From the XRD results it was found that the aged solution samples are having hexagonal structure with large grain size and this leads to the improved quality of the ZnO nanostructure. The aging conditions are such that the systems will tend to reach their thermodynamic colloidal stability and this stability leads to the variations of morphology, size, or structural properties as a result of the aging mechanism [19].

Figures 4 and **5(a)-(d)** show the SEM images of ZnO nanostructure grown on glass substrates prepared from fresh and two days aged solution annealed at 300°C, 400°C, 500°C and 600°C respectively. At below 300°C uniform crystal structure was not grown completely and poor alignment of the nanostructures was observed. This was confirmed by X-ray diffraction pattern by showing weak orientation peak at below 300°C. When increasing the annealing temperature from 300°C to 600°C the nanostructure grows along the (101), (100) and (002) directions and also improved the structural alignment. From **Figures 4(c)** and **5(c)**, it is observed that the crystal started to grow on the surface which is perpendicular to the substrate. The ZnO films prepared from fresh solution and annealed at 600°C shows that the grains sizes are large and no evidence for hexagonal structure of the ZnO nanostructure as shown in **Figure 4(d)**. In **Figure 5(d)**, the sample shows hexagonal structure and it confirms that the films are indexed to the hexagonal phase of ZnO. These results agree well with the X-ray results. The samples annealed at 600°C using two days aged solution

Figure 4. SEM images of the ZnO nanostructure prepared from fresh solution with different annealing temperature of (a): 300°C; (b): 400°C; (c): 500°C; (d): 600°C.

Figure 5. SEM images of the ZnO nanostructure prepared from two days aged solution with different annealing temperature of (a): 300°C; (b): 400°C; (c): 500°C; (d): 600°C.

gives the same wurtzite ZnO patterns with high intensity of diffraction peaks which provides further evidence that the aged solution has little influence on the orientation of the ZnO nanostructure.

Raman spectrum is an effective method to investigate the crystallization, structure and defects in the nanostructure materials. Wurtzite ZnO belongs to the C^4_{6V} space group with two formula units per primitive cell. **Figures 6** and **7** show the Raman spectra of ZnO nanostructure grown at temperatures ranging from 100°C to 600°C using fresh and two days aged solution. There is no variation observed for the samples annealed at 100°C to 300°C as shown in **Figures 6** and **7**. When samples are annealed

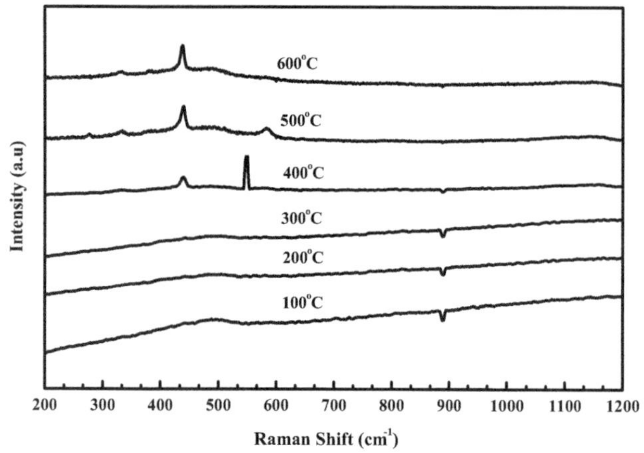

Figure 6. Raman spectra of ZnO nanostructure prepared from fresh solution with different annealing temperatures.

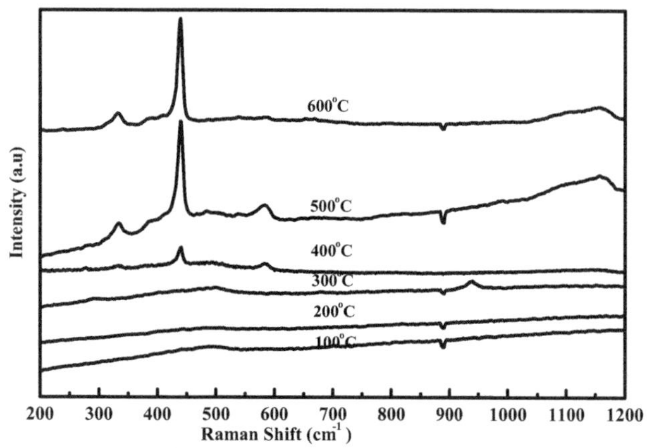

Figure 7. Raman spectra of ZnO nanostructure prepared from two days aged solution with different annealing temperatures.

at 400°C to 600°C, the peak at 437 cm^{-1} was observed which leads to the ZnO nonpolar optical phonon (E_2) modes and it is an assigned Raman peak of ZnO bulk [20]. The peak at 330 cm^{-1} assigned to the Raman spectrum arising from zone boundary phonons $3E_{2H}$-E_{2L} (multiphonon), while the peak at 561 cm^{-1} leads to the E_1 (LO) mode of ZnO associated with oxygen deficiency [21]. The samples grown at higher temperature exhibit a very low oxygen vacancy and this may be due to influence of change in particle size annealed at high temperature [22]. However when samples are annealed in air, excess zinc in the film becomes oxidized, the impurity states disappear and the magnitude of resonance Raman enhancement of the E_1 (LO) mode is diminished. The peak intensity of optical phonon mode E_2 is increased slightly when increasing the annealing temperature from 400°C to 600°C for the samples prepared from fresh solution but in two days aged solution the intensity of E_2 mode is largely increased when compared with fresh solution. **Figure 8** shows the Raman spectra of ZnO films obtained from (a)

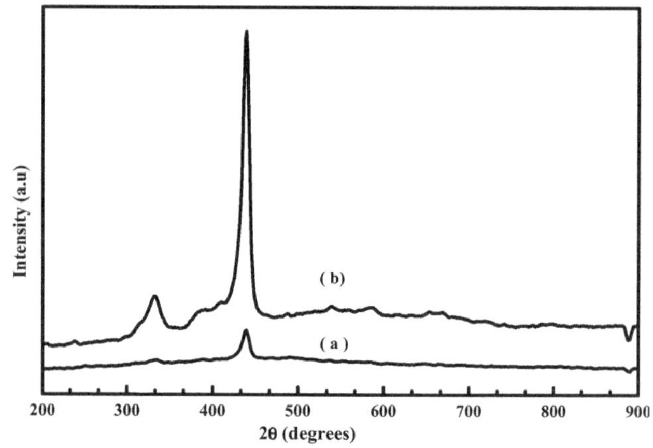

Figure 8. Raman spectra of ZnO nanostructure prepared from (a) fresh and (b) two days aged solution annealed at 600°C.

fresh and (b) two days aged solution annealed at 600°C. Samples prepared from aged solution indicate the most intense peak at 437 cm^{-1} when compared with fresh solution, which corresponds to the vibrational mode of E_2 (high). The E_2 mode corresponds to the band characteristics for the wurtzite hexagonal plane of ZnO. The highest intensity and narrower spectral width of the Raman active E_2 mode indicated that the as-grown ZnO have good quality with a hexagonal wurtzite crystal phase. From this result, it is confirmed that the ZnO nanocrystals possibly gives the better crystal quality and less impurity when using the two days aged solution instead of using a fresh solution.

Figures 9 and **10** show the PL spectra of the samples prepared from fresh and two days aged solution annealed at 400°C, 500°C and 600°C. In **Figure 9** the UV emission was not observed clearly for samples annealed from 400°C and 500°C. When samples annealed at 600°C the broad luminescence band was observed around 470 nm other than the weak UV emission. However the aged solutions do not exhibit this luminescence band. Detlef *et al.*, [23] reported that the freshly prepared ZnO colloidal suspension exhibit a single broad fluorescence band with initial maximum around 460 nm. From this we can understand that the broad band around 470 nm may be due to the blue shift of fresh colloidal suspension of ZnO. The results of PL spectra may prove that the solution condition has a little influence on the ZnO nanostructure. All the spectra from aged solution have a UV emission around 383 nm and visible emission around 530 nm, which is generally discussed with respect to oxygen vacancies at the surface. The UV band emission of ZnO has been well demonstrated to the near band-edge transition, namely the recombination of free excitons through exciton-exciton emission process [24]. In **Figure 10** the samples annealed at 400°C shows that the UV emission peak which is almost equal to the green band emission. This

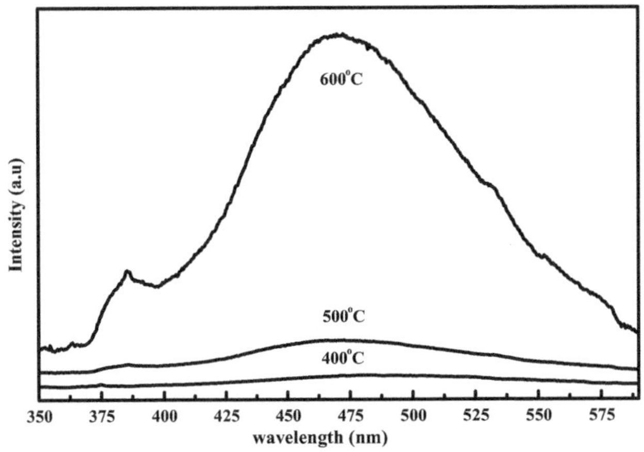

Figure 9. PL spectra of the ZnO nanostructure prepared from fresh solution with different annealing temperatures measured at room temperature.

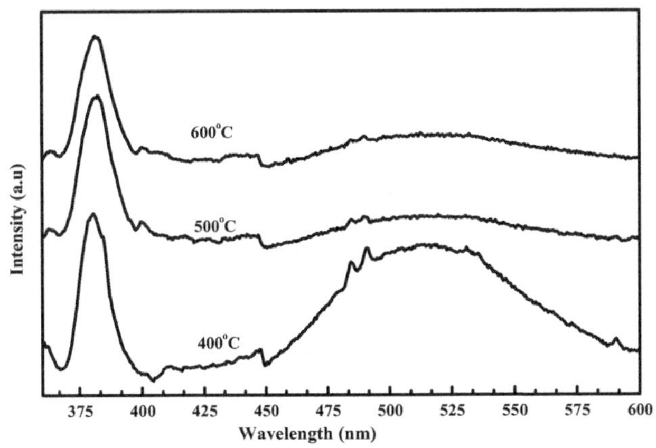

Figure 10. PL spectra of the ZnO nanostructure prepared from two days aged solution with different annealing temperatures measured at room temperature.

proves that these samples do not have good crystalline quality as they have many oxygen vacancies. But in the case of sample annealed at 500°C and 600°C much weaker visible emission and sharper UV peak helps to demonstrate much better optical properties. The green PL has been observed in the ZnO nanocrystals grown by hydrothermal, CVD and electrochemical methods and its origin are commonly attributed to the photo generated holes with singly ionized oxygen vacancies. However the PL spectrum of ZnO nanocrystals array is depending on the growth temperature. The band edge PL intensity increases greatly and the intensity of green PL decreases a lot when increasing the annealing temperature from 500°C to 600°C. Increasing the growth temperature will improve the crystal quality of ZnO nanostructure to a certain extent [25]. Although the reason for the growth temperature dependent PL is not quiet clear when related to the formation mechanism of oxygen vacancies. From the PL results it was observed that samples annealed at 600°C gives a strong UV peak and much lower green

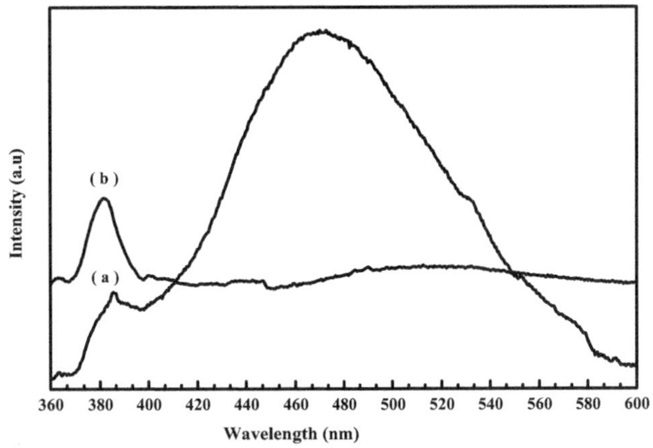

Figure 11. PL spectra of the ZnO nanostructure prepared from (a) fresh and (b) two days aged solution annealed at 600°C.

peak when compared with samples annealed at 400°C and 500°C. **Figure 11** shows the PL spectra measured at room temperature for (a) fresh and (b) two days aged solution annealed at 600°C. From aged solution a strong and high intensity peak around 383 nm in the UV region and week band around 530 nm in the visible region was observed. The strong room temperature UV emission of aged solution should be attributed to the high purity, good crystallinity of the as synthesized ZnO nanostructure. It has been reported that the improvement in the crystal quality may cause the appearance of a sharp and strong UV emission and a suppressed and week green emission [26]. The PL result indicates that the optical properties of ZnO nanostructure were improved by the aged solution when compared with fresh solution.

4. Conclusion

The success of this work is that the nanostructure of ZnO has been prepared at low cost by a simple solution method. This method is based on a simple procedure and doesn't need template supporting which leads to the formation of ZnO nanostructure. A systematic study of this work reveals that the use of aged sol-gel solution along with growth parameters will lead to the formation of ZnO nanostructure with low oxygen vacancy defect density, which might find potential applications in nanodevices. The annealing temperature plays a crucial role for the improvement of ZnO nanostructure. For the samples annealed below 300°C, the crystal growth is non-crystalline and incomplete with poor orientations. The samples prepared by two days aged solution exhibit stronger UV emission than broad green emission, which indicates that the high optical quality of the ZnO nanostructures is prepared from aged solution. It is worthy to note that the aging mechanisms have a great influence on the formation of good quality ZnO nanostructure.

5. Acknowledgements

This work was supported by the Center for Social Contribution and Collaboration (C-SoCC), Nagoya Institute of Technology, Nagoya, Japan and FONDECYT No: 1130916, Government of Chile, Santiago, Chile.

REFERENCES

[1] A. M. Morales and C. M. Lieber, "A Laser Ablation Method for the Synthesis of Crystalline Semiconductor Nanowires," *Science*, Vol. 279, No. 5348, 1998, pp. 208-211.

[2] D. C. Look, "Recent Advances in ZnO Material and Devices," *Materials Science and Engineering: B*, Vol. 80, No. 1, 2001, pp. 383-387.

[3] Y. Li, G. W. Meng, L. D. Zhang and F. Phillip, "Ordered Semiconductor ZnO Nanowires Arrays and Their Photoluminescence Properties," *Applied Physics Letters*, Vol. 76, No. 15, 2000, pp. 2011-2013.

[4] R. S. Wagner and W. C. Ellis, "Vapor-Liquid-Solid Mechanism of Single Crystal Growth," *Applied Physics Letters*, Vol. 4, No. 5, 1994, pp. 89-90.

[5] Y. Xia, P. Yang, Y. Sun, Y. Wu, B. Mayers, B. Gates, Y. Yin, F. Kim and H. Yan, "One Dimensional Nanostructures; Synthesis, Characterization, and Applications," *Advanced Materials*, Vol. 15, No. 5, 2003, pp. 353-389.

[6] Z. S. Wu, S. Z. Deng, N. S. Xu, J. Chen, J. Zhou and J. Chen, "Single Wall Carbon Nanotube Atomic Force Microscope Probes," *Applied Physics Letters*, Vol. 80, No. 20, 2002, pp. 3829-3831.

[7] J. Liu, X. M. Meng, Y. Jiang, C. S. Lee, I. Bello and S. T. Lee, "Gallium Nitride Nano Wires Doped with Silicon," *Applied Physics Letters*, Vol. 83, No. 20, 2003, pp. 4241-4243.

[8] Y. F. Chan, X. F. Duan, S. K. Chan, I. K. Sou, X. X. Zhang and N. Wang, "ZnSe Nanowires Epitaxially Grown on Gap (111) Substrates by Molecular Beam Epitaxy," *Applied Physics Letters*, Vol. 83, No. 13, 2003, pp. 2665-2667.

[9] M. H. Huang, S. Mao, H. Feick, J. Q. Yan, Y. Y. Wu, H. Kind, E. Weber, R. Russo and P. D. Yang, "Room-Temperature Ultraviolet Nanowires Nanolasers," *Science*, Vol. 292, No. 5523, 2001, pp. 1897-1899.

[10] T. Xu, C. S. Xie and Prog, "Tetrapod–Like Nano-Particle ZnO/Acrylic Resin Composite and Its Multi-Function Property," *Progress in Organic Coatings*, Vol. 46, No. 4, 2003, pp. 297-301.

[11] B. L. Zhu, D. W. Zeng, J. Wu, W. L. Song and C. S. Xie, "Synthesis and Gas Sensitivity of In-Doped ZnO Nanoparticles," *Journal of Materials Science: Materials in Electronics*, Vol. 14, No. 8, 2003, pp. 521-526.

[12] A. Sekar, S. H. Kim, A. Umar and Y. B. Hahn, "Catalyst Free Synthesis of ZnO Nanowires on Si by Oxidation of Zn Powder," *Journal of Crystal Growth*, Vol. 277, No. 1-4, 2005, pp. 471-478.

[13] W. L. Hughes and Z. L. Wang, "Nanobelts as Nanocantilevers," *Applied Physics Letters*, Vol. 82, No. 14, 2003, pp. 2886-2888.

[14] J. Yu. Lao, J. G. Wen and Z. F. Ren, " Hierarchical ZnO Nanostructure," *Nano Letters*, Vol. 2, No. 11, 2002, pp. 1287-1291.

[15] S. Lib, Y. S. Chu, D. Z. Shen, Y. M. Lu, Y. J. Zhang and X. W. Fan, "High Quality ZnO Thin Films Grown," *Journal of Applied Physics*, Vol. 91, No. 1, 2002, pp. 501-505.

[16] L. L. Wu, Y. S. Wu and W. Lu, "Preparation of ZnO Nanorods and Optical Characterizations," *Physica E*, Vol. 28, No. 1, 2005, pp. 76-82.

[17] Y.-H. Ni, X.-W. Wei, J.-M. Hong and Y. Ye, "Hydrothermal Preparation and Optical Properties of ZnO Nanorods," *Materials Science and Engineering: B*, Vol. 121, No. 1-2, 2005, pp. 42-47.

[18] J.-H. Lee, B.-W. Yeo and B.-O. Park, "Effects of the Annealing Treatment on Electrical and Optical Properties of ZnO Transparent Conduction Films by Ultrasonic Spraying Pyrolysis," *Thin Solid Films*, Vol. 457, No. 12, 2004, pp. 333-337.

[19] E. Hosono, S. Fujihara and T. Kimura, "Synthesis, Structure and Photoelectrochemical Performance of Micro/Nano Textured ZnO/Eosin y Electrodes," *Electrochimica Acta*, Vol. 49, No. 14, 2004, pp. 2287-2293.

[20] X. L. Xu, S. P. Lau, J. S. Chen, G. Y. Chen and B. K. Tay, "Polycrystalline ZnO Thin Films on Si (100) Deposited by Filtered Cathodic Vaccum Arc," *Journal of Crystal Growth*, Vol. 223, No. 1-2, 2001, pp. 201-205.

[21] G. J. Exarhos and S. K. Sharma, "Influence of Processing Variables on the Structure and Properties of ZnO Films," *Thin Solid Film*, Vol. 270, No. 1-2, 1995, pp. 27-32.

[22] A. K. Pradhan, K. Zhang, G. B. Louttus, U. N. Roy, Y. Cui and A. Burger, "Structural and Spectroscopic Characteristics of ZnO and ZnO: Er3+ Nanostructure," *Journal of Physical and Condensed Matterials*, Vol. 16, No. 39, 2004, pp. 7123-7130.

[23] Y. C. Kong, D. P. Yu, B. Zhang, W. Fang and S. Q. Feng "Ultraviolet-Emitting ZnO Nanowires Synthesized by a Physical Vapor Deposition Approach," *Applied Physics Letters*, Vol. 78, No. 4, 2001, pp. 407-409.

[24] W. Detlef, Bahnemann, C. Kormann and M. R. Hoffmann, "Preparation and Characterization of Quantum Size Zinc Oxide: A Detailed Spectroscopic Study," *Physical Chemistry*, Vol. 91, No. 14, 1987, pp. 3789-3798.

[25] C.-H. Hung and W.-T. Whang, "A Novel Low-Tempera-

ture Growth and Characterization of Single Crystal ZnO Nanorods," *Materials Chemistry and Physics*, Vol. 82, No. 3, 2003, pp. 705-710.

[26] D. M. Bagnall, Y. F. Chen, Z. Zhu, T. Yam, S. Koyama, M. Y. Shen and T. Goto, "High Temperature Excitonic Stimulated Emission from ZnO Epitaxial Layers," *Applied Physics Letters*, Vol. 73, No. 81, 1998, pp. 1038-1040.

Elucidation of Active Species over Ru/MnO Catalyst on Co_2/H_2 Methanation Reaction

Wan Azelee Wan Abu Bakar*, Susilawati Toemen, Rusmidah Ali, Hazwan Faiz Abd Rahim
Department of Chemistry, Faculty of Science, Universiti Teknologi Malaysia, Johor Bahru, Malaysia

ABSTRACT

The information regarding phase changes and structure transformation, particle sizes as well as active species of the catalyst are briefly discussed towards alumina supported Ru/MnO catalysts according to their various parameters of calcination temperatures and Mn loading. The Ru/Mn-75/Al_2O_3 catalysts calcined at 400˚C, 700˚C, 900˚C, 1000˚C and 1100˚C, Ru/Mn-65/Al_2O_3 and Ru/Mn-85/Al_2O_3 catalyst calcined at 1000˚C were synthesized by the wetness impregnation method. All the prepared catalysts exhibited crystallite size in the range of 95 nm to 114 nm. It was found that the catalyst with Ru/Mn-75/Al_2O_3 calcined at 1000˚C showed the highest 60.21% CO_2 conversion with 57.84% formation of CH_4 at the reaction temperature 200˚C. The expected active species that assist the CO_2 methanation activity over this catalyst was Mn_3O_4.

Keywords: X-Ray Diffraction; CO_2 Methanation; Catalysts; Active Species

1. Introduction

X-Ray Diffraction (XRD) analysis is one of the most important methods in determining the structure of complex natural products such as steroids, vitamins and antibiotics. Meanwhile, XRD analysis on the supported catalysts was done to reveal information about the crystallographic structure, chemical composition and physical properties of the catalyst [1-3]. Kizling and Regali [4] found that zirconium oxide analyzed by XRD exists in two crystalline structures, tetragonal and monoclinic. The synthesized material remained amorphous when calcined at 400˚C, while at 500˚C it showed crystalline structure consisting mainly of the tetragonal zirconia. From this result, it proved that the XRD technique can provide the qualitative identification of crystalline materials [5]. However, to obtain qualitative and quantitative information of the compound, the sample must be in solid form. The energetic X-rays can penetrate deep into the materials and provide information about the bulk structure. The operation follows Bragg's law as shown in Equation (1).

$$2d\ \sin\theta = n\lambda \tag{1}$$

where d = distance between two lattice points, θ = scattering angle, n = integer representing the order of the diffraction peak, λ = wavelength of the X-ray.

The output of this analysis is in the form of a diffractogram which is the plot of intensity versus 2θ. Peak position, width and intensity were then identified by a comparison with standards data Powder Diffraction Files (PDF) provided by the software used in this technique. Crystallographic phase present in the catalyst produces well-defined diffraction peaks whose widths are related to the crystalline quality. Crystalline phases are represented by sharp and narrow peaks, while amorphous phases are represented by broad and weak peaks [5]. Moreover, it can supply information regarding the crystallite size that can be calculated by Scherrer equation as shown in Equation (2).

$$d = k\lambda/\beta\cos\ \theta \tag{2}$$

where d = Crystallite Size (nm), λ = wavelength of the X-ray, θ = scattering angle, β = FWHM value.

The knowledge of catalyst structure is essential in order to understand the properties of the catalyst's surface. Therefore, most of the researchers had used the XRD analysis in their research. Nasr-Allah [6] who had examined the XRD analysis on pure and doped Mn/Al mixed solids calcined at 500˚C, 800˚C, and 900˚C revealed a nanocrystalline nature for the bulk structure of the catalysts

*Corresponding author.

prepared. Though in heterogeneous catalysis the active sites are usually located on the solid surface, the bulk structure also plays an important role, because many of the catalyst characteristics depend on it [7]. Therefore, this paper will focus on the application of XRD analysis technique over alumina supported Ru-MnO catalysts with particular emphasis on their catalytic performance towards CO_2 methanation reaction.

2. Experimental

2.1. Preparation of Catalysts

$MnCl_2 \cdot 2H_2O$ (5 g) obtained from Sigma Aldrich was dissolved with 2 mL of triply distilled water. Mixed catalysts solution was prepared by mixing an appropriate amount of $RuCl_3 \cdot xH_2O$ (Sigma Aldrich) with $MnCl_2 \cdot 2H_2O$ solution according to its ratio. The catalysts were prepared by impregnating the catalyst solution on 10 g of Al_2O_3 beads (Sigma Aldrich) support for 15 minutes and then it was took out and dried at ambient temperature. The impregnation and drying were repeated for three times. Then the impregnated Al_2O_3-bead was washed-dispensed with triply distilled water to remove chloride ion. This process was repeated until no colour change of $AgNO_3$ reagent. It was then aged at 80˚C for 24 hours and calcined in the furnace over various calcination temperatures.

2.2. Characterization

XRD analysis was conducted by using Siemens D5000 Crystalloflex X-Ray Diffractometer equipped with Cu target ($\lambda_{CuK\alpha}$ = 1.54 Å) radiation. Using a step scan of 0.050˚ and a step time of a second per step, data were collected over the range of 2θ from 20˚ to 80˚.

2.3. Catalytic Performance

All the prepared catalysts were tested its catalytic activity towards carbon dioxide methanation reaction by using in-house-built micro reactor. The analysis was conducted using continuous flow of CO_2, H_2 and compressed air (10:40:50) mixture with flow rate of 100 mL/min. This reactor was coupled with FTIR Nicolet Avatar 370 DTGS. The formation of methane was detected by Hewlett-Packard 6890 Series GC System (Ultra 1) with 25.0 m × 200 μm × 0.11 μm nominal column, helium gas as the carrier gas with flow rate of 20 mL/min at 75 kPa, and using Flame Ionization Detector (FID). The measurements were conducted at reaction temperatures of 100˚C to 400˚C.

3. Results and Discussion

3.1. Characterization of the Catalysts

All the prepared catalysts presented in **Table 1** were analyzed by XRD analysis in order to determine the pattern of diffractogram as well as to identify the active oxide phases which contribute to the good catalytic activity. All the XRD patterns towards the fresh $Ru/Mn/Al_2O_3$ catalysts over various calcination temperatures is shown in **Figures 1** and **2** while over various loading of Mn in **Figure 3**. The XRD pattern over fresh and used catalyst is displayed in **Figure 4**. The peak phases, 2θ values and d spacing value obtained from XRD are clearly discussed. The average crystallite sizes of the prepared catalysts are calculated from XRD peak broadening using the Scherrer equation (Equation (2)). The crystallite size observed was not significantly different among the prepared catalysts which around 95 nm - 114 nm.

From **Figures 1** and **2**, it can be seen that the phase-changes with the increasing of calcination temperatures from 400˚C to 700˚C and 900˚C to 1100˚C. XRD diffractogram of fresh (**Figure 1(a)**) $Ru/Mn\text{-}75/Al_2O_3$ catalyst for calcination temperature of 400˚C showed very low degree of crystallinity. All the peaks could be attributed

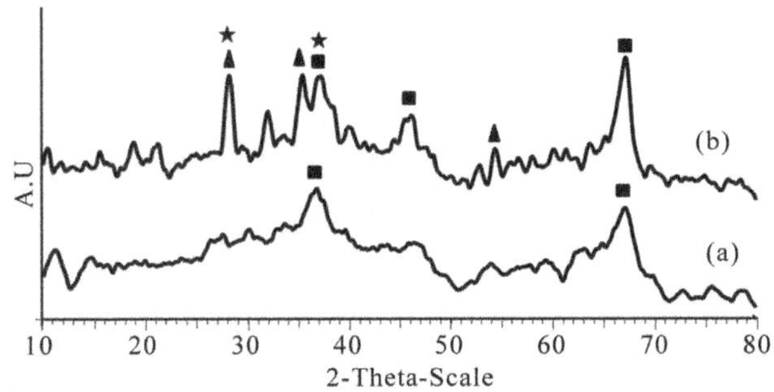

Figure 1. The overlay of XRD diffractograms of fresh $Ru/Mn\text{-}75/Al_2O_3$ catalysts calcined at (a) 400°C and (b) 700°C for 5 h.

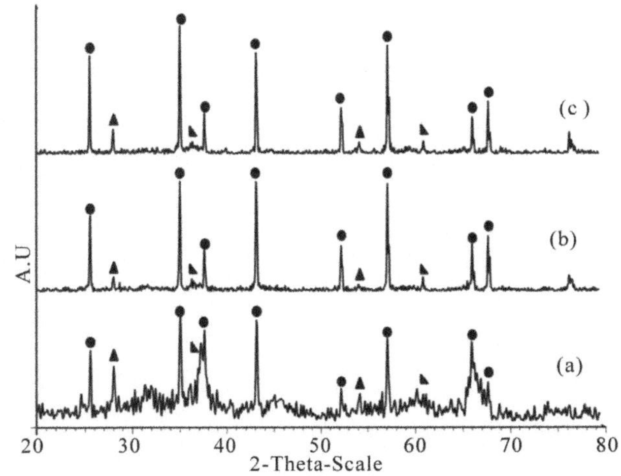

Figure 2. The overlay of XRD diffractograms of fresh $Ru/Mn\text{-}75/Al_2O_3$ catalysts calcined at (a) 900 °C, (b) 1000°C and (c) 1100°C for 5 h.

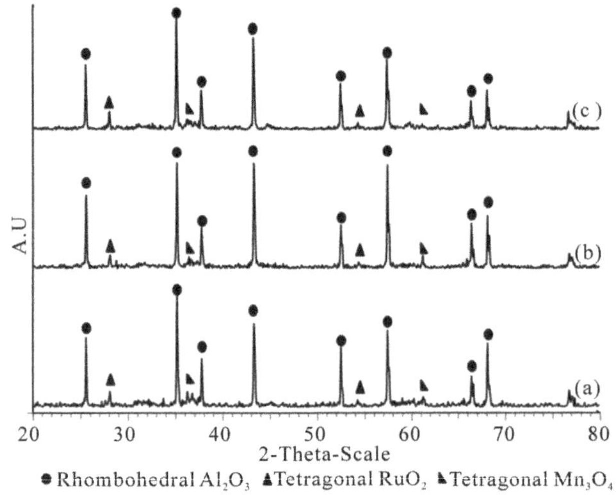

Figure 3. The overlay of XRD diffractograms of fresh (a) Ru/Mn-65/Al_2O_3; (b) Ru/Mn-75/Al_2O_3; and (c) Ru/Mn-85/Al_2O_3 catalysts calcined at 1000°C for 5 h.

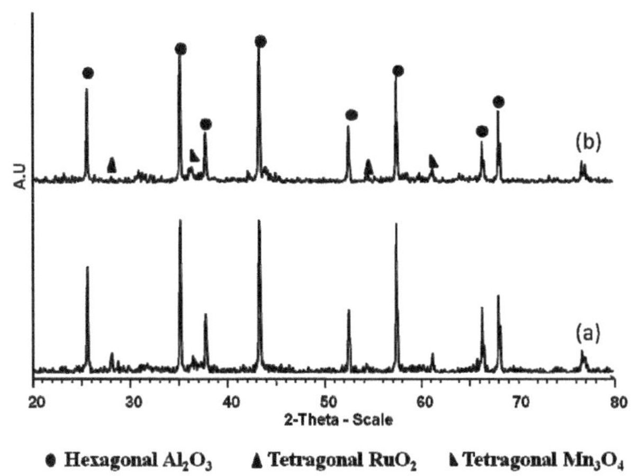

Figure 4. The overlay of XRD diffractograms of (a) fresh and (b) used Ru/Mn-75/Al_2O_3 catalyst calcined at 1000 °C for 5 h

Table 1. Ru/Mn/Al_2O_3 catalyst prepared by wetness impregnation method.

Catalysts	Calc. Temp. (°C)	Mn loading (wt%)	Ru loading (wt%)	Cryst. Size (nm)
Ru/Mn-65/Al_2O_3	1000	16.92	0.46	102
Ru/Mn-75/Al_2O_3 (a)	900	16.80	5.44	95
Ru/Mn-75/Al_2O_3 (b)	1000	16.47	7.41	106
Ru/Mn-75/Al_2O_3 (b)-used	1000	16.68	5.40	104
Ru/Mn-75/Al_2O_3 (c)	1100	16.05	3.19	114
Ru/Mn-85/Al_2O_3	1000	16.52	2.28	111

to the cubic alumina at 2θ around 37° and 67°. At such low temperature, the crystallinity of the sample could not be observed and no peaks can be assigned. This result is in line with the result obtained by Tang *et al.* [8] who claimed that the support only exhibited broad diffraction peaks when calcined at below 500°C and no other crystalline peaks was observed at low calcination temperature [9].

By increasing the calcination temperature of the catalyst to 700°C (**Figure 1(b)**), the intensity of those alumina peaks was slightly increased and became more profound [1]. This revealed that the intermediate crystallinity was also observed in this catalyst. The dominated Al_2O_3 cubic phase can be seen at similar position as calcination temperature of 400°C at 2θ values of 46.041° (I_{100}), 67.035° (I_{93}) and 37.641° (I_{47}) with d^*_{obs} values (Å) of 1.970, 1.394 and 2.387 (d^*_{ref} values (Å): 1.976, 1.397 and 2.383). Interestingly, new peaks attributable to RuO_2 species were observed at calcination temperature of 700°C. The three peaks of RuO_2 tetragonal phase were appeared at 2θ values of 28.124° (I_{100}), 35.112 (I_{72}) and 54.386° (I_{54}) with d^*_{obs} values (Å) of 3.170, 2.553 and 1.685 (d^*_{ref} values (Å): 3.177, 2.555 and 1.687). However, RuO_2 tetragonal phase at 28.124° was more intense due to the sharing of peaks with tetragonal MnO_2 which was also observed at 28.680° (I_{100}). Another one peak of $MnO_2$was assumed in the enveloped of Al_2O_3 peaks at 2θ of 37.441° (I_{50}) or d spacing values of 2.394 Å (d^*_{ref} values (Å): 2.40).

When the calcination temperature was increased to 900°C, the polycrystalline state of Ru/Mn-75/Al_2O_3 catalyst was detected. The Al_2O_3 cubic phase which was observed earlier in catalyst calcined at 400°C and 700°C had transformed into Al_2O_3 rhombohedral phase. The peaks were observed at 2θ values of 35.131° (I_{100}), 43.329°(I_{96}), 57.410° (I_{91}), 25.550° (I_{68}), 68.089° (I_{52}), 52.475° (I_{47}), 37.323° (I_{46}) and 66.398° (I_{35}) with d_{obs} values of 2.552, 2.087, 1.604, 3.484, 1.376, 1.742, 2.407 and 1.407 Å (d_{ref} values (Å): 2.552, 2.086, 1.602, 3.482, 1.375, 1.741, 2.381 and 1.405). The peaks assigned to the tetragonal RuO_2 remained similar and appeared sharply at 2θ values of 28.076° (I_{100}) and 54.965° (I_{54}) with d^*_{obs} values (Å) of 3.176 and 1.684 (d^*_{ref} values (Å): 3.177 and 1.687) as in catalyst calcined at 700°C. The tetragonal phase of$RuO_2$is thermostable and it can prevent the phase changing to occur even when calcined at higher temperature as suggested by Richardson *et al.* [10].

Furthermore, the manganese oxide species was appeared in the form of tetragonal Mn_3O_4 and detected at 2θ values of 61.872° (I_{60}) with d^*_{obs} values (Å) of 1.512 (d^*_{ref} values (Å): 1.542). However, the pattern shows a very small peak which was hardly distinguished from the background noise. The overlapping of Mn_3O_4diffraction peaks with those of the support peaks at 2θ of 36.466° (I_{100}) with d^*_{obs} values (Å) of 2.484 was also occurred. The transition of manganese oxide to another manganese oxide species depends on the certain temperature and atmosphere. Therefore, the Mn_3O_4 species is the most stable form within this temperature.

Upon calcination at 1000°C for 5h, well-definedsharp peaks were obtained representing the growth of the particle size for each species as shown in **Table 1**. The average crystallite size over Ru/Mn-75/Al_2O_3 catalyst calcined at 1000°C was about 106 nm which is 10% larger than the one that was calcined at 900°C. Besides, no overlapping peaks are observed indicating that the calcination temperature of 1000°C is enough for preparing highly pure crystalline of Ru/Mn-75/Al_2O_3 catalyst. On increasing the calcinations temperature to 1100°C, no phase changes were detected. Similar peaks that appeared on Ru/Mn-75/Al_2O_3 catalyst calcined at 1000°C can also be observed at Ru/Mn-75/Al_2O_3 catalyst calcined at 1100°C. The crystallinity for both catalysts seems to be similar to each other. However, the intensity of tetragonal RuO_2 especially at $2\theta = 28°$ was slightly increased.

Therefore, it can be summarized that the degree of crystallinity of all the catalysts improved as the calcination temperature was increased from 400°C to 1100°C This results suggested that the calcination temperature has an effect in rearranging the crystal lattice to be in ordered periodic arrangement to produce narrower and sharper peaks attributable to the crystalline phase. Jones *et al.* [11] claimed that catalysts calcined at temperatures above 500°C showed high crystallinity. However, high crystallinity phase was only observed for catalyst calcined at 1000°C and 1100°C, whereby catalyst calcined at 400°C showed amorphous phase and those calcined at 900°C showed polycrystalline phase.

The Ru/Mn-75/Al_2O_3 catalyst calcined at 1000°C was selected to be further analyzed by using XRD analysis towards its Mn loading since the intensity of RuO_2 over this catalyst is lower (**Figure 2(b)**) compared to other calcination temperature of 900°C (**Figure 2(a)**) and 1100°C (**Figure 2(c)**). Therefore, the XRD patterns towards various loading of Mn over Ru/Mn-75/Al_2O_3 catalyst calcined at 1000°C is illustrated in **Figure 3**. The loading used in this research are 65 wt%, 75 wt% and 85 wt%. The species obtained from Ru/Mn-75/Al_2O_3 catalyst was described above. Overall, the effect of Mn loading on the change of phase transformation over XRD analysis is not clearly distinguished. All the species are in stable phase even when Mn loading was increased or decreased by 10 wt% as well as the peaks position which remained unchanged. However, the intensity of the peaks makes some differences between the catalysts prepared. Obviously, the intensity of tetragonal Mn_3O_4 especially at 2θ of 61.72 (I_{100}) is higher (**Figure 3(b)**) compared to the catalysts with Mn loading of 85 wt% (**Figure 3(c)**) and 65 wt% (**Figure 3(a)**), respectively.

From these results, it can be concluded that there is some increment and reduction of species intensity with Mn loading variation. Therefore, the selection of metal loading is tremendously important in order to produce a catalyst that could enhance the CO_2 methanation reaction. The expected active species obtained from XRD analysis (**Figures 1-3**) are Mn_3O_4 in the form of tetragonal phase. It was also noted that the Al_2O_3 phase should be rhombohedral for the catalyst to possess catalytic activity towards CO_2 methanation reaction. Therefore, it can be suggested that the potential catalyst from these results is Ru/Mn-75/Al_2O_3 catalyst calcined at 1000°C.

The XRD spectra of fresh and used Ru/Mn-75/Al_2O_3 catalyst calcined at 1000°C is shown in **Figure 4**. From comparison, it can be suggestedthat there was a major differences in the diffraction patterns of fresh and used Ru/Mn-75/Al_2O_3 catalysts calcined at 1000°C. The two peaks of RuO_2 at 2θ of 28.076° (I_{100}) and 54.965° were disappeared. This may be due to the modification of the species that occurred in the matrix after the catalytic testing, whereby the active species have migrated into the bulk matrix of the catalyst or been adsorbed into the pores of the support due to the high porosity of alumina support. It is in good agreement with our results taken from EDX analysis (**Table 1**). About 7.41 wt% of Ru was detected on the surface of the fresh catalyst compared to the 5.40 wt% over used Ru/Mn-75/Al_2O_3(b) catalyst.

The reduction of some alumina peaks especially at 2θ of 57.410° over used Ru/Mn-75/Al_2O_3(b) catalyst can also be observed on the diffractogram. No significant difference can be noticed among the other species. From this XRD analysis results, the deactivation over this catalyst might occurred after catalytic testing (used catalyst) towards CO_2 methanation reaction since some species was no longer observed after exposing to high temperature during catalytic testing.

3.2. Catalytic Measurement

All the prepared catalysts were then tested on its catalytic activity using a fixed bed micro reactor coupled with FTIR and GC to determine the potential catalyst in this research. Several parameters were used in preparing the catalysts for the catalytic activity measurement such as different calcination temperatures and loading of Mn.

3.2.1. Effect of Calcination Temperatures

Ru/Mn-75/Al_2O_3 catalyst was prepared via impregnation method and aged in an oven for 24 hours before calcined the catalyst atfive different temperatures of 400°C, 700°C, 900°C, 1000°C and 1100°C. **Figure 5** shows the trend of CO_2 conversion catalyzed by the manganese oxide based catalyst calcined at different calcination temperaturesand the data of the catalytic activity is shown in **Table 2**. It can be seen that the CO_2 conversion over Ru/Mn-75/Al_2O_3 catalyst calcined at 400°C achieved 13.75% at reaction temperature of 200˚C. The percentage conversion keeps on increasing until it reaches its maximum point at 400˚C, whereby the percentage of CO_2 conversion is at the highest which is 90.53%.

Similar pattern of CO_2 conversion was also observed for Ru/Mn-75/Al_2O_3 catalyst calcined at 700˚C. Initially, the conversion of CO_2 gave only 15.87% at 200˚C but continuously increased until achieved 91.77% at 400˚C reaction temperature. However, there was no methane gas detected over the converted CO_2 from both catalysts at reaction temperature below 200˚C. The conversion of CO_2 was then slightly increased after calcined the similar catalyst at 900˚C. The maximum conversion achieved was approximately 99% but only 55.43% of methane gas was detected at 400˚C reaction temperature.

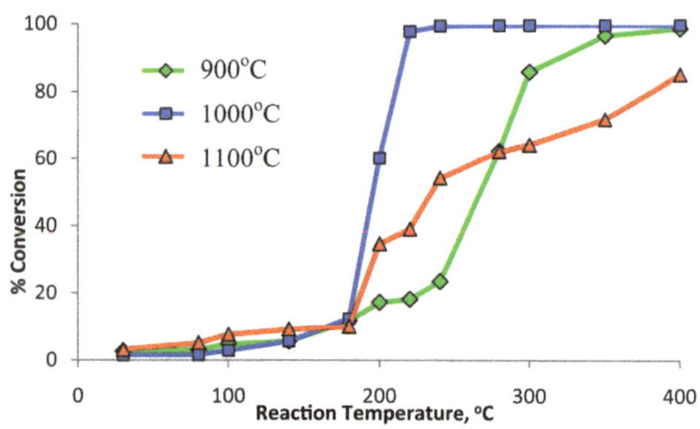

Figure 5. Percentage conversion of CO_2 over Ru/Mn-75/Al_2O_3 catalyst by using various calcination temperatures.

Table 2. Comparison of CO_2 conversion from methanation reaction over Ru/Mn-75/Al_2O_3 catalyst calcined at different temperatures.

Calcination Temperatures	Reaction Temp. (˚C)	% CO_2 Conversion	% CH_4 Formation
400	100	9.84	0
	200	13.75	0
	300	54.76	23.21
	400	90.53	44.23
700	100	7.22	0
	200	15.87	0
	300	67.88	34.76
	400	91.77	48.97
900	100	4.82	0
	200	17.29	10.33
	300	85.93	45.72
	400	98.88	55.43
1000	100	2.92	1.14
	200	60.21	57.84
	300	99.73	60.48
	400	99.91	67.62
1100	100	7.65	0
	200	34.71	18.71
	300	64.22	31.00
	400	85.23	44.67

This catalyst was then further subjected to higher calcination temperatures of 1000˚C. At reaction temperature of 200˚C, this catalyst already converting 60.21% of CO_2 and yielded 57.84% of methane. The maximum conversion up to 100% was observed for this catalyst at maximum reaction temperature of 400˚C with 67.62% of methane was formed. Further increased the calcination temperature to 1100˚C, the methane formation of Ru/Mn-75/Al_2O_3 catalyst was dropped to only 44.67% over 85.23% of CO_2 conversion at similar reaction temperature.

Therefore, it can be concluded that the percentage of CO_2 conversion increases as the calcination temperature increases up to 1000˚C. XRD analysis in **Figures 1** and **2** shows that the increasing calcination temperature of the catalyst had transformed the amorphous state of the catalyst to crystalline state. All the peaks become more profound, shaper and intense at their location. No overlapping peak was observed meaning that the pure metal oxide had obtained. It is indicate that there are no remaining impurities on the surface of the catalyst which will prevent the chemisorption of CO_2 and H_2 hence will increase the CO_2 methanation activity.

3.2.2. Effect of Mn Loadings

Table 3 compares the amount of Mn loading towards the percentage CO_2 conversion by the Ru/Mn/Al_2O_3 catalyst. All the catalysts were calcined at 1000˚C. The detailed trend plot of the catalytic activity is as shown in **Figure 6**. The performance of Ru/Mn-75/Al_2O_3 catalyst calcined at 1000˚C had been discussed before in Section 3.2.1. At reaction temperature of 200˚C, about 57.84% of methane was formed from 60.21% of CO_2 conversion.

By decreasing the Mn content to 65 wt%, the CO_2 conversion was decreased to 30.72% with 99% selectivity towards methane at reaction temperature of 200˚C. Over 100% CO_2 conversion, only 34.67% of methane was formed from the reaction at maximum temperature. At higher Mn loading of 85 wt%, about 13.63% of CO_2 was converted to methane at reaction temperature of 200˚C. Overall, there is no significant difference in CO_2 conversion throughout the reaction process except at reaction temperature of 200˚C (**Figure 6**). Therefore, Ru/Mn-75/Al_2O_3 (b) catalyst was selected as potential catalyst in this research since the CO_2 conversion of Ru/Mn-65/Al_2O_3 and Ru/Mn-85/Al_2O_3 catalysts did not exceed the conversion of Ru/Mn-75/Al_2O_3 catalyst at reaction

Table 3. Comparison of CO_2 conversion from methanation reaction over $Ru/Mn/Al_2O_3$ catalyst prepared by wetness impregnation method over various Mn loading and calcined at 1000°C for 5 h.

Catalysts	Reaction Temp. (°C)	% CO_2 Conversion	% CH_4 Formation
$Ru/Mn-65/Al_2O_3$	100	9.51	0
	200	30.72	30.52
	300	100	37.41
	400	100	34.67
$Ru/Mn-75/Al_2O_3$	100	2.92	1.14
	200	60.21	57.84
	300	99.73	60.48
	400	99.91	67.62
$Ru/Mn-85/Al_2O_3$	100	4.63	0
	200	15.88	13.63
	300	100	55.72
	400	100	50.43

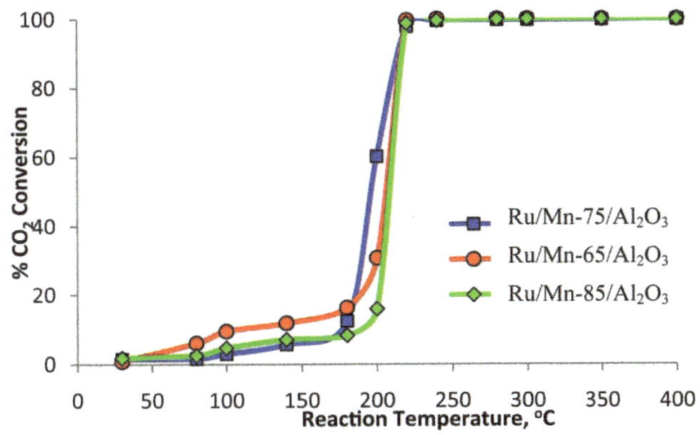

Figure 6. Percentage conversion of CO_2 over $Ru/Mn/Al_2O_3$ catalyst calcined at 1000°C for 5 hours with various compositions.

temperature 200°C.

It has been shown by XRD analysis thatthe catalyst with various Mn loading had similar pattern of diffractogram with similar phases and peaks position. However, the changes in intensity of Mn_3O_4 (**Figure 3(b)**) was observed over Mn based catalyst with various loading whereby the intensity of Mn_3O_4 species over $Ru/Mn-75/Al_2O_3$ catalyst is slightly higher compared to other catalyst. Therefore, it can be suggested that Mn_3O_4 is one of the active species which plays an important role to assist the methanation reaction on the catalyst surface.

4. Conclusion

XRD analysis can provide the information regarding phase changes and structure transformation, particle sizes as well as active species of the catalyst. From the results obtained, the $Ru/Mn-70/Al_2O_3$ calcined at 1000°C can be categorized as the potential catalyst among all the prepared catalysts. It can give 86.82% CO_2 conversion with 61.94% of methane formation. The structural study showed that the Mn_3O_4 compound in tetragonal structure was the active species which enhance the catalytic activity.

5. Acknowledgements

The authors are gratefully acknowledged to the Ministry of Science, Technology and Innovation, Malaysia for E-Science Fund Vote 4S002, Ministry of Higher Education, Malaysia for GUP Fund vote 01H57 and Universiti Teknologi Malaysia for financial support.

REFERENCES

[1] W. A. W. Abu Bakar and R. S. Toemen, "Catalytic Methanation Reaction over Supported Nickel-Ruthenium Oxide Base for Purification of Simulated Natural Gas," *Scientia Iranica*, Vol. 19, No. 3, 2012, pp. 525-534.

[2] W. A. W. A. Bakar, R. Ali and S. Toemen, "Catalytic Methanation Reaction over Supported Nickel-Rhodium Oxide for Purification of Simulated Natural Gas," *Journal of Natural Gas Chemistry*, Vol. 20, No. 6, 2011, pp. 585-594.

[3] W. A. Wan Abu Bakar, M. Y. Othman, R. Ali, C. K. Yong and S. Toemen, "The Investigation of Active Sites on Nickel Oxide Based Catalysts towards the *in-Situ* Reactions of Methanation and Desulfurization," *Modern Applied Science*, Vol. 3, No. 2, 2009, pp. 36-44.

[4] M. B. Kizling and F. Regali, "Preparation of Zirconium Oxide Particles for Catalyst Supports by the Microemulsion Technique. Characterization by X-Ray Diffraction, BET, SEM-EDX, FT-IR and Catalytic Tests," *Studies in Surface Science and Catalysis*, Vol. 118, 1998, pp. 495-504.

[5] D. A. Skoog, E. J. Holler and S. R. Crouch, "Principles of Instrumental Analysis," Thomson Brooks/Cole Corporation, Belmont, 2007.

[6] M. D. Nasr-Allah, "Physicochemical, Surface, and Catalytic Properties of Pure and Ceria-Doped Manganese/Alumina Catalysts," *Chinese Journal of Catalysis*, Vol. 29, No. 8, 2008, pp. 687-695.

[7] G. Perego, "Characterization of Heterogeneous Catalysts by X-Ray Diffraction Techniques," *Catalysis Today*, Vol. 41, No. 1-3, 1998, pp. 251-259.

[8] Q. Tang, X. Huang, C. Wu, P. Zhao, Y. Chen and Y. Yang, "Structure and Catalytic Properties of K-Doped Manganese Oxide Supported on Alumina," *Journal of Molecular Catalysis A: Chemical*, Vol. 306, No. 1-2, 2009, pp. 48-53.

[9] Q. Tang, X. Gong, C. Wu, Y. Chen, A. Borgna and Y. Yang, "Insights into the Nature of Alumina-Supported MnOOH and Its Catalytic Performance in the Aerobic Oxidation of Benzyl Alcohol," *Catalysis Communications*, Vol. 10, No. 7, 2009, pp. 1122-1126.

[10] J. T. Richardson, M. Garrait and J.-K. Hung, "Carbon Dioxide Reforming with Rh and Pt-Re Catalysts Dispersed on Ceramic Foam Supports," *Applied Catalysis A: General*, Vol. 255, No. 1, 2003, pp. 69-82.

[11] C. Jones, K. J. Cole, S. H. Taylor, M. J. Crudace and G. J. Hutchings, "Copper Manganese Oxide Catalysts for Ambient Temperature Carbon Monoxide Oxidation: Effect of Calcination on Activity," *Journal of Molecular Catalysis A: Chemical*, Vol. 305, No. 1-2, 2009, pp. 121-124.

Preparation of Olive Oil Doped SiO_2 via Sol-Gel Technique

Fawaz Nasro Saiof[1], Mohammed Alwan Hamza[2]
[1]Higher Institute for Laser Research, Damascus University, Damascus, Syrian
[2]College of Science, Baghdad University, Baghdad, Iraq

ABSTRACT

Olive oil doped Nanosilica is successfully prepared by wet chemical synthesis method. Samples were analyzed by a variety of techniques, including X-ray diffraction, FTIR, absorption and emission spectrometers to report the capability of Sol-Gel technology on preparing of silica monolith as a host material for Olive oil, and investigate the effect of converting Olive oil to a solid state on its optical properties. Some absorption bands of Olive oil are disappearing from absorption spectrum of doped silica. Only single strong fluorescence peak was appearing in fluorescence spectrum of pure Olive oil at wavelength around 678 nm which attributed to chlorophylls, while fluorescence spectrum to doped sample shows two strong fluorescence peaks at the wavelength 681 nm and 585 nm which attributed to chlorophylls and Vitamin E respectively. The doping process enhances fluorescence activity of Olive oil through enhancing intensity of the fluorescence peak corresponding to Vitamin E. Absorption and fluorescence spectrums to doped silica sample give a good indication in direction of using Sol-Gel technique to prepare for Olive oil doped SiO_2 as an optical active material.

Keywords: Sol-Gel; SiO_2; Olive Oil

1. Introduction

Virgin olive oil, due to its high price, is a target for adulteration with low price/quality oils. Fluorescence spectroscopy has been used to determine the authenticity of olive oils [1]. Many papers have been published in recent years on the use of fluorescence in vegetable oils [2-9]. Sayago *et al.* [7] applied fluorescence spectroscopy for detecting hazelnut oil adulteration in virgin olive oils. Kyriakidis and Skarkalis [8] showed that emission fluorescence spectra of virgin olive oils between 400 and 700 nm measured at excitation wavelength 365 nm have clear differences compared with the spectra of other vegetable oils. Virgin olive oils present two low peaks at 445 and 475 nm (related to conjugated hydroperoxides), one intense peak at 525 nm (due to Vitamin E) and another peak at 681 nm (due to chlorophylls) [8]. The very low intensity of the peaks at 445 and 475 nm is due to their large content on monounsaturated fatty acids and phenollic antioxidants, which provide more stability against oxidation. All refined oils show only one intense peak at 445 nm. It is due to fatty acid oxidation products formed as a result of the large percentage of polyunsaturated fatty acids present in these oils.

F. Guimet Vila [5,9] study Excitation-Emission Matrix EEMs of virgin and pure olive oils in the range λ_{ex} = 300 - 400 nm, λ_{em} = 400 - 695 nm. He shows that emission peak at the range 600 - 695 nm is attributed to chlorophylls (**Figure 1**), while emission peak at the range 400 - 600 nm is attributed to oxidation products and Vitamin E (**Figure 2**).

According to **Figures 1** and **2**, we can note that the two fluorescence peaks which attributed to chlorophylls and Vitamin E have a maximum intensity when excitation wavelength around 400 nm and 350 nm respectively. Also we can note that the second fluorescence peak is weaker in comparison with the first fluorescence peak.

The purpose of this paper is to report the capability of Sol-Gel technology on preparing of silica monolith as a host material for Olive oil, and investigate the effect of converting Olive oil to a solid state on its optical properties. Major optical properties are determined from analysis of absorption and emission spectra to both of pure Olive oil and prepared doped silica sample.

2. Experimental

2.1. Samples Preparation

Sol-Gel process was used to preparing Silicon nanostructures monolith because it's capable to producing highly

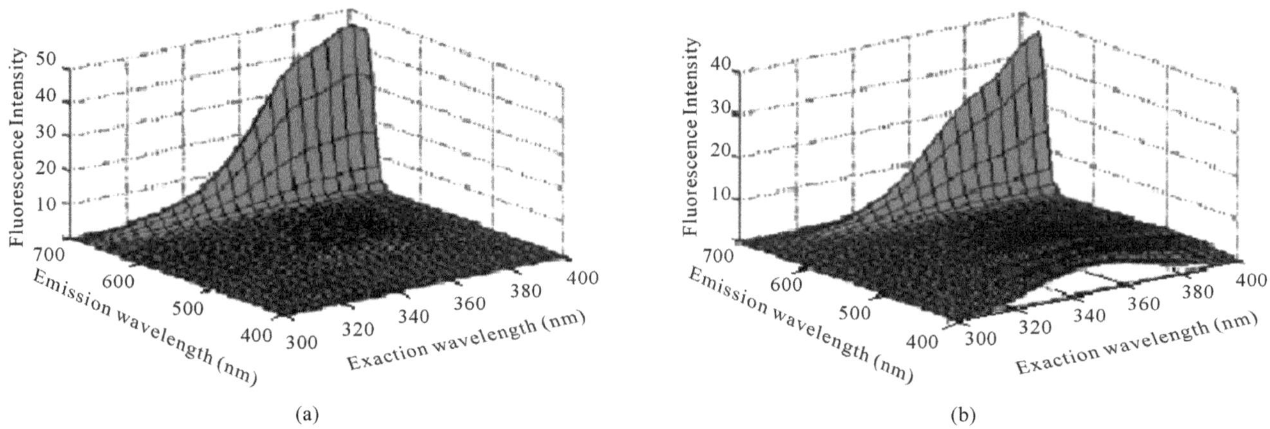

Figure 1. EEMs of a virgin (a) and a pure (b) olive oil between λ_{ex} = 300 and 400 nm, λ_{em} = 400 and 695 nm [5].

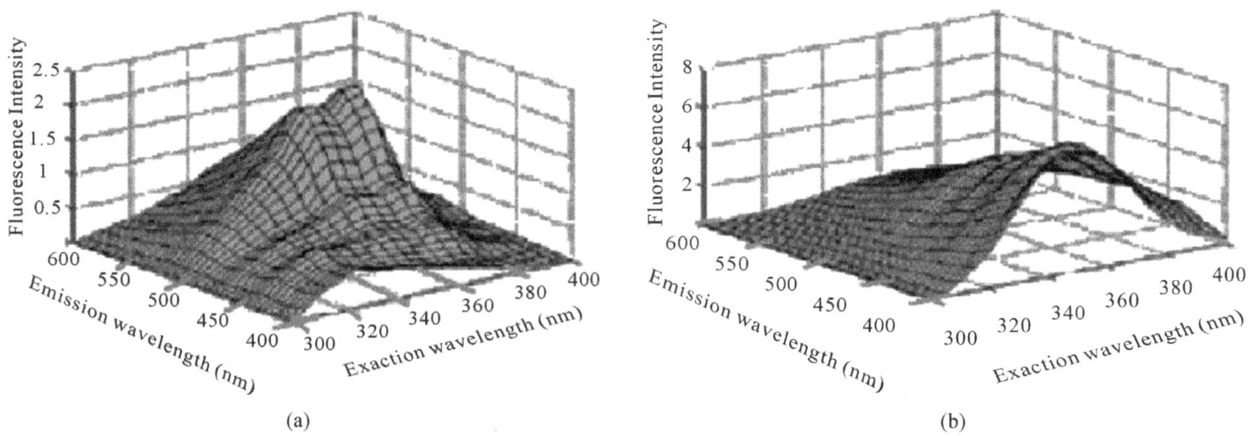

Figure 2. EEMs of a virgin (a) and a pure (b) olive oil between λ_{ex} = 300 and 400 nm, λ_{em} = 400 and 600 nm [5].

homogeneous metal oxide nanoparticles as well as its facility for dopant addition [10,11].

The doped and un-doped samples were synthesized by sol-gel method from tetraethylorthosilicate (TEOS) (Aldrich 98%), Ethanol (EtOH 99.9%) from GCC, hydrochlor ic acid (HCl, 34.5%) from BDH. Deionized water was used for the hydrolysis of (TEOS) and preparation of pure and doped SiO_2 sol. The reaction was performed at a cooler water jacket (*i.e.* the solution was cooled by water at temperature (10˚C - 15˚C). The amount of each chemical in this procedure was TEOS:H_2O:EtOH:HCl = 1:1:10:0.1 in molar ratio. All solution was prepared as follows: 1 mole of tetraethylorthosilicate (TEOS) and 1 moles of ethanol (EtOH) were mixed and stirred for 10 min. While stirring, 0.1 M catalysts in water were added dropwise to the solution until water to TEOS molar ratio of $R = 2$ are attained. For dopand samples, a 0.6 ml commercial Syrian Olive oil was used and added with ethanol before mixing with TEOS. All solutions were then stirred at room temperature for 2 hours further and aged for 24 hours before use.

After aging for 24 hour, first drying occurred for 2 hour at temperature 60˚C. Then samples left in room temperature without covers in order to permit solvent evaporation through the drying process. **Figure 3** shows a Snapshot for prepared doped sample.

2.2. Samples Characterization

Structural characterizations of the doped and un-doped SiO_2 samples were done by X-ray diffraction (XRD). θ - 2θ scans were recorded using ITAL-STRUCTURE diffractometer equipped. While Mid-IR spectra were obtained for the prepared samples using FT-IR spectrometer, Shimadzu, on KBr pellets of the samples.

Absorption spectra were measured at room temperature with TupCen UV-VIS Spectrometer. Emission spectra were measured at room temperature by using Solar-Lab mono chromatore. As excitation source we used the 405 nm/250 mW Laser diode. For comparison, further more absorption and emission spectra were measured to pure Olive oil.

3. Result and Discussion

Crystalline phases were not observed in the olive oil doped samples in agreement with the X-ray diffraction measurements as shown in **Figure 4.** The broad peaks presented of doped monoliths at about $2\theta \sim 24.2°$, show that the Olive oil can be randomly distributed in the SiO_2 network.

Fourier transform infrared (FTIR) absorption spectra in the range of 4000 - 400 cm^{-1} were taken. The FTIR spectrum of doped and un-doped samples was illustrated in **Figure 5**. The characteristics vibrational bands of silica were found in the FTIR spectra. Where; the absorption bands at about 472 cm^{-1}, 808 cm^{-1} and 1100 cm^{-1} which were due to bending, symmetric stretching and asymmetric stretching vibrations of Si-O-Si groups respectively. The band at around 960 cm^{-1} which is ascribed to stretching vibration of silanol (Si-OH) groups [12,13], indicates the amount of these groups in the silica network [12]. The amount of silanol groups is proportional with the ratio of condensation reaction of Sol-Gel method, and indicates that condensation reaction dose not completely finished.

Another two bands were appeared at around 1600 cm^{-1} and 3400 cm^{-1}. These two bands were the characteristics vibration of O-H bond in water molecules [12], and indicating that the drying process at 60°C dose not com-

Figure 3. Snapshot for prepared silica sample doped with Olive oil.

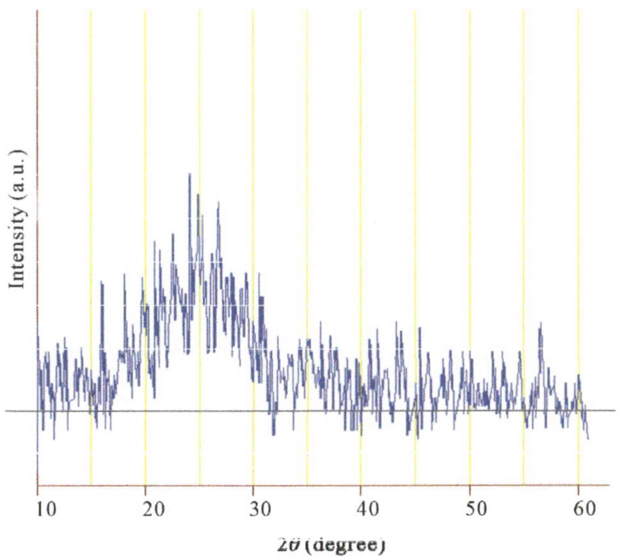

Figure 4. X-ray diffraction patterns of SiO_2 doped with Olive oil.

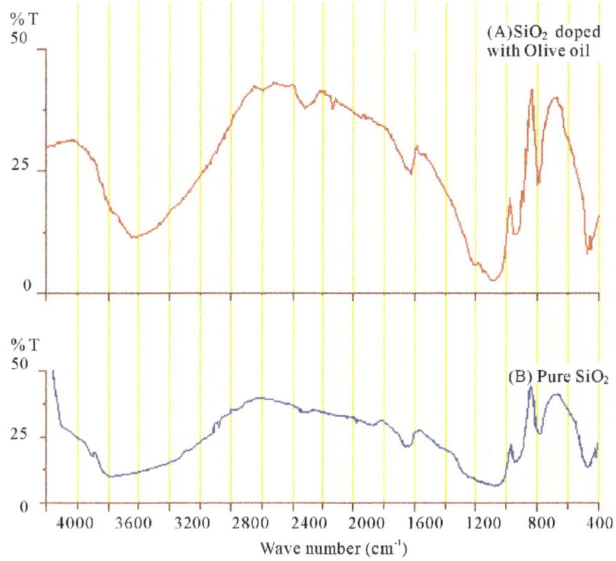

Figure 5. FTIR spectrums for samples: (A) SiO_2 doped with Olive oil and (B) Pure SiO_2.

pletely trap the water molecules from the pores of silica Xerogel network.

UV/Vis absorption spectrums in the range of 250 - 850 nm were taken at room temperature to pure olive oil and to prepared doped silica and presented in **Figure 6**. Absorption spectrum of pure olive oil has a strong two peak around wavelength at 409 nm and 668 nm, with three other weak peaks appears at wavelength around 506 nm, 534 nm and 610 nm [2,3]. All these peaks are disappearing from absorption spectrum of doped silica, in which it could see only wide absorption peak cover most of UV region. That's mean silica matrix effect on optical properties of Olive oil through changing of its absorption bands.

To determine the optical characteristics of the samples, photoluminescence measurements were carried out using the 405 nm/250 mW Laser Diode for excitation. Fluorescence spectrums in the range of 450 - 900 nm were taken at room temperature to pure Olive oil and to prepared silica doped with Olive oil and presented in **Figure 7**. The fluorescence spectrum of pure Olive oil in range 600 - 695 nm have a peak around the wavelength 678nm which attributed to chlorophylls [5,9]. The fluorescence peak at the range 450 - 600 nm does not appears which could be attributed that the fluorescence peak is so weak in comparison with other fluorescence peak.

The fluorescence spectrum of doped sample show two fluorescence peaks, first one in range of 600 - 695 nm appear at wavelength 681 nm which attributed to chlorophylls. The seconds peaks in range of 450 - 600 nm appear at wavelength 585 nm which attributed to Vitamin E [5,8,9]. Unlike with fluorescence spectrum of pure Olive oil, the intensity of the second peak becomes strong with

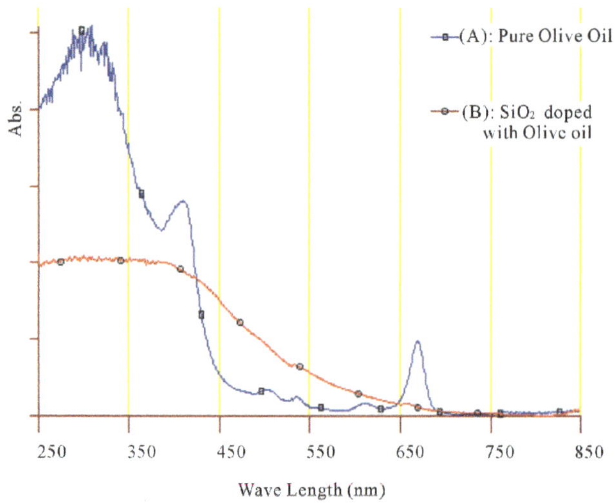

Figure 6. Absorption spectrums for: (A) Pure Olive oil and (B) SiO_2 doped with Olive oil.

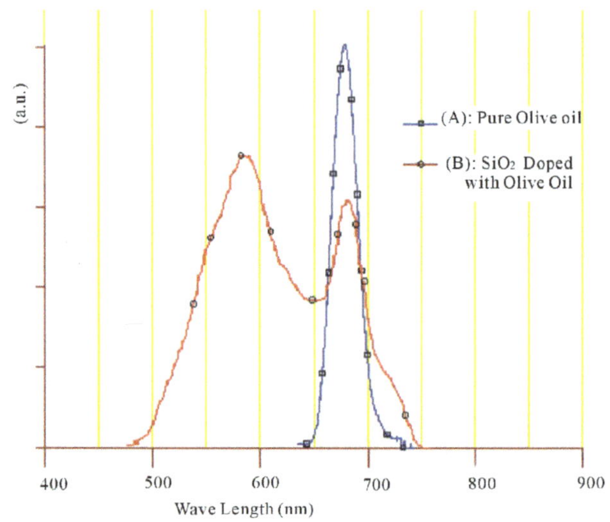

Figure 7. Fluorescence spectrums for: (A) Pure Olive oil and (B) SiO_2 doped with Olive oil.

wide bandwidth and somewhat larger than the intensity of first peak. That's mean the doping process enhanced of fluorescence activity of Olive oil especially in the range of 450 - 600 nm. According to **Figure 2**, it can note that SiO_2 host medium effect on the second fluorescence peak (attributed to Vitamin E) and shift the peak location intensity to wavelength around 585 nm.

4. Conclusions

The Olive oil doped Nanosilica is successfully prepared by wet chemical synthesis method. The doped prepared Nanosilica sample has amorphous structures, and seems to be transparent and unbroken which due to the suitable Sol-Gel parameters used. Some absorption bands of Olive oil are disappearing from doped silica absorption spectrum. Fluorescence spectrum of pure olive oil shows only a single strong fluorescence peak at 678 nm which attributed to chlorophylls, while fluorescence spectrum of doped sample shows two strong fluorescence peaks at the wavelength 681 nm and 585 nm which attributed to chlorophylls and Vitamin E respectively. A peak shift was recorded on the Vitamin E fluorescence peak.

The doping process enhances of fluorescence activity of Olive oil especially at the range of 450 - 600 nm through enhancing intensity of the fluorescence peak corresponding to Vitamin E.

Absorption and fluorescence spectrums to doped silica sample give a good indication in direction of using Sol-Gel technique to prepare of Olive oil doped SiO_2 as an optical active material.

5. Acknowledgements

This work was financially supported by High institute for laser technology and their applications/Damascus University.

REFERENCES

[1] J. Gracian, "Analysis and Characterization of Oils, Fats and Fat Products," Vol. 2, John Wiley & Sons, London, 1968.

[2] S. Fawaz, M. A. Gafary and L. A. Mamouly, "Determination of the Elements in the Olive Oil Responsible for the Luminescence Spectra Using a Green Laser," *AIP Conference Proceedings*, Vol. 1172, Cairo, 17-21 May 2009, pp. 20-23.

[3] F. Saiof, "Spectral Characteristic of Olive Oil and Some Other Vegetable Oils in Range 200 - 2000 nm," 2009.

[4] F. Guimet, R. Boqué and J. Ferré, "Cluster Analysis Applied to the Exploratory Analysis of Commercial Spanish Olive Oils by Means of Excitation-Emission Fluorescence Spectroscopy," *Journal of Agricultural and Food Chemistry*, Vol. 52, No. 22, 2004, pp. 6673-6679.

[5] F. Guimet, J. Ferré, R. Boqué and F. X. Rius, "Application of Unfold Principal Component Analysis and Parallel Factor Analysis to the Exploratory Analysis of Olive Oils by Means of Excitation-Emission Matrix Fluorescence Spectroscopy," *Analytica Chimica Acta*, Vol. 515, No. 1, 2004, pp. 75-85.

[6] F. Guimet, R. Boqué and J. Ferré, "Study of Oils from the Protected Denomination of Origin "Siurana" Using Excitation Emission Fluorescence Spectroscopy and Three-Way Methods of Analysis," *Grasas y Aceites*, Vol. 56, No. 4, 2005, pp. 292-297.

[7] A. Sayago, M. T. Morales and R. Aparicio, "Detection of Hazelnut Oil in Virgin Olive Oil by a Spectrofluorimetric Method," *The Journal European Food Research and Technology*, Vol. 218, No. 5, 2004, pp. 480-483.

[8] N. B. Kyriakidis and P. Skarkalis, "Fluorescence Spectra Measurement of Olive Oil and Other Vegetable Oils,"

Journal of AOAC International, Vol. 83, No. 6, 2000, pp. 1435-1439.

[9] F. Guimet "Olive Oil Characterization Using Excitation-Emission Fluorescence Spectroscopy and Three-Way Methods of Analysis," PH.D. Thesis, University of Rovira i Virgili, Tarragona, 2005.

[10] J. Livage, M. Henry and C. Sanchez, "Sol-Gel Chemistry of Transition Metal Oxides," *Progress in Solid State Chemistry*, Vol. 18, No. 4, 1988, pp. 259-342.

[11] J. P. Boilot, F. Chaput, T. Gacoin, *et al.*, "Organic-Inorganic Solids by Sol-Gel and Optical Applications," *Comptes Rendus de l'Académie des Sciences*, Vol. 322, No. 1, 1996, pp. 27-43.

[12] L. Yang, "Fabrication and Characterization of Microlasers by Sol-Gel Method," Ph.D. Thesis, California Institute of Technology, California, 2005.

[13] M. N. Muralidaran, C. A. Rasmitha and R. Ratheesh, "Photoluminescence and FTIR Studies of Pure and Rare Earth Doped Silica Xerogel and Aerogel," *Journal of Porous Materials*, Vol. 16, No. 6, 2009, pp. 635-640.

23

Effect of Switching on Metal-Organic Interface Adhesion Relevant to Organic Electronic Devices

Babaniyi Babatope[1*], Akogwu Onobu[2], Olusegun O. Adewoye[3], Winston O. Soboyejo[2,3]
[1]Department of Physics, Obafemi Awolowo University, Ile-Ife, Nigeria
[2]Department of Mechanical and Aerospace Engineering, Princeton University, Princeton, USA
[3]Materials Science and Engineering Department, African University of Science and Technology, Abuja, Nigeria

ABSTRACT

Considerable efforts are currently being devoted to investigation of metal-organic, organic-organic and organic-inorganic interfaces relevant to organic electronic devices such as organic light emitting diode (OLEDs), organic photovoltaic solar cells, organic field effect transistors (OFETs), organic spintronic devices and organic-based Write Once Read Many times (WORM) memory devices on both rigid and flexible substrates in laboratories around the world. The multilayer structure of these devices makes interfaces between dissimilar materials in contact and plays a prominent role in charge transport and injection efficiency which inevitably affect device performance. This paper presents results of an initial study on how switching between voltage thresholds and chemical surface treatment affects adhesion properties of a metal-organic (Au-PEDOT:PSS) contact interface in a WORM device. Contact and Tapping-mode Atomic Force Microscopy (AFM) gave surface topography, phase imaging and interface adhesion properties in addition to SEM/EDX imaging which showed that surface treatment, switching and surface roughness all appeared to be key factors in increasing interface adhesion with implications for increased device performance.

Keywords: AFM; Interface; Adhesion Force; Organic Electronics; Voltage Switching; Organic Memory Devices; Surface Treatment

1. Introduction

The investigation of interfaces between dissimilar organic-metal, organic-organic and organic-inorganic materials which are inherent in the devices made from them has been intensified in recent times. The interface phenomena are thus crucial towards the development, understanding and improvement of organic-based semiconductor electronic device [1-14] applications such as organic light emitting devices (OLEDs) [15-18], organic photovoltaic devices [19-21], organic thin film transistor devices [22-27] and organic spin electronic devices in which the transport and control of spin polarized information are represented [28-30]. The interface between the different materials that make them up determines the charge transport and charge injection efficiency, with implications for the performance of the devices.

Interfacial phenomena are particularly crucial towards the development and improvement of applications of these devices and in order to effectively investigate multilayer structured devices, the overall flexibility becomes very critical. This, in addition to their molecular nature, makes the study of organic thin films interfaces to be more intensified compared to inorganic semiconductors.

The type of interaction at the interface is either physical or chemical and progress in organic electronics requires their detail understanding [31]. Investigations of the chemical nature of interfaces are common in thin film characterization, but not much attention has been directed at measuring physical interaction until recently when advanced characterization tools are becoming more widely available. However, in a work reported on the nanoscale adhesion between organic-organic, organic-inorganic, and inorganic-inorganic thin film interfaces [32], the AFM technique was used in quantifying the interfaces, though with some limitations. The pull-off forces and surface parameters were measured and incorporated into theo-

*Corresponding author.

retical models for the estimation of surface energies. Obviously, therefore, the improvement of performance of organic electronic devices depends on a clear understanding of the principles underlying organic-film/metal-electrode interfaces.

Data-storage and switching applications of conjugated polymer-based devices in which, depending on the voltage-sweep direction, two different current-voltage (*I-V*) characteristics, and hence low and high conducting states are used, have been observed and reported (e.g. [33-39]). These devices have an associated memory effect for data-storage applications especially the bistable architectures (e.g. [21,26,38,40-47]). The bistable and multiple-layer stacking structures have therefore emerged as a viable technology for flexible, ultrafast, and ultrahigh-density memory devices in the field of organic electronics [48].

In a multilayer organic electronic device structure (**Figure 1**), charges often move from one layer to another by either injecting holes and/or electrons or transporting the same in one direction or the other. In the emissive layers, injected charges produce localized charge carrying species which move through the device under the influence of external field (bias) across the interface [49]. Charge injection transport efficiency through the interfaces is thus critical to device performance. However, there are problems of adhesion, morphological inhomogeneities and conductivity anisotropy at these interfaces which are not yet properly understood [50].

Charge transport efficiency could be enhanced by reducing the number of interfaces in organic devices. This is often done using the standard polymer processing method of blending in which organic materials, forming individual layers, are innovatively blended, thereby reducing the number of layers. This has been recognized as one of the possible ways forwarded for improved charge transport efficiency [51], as illustrated in **Figure 2**.

This investigation was therefore motivated by the need to explore additional interface reengineering techniques

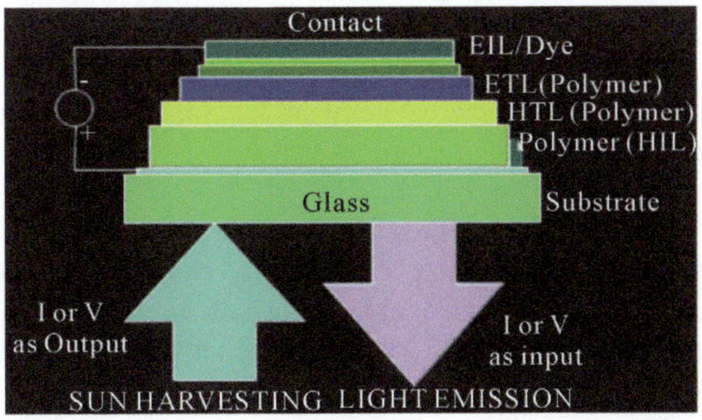

Figure 1. Typical device structure for sun harvesting (Polymer Solar Cells) and light emission (OLEDs). The figure shows the contacts and the emissive layers made up of different carefully selected conducting (conjugated) polymers.

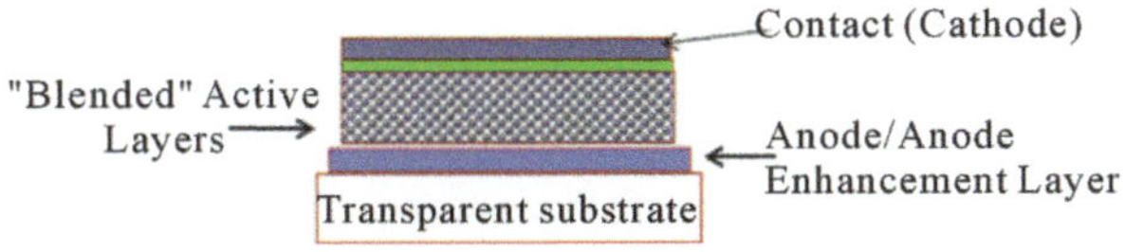

Figure 2. Device structure that reduced the number of interfaces using conventional polymer blending technique in which each phase would be expected to independently their different roles of charge transportation and injection.

using a combination of memory switching, adhesion and surface treatment. Also, this type of study could be incorporated into device performance improvement studies in view of recent reported improving power conversion efficiency of conjugated polymer-based solar cells [52], which remains largely unsatisfactory for large-scale commercial production in competition with its amorphous silicon-based devices. There are indications that charge transportation at the interface could be improved by a better understanding of the physical phenomena at the interface [32]. This has brought into the fore, an intensified investigation of the physical interaction between the dissimilar materials in direct contact, which is vital to sustainable device performance.

2. Experimental Details

2.1. Materials and Equipment

The materials used in this investigation include AL 4083 PEDOT:PSS suitable for OLED fabrication (H. C. Starck, MA, Newton, USA) as the organic layer, n-doped silicon wafer (Eagle-Picher, Miami, OK, USA) as the substrate, gold (Au) as the metal contact layer (Alfa Aesar, USA), hydrochloric acid (Aldrich, USA) for the surface treatment of the organic layer, Hydrofluoric acid (Aldrich, USA) and standard assorted reagents for substrate cleaning. The thermal deposition of the gold contact on the organic layer was carried out using an Edwards E306A deposition system (Edwards, Sussex, UK), the switching was done with an HP semiconductor parameter analyzer (HP 4145B) for the current-voltage measurement, etched silicon contact AFM tips were purchased from Veeco Instruments (Woodbury, NY, USA), Digital Instruments Dimension 3000 Atomic Force Microscope (AFM) (Digital Instruments, Plainview, NY), Scanning Electron Microscope (SEM-EDX) (Philips FEI XL30 FEG-SEM, Hillsboro, USA), a standard Fishers ultrasonic bath, an ultra violet ozone cleaning system and a drying chamber were all used in the course of this study.

2.2. Experimental Procedure

A Write Once Read Many times (WORM) device was fabricated by direct spin-casting of PEDOT:PSS on a suitably cleaned n-Si substrate followed by vacuum deposition of Au Contact. The cleaning procedure for the substrate includes the use of a combination of different solvents including acetone, trichloroethylene (TCE), isopropanol, hydrofluoric acid (HF), a detergent and deionized water under different temperature conditions in an ultrasonic bath. The substrate was then transferred to an ultra violet ozone cleaning system for 5 min for further cleaning after which it was transferred to a glove box for safe keeping, ready for spin coating of the organic layer.

A thin layer of PEDOT:PSS (about 500 nm) determined based on standard calibration curves obtained by plotting series of spinning speeds, accelerations and thicknesses and after drying (H. C. Starck, Newton, MA USA suppliers of PEDOT:PSS) was then deposited on the substrate by spin coating in ambient air condition followed by baking in an oven at 200˚C for about 60 min. The sample was again transferred to a glove box ready for the Au deposition. Some of the spin-coated samples were chemically treated with dilute HCl (0.1 N) prior to deposition of the Au contact (approximately 100 nm, preceded by a thin layer, about 5 nm, of Cr) which was carried out by thermal evaporation under a vacuum of 1×10^{-6} Torr (1 Torr = 133.32 Pa).

A selection of the Au/PEDOT:PSS/Si device was quasistatically switched or "blown" using a pulsed voltage ramp with 10 ms long 100 mV steps and applied for 0.5 - 4 ms as necessary The switching was carried out by use of current transients to change the polymer fuse (PEDOT:PSS) from a conducting ("1" or "ON") to a nonconducting ("0" or "OFF") state. Quasistatic (continuous) condition (J~10 A/cm^2): 0 V - 10 V - 0 V and rapid voltage pulsed transient condition (J~1 kA/cm^2): 10 V, 2 μs were used for this purpose [53,54].

The Au contact was deposited either before or after the polymer surface treatment and was carefully removed by a peeling process before and after switching. The same was done before and after polymer surface treatment. Phase imaging, surface topography and force calibration curves were obtained for the different surfaces by AFM in contact and tapping modes to quantify the physical interaction at the polymer interface using Au-coated cantilever tips. Interaction force response between the sample and cantilever tip [32,55] was measured for each sample. The surface roughness for each surface was recorded and related to adhesion data as measured by the deflection as recorded from the interaction between Au-coated cantilever tip and the polymer surface.

The differently deposited, treated or untreated, switched or unswitched surfaces before or after peeling the Au-contact, were separately investigated by SEM/EDX to qualitatively correlate the effectiveness of the peeling process, and the relative degree of polymer-metal interface adhesion. These tips were coated with Au the complementary materials that make direct contact with the surfaces. The purpose is also to investigate whether part of the polymer peeled with the gold contact.

3. Result and Discussion

Figures 3 and **4** show the surface topography (**Figures 3(a)-(d)**) as determined by the degree of roughness obtained in the tapping mode; and the force calibration plots (**Figures 3(a)-(d)**) obtained in contact mode for the unswitched (**Figures 4(a)** and **(c)**) and switched (**Figures 3(b)** and **(d)**) devices when untreated (**Figure 3**) and treated (**Figure 4**). The values for the respective roughness are as shown in **Table 1**.

Analysis of the force calibration curves shows that in the unswitched mode, when untreated (**Figure 3(b)**), the Au-coated cantilever tips dragged on the polymer surface prior to disengagement whereas, after switching, the disengagement of the tip was without dragging (**Figure 3(d)**). This was accompanied with a reduced adhesion force by almost 50%. Upon surface treatment, this effect combined with that of switching to give the highest adhesion force in the samples under consideration with a much greater dragging prior to cantilever tip disengagement. This appears to suggest that surface treatment and switching are significant to improved interface adhesion between the Au contact and the polymer. This is expected to lead to an improvement in device performance; the next in the on-going work.

The summary of the deflection on the force calibration curve (a direct measure of adhesion force) as estimated directly from the calibration plots under different surface treatment and switching conditions, is shown in **Table 2**.

Figures 5 and **6** show the SEM/EDX images for an untreated surface in the unswitched state (**Figure 5**) and in the treated and switched state (the two extremes). **Figure 5** indicates that peeling the Au contact from the PEDOT:PSS surface was "harder" with more of the Au remaining adhered to the polymer under similar peeling conditions.

This qualitatively suggests that adhesion force would be high. However, upon surface treatment (which is expected to make the surface to be more even, due to removal of surface peat, hence minimizing conductivity anisotropy and heterogeneity at the interface), the Au peeled off neatly leaving a smooth polymer surface, (**Figure 6**).

This study is significant in the quest to finding ways of improving the lifetimes of organic electronic devices through interface reengineering which is vital to better device performance. It has also qualitatively and quantitatively revealed the possibility of using this technique in combination with established packaging (or encapsulating methods to significantly contribute to general improvement in device performance). It was observed that peeling the gold contact in the untreated, unswitched

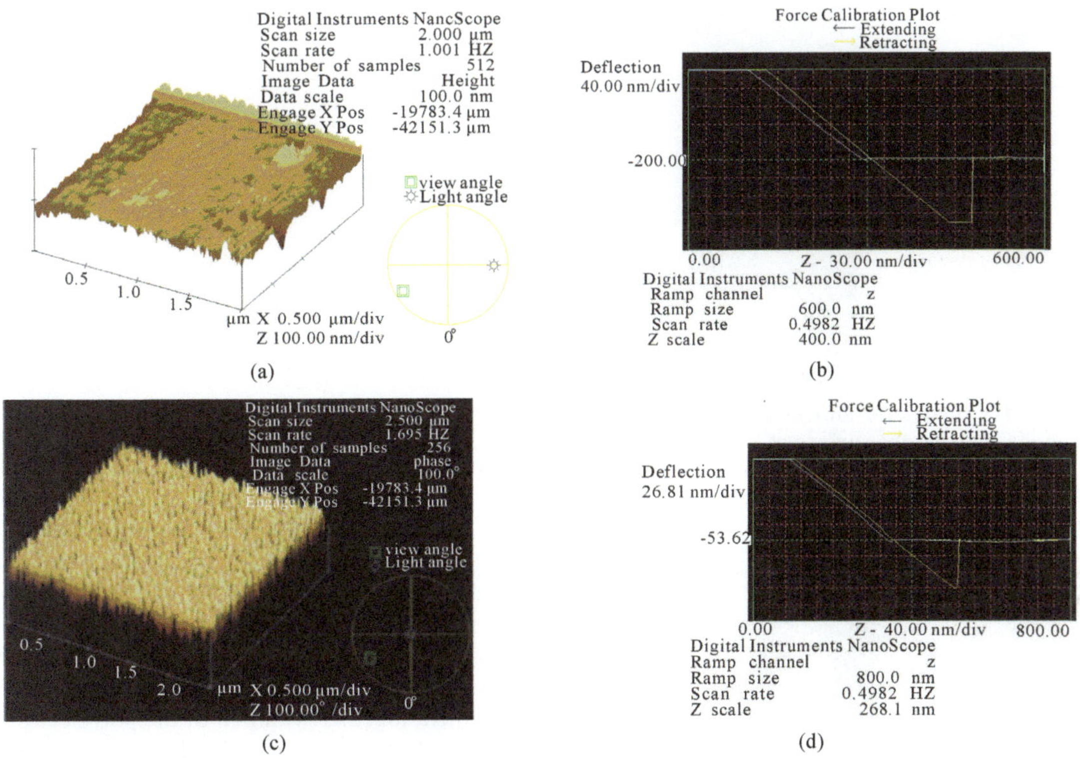

Figure 3. Surface roughness ((a) and (c)) and force displacement curves ((b) and (d)) for untreated PEDOT:PSS on silicon substrate when in the unswitched ((a), (b)) and switched ((c), (d)) modes. (a) Surface roughness in the untreated and unswitched mode; (b) Force-displacement curve in untreated and unswitched mode; (c) Surface roughness in the untreated and switched mode; (d) Force-displacement curve in the untreated and switched mode.

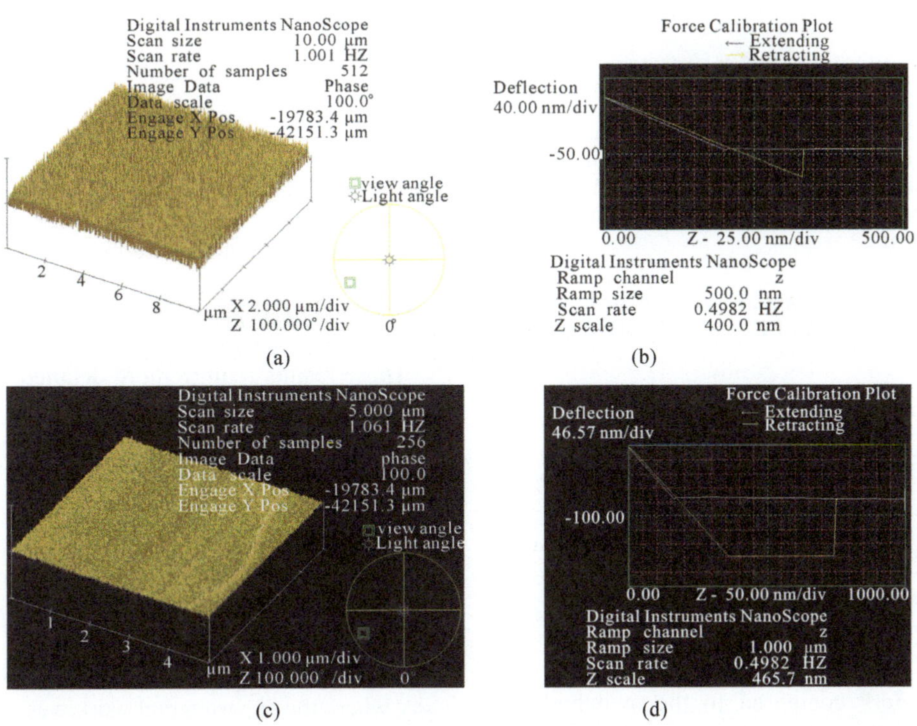

Figure 4. Surface roughness ((a) and (c)) and force displacement curves ((b) and (d)) for untreated PEDOT:PSS on silicon substrate when in the unswitched ((a), (b)) and switched ((c), (d)) modes. (a) Surface roughness in the treated and unswitched mode; (b) Force-displacement curve in the treated and unswitched mode; (c) Surface roughness in the treated and switched mode; (d) Force-displacement curve in the treated and switched mode.

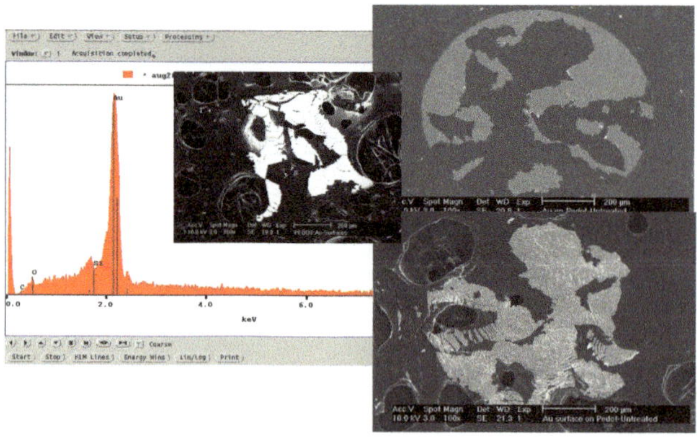

Figure 5. SEM/EDX images for an Untreated sample in the unswitched mode showing the surface of PEDOT:PSS after peeling off the Au contact (top and bottom right), the Au surface (inset) and the EDX for the contact showing that no polymer adhered to it.

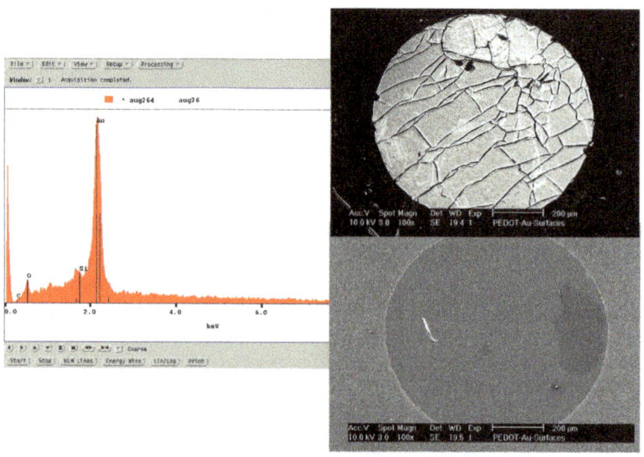

Figure 6. SEM/EDX images for an treated sample in the switched mode showing the surface of PEDOT:PSS after peeling off the Au contact (bottom right), the Au surface (top right) and the EDX for the contact showing that no polymer adhered to it.

Table 1. Summary data showing the effect of surface treatment and switching on the Au-PEDOT:PSS interface adhesion.

	Unswitched	Switched	Adhesion Force
Untreated	Rough (13.667 nm)	Smooth (5.609 nm)	Increased
Treated	Rough (9.739 nm)	Smooth (1.434 nm)	Increased
Adhesion Force	Not Significant	Significant Increase	

Table 2. Summary of adhesion data showing deflection values.

Untreated		Treated	
Unswitched	140.00 nm	Unswitched	70.00 nm
Switched	78.00 nm	Switched	185.50 nm
Unswitched		**Switched**	
Treated	70.00 nm	Treated	185.50 nm
Untreated	140.00 nm	Untreated	78.00 nm

sample was "harder" compared to the switched and treated samples that peeled more easily with the highest interface adhesion. Since this is more or less an exploratory study, the concept of switching especially the effect of treating the surface with different reagents requires further detailed investigation.

The quantitative analysis of the micrographs consistently showed convincingly that there was no evidence of the presence of the polymer, PEDOT:PSS, on the gold surface as all the specimens investigated showed that each time the Au contact was peeled, no polymer adhered to it (**Figures 5** and **6**). The adhesive force at the interface is thought to be a combination of Van Der Waal and electrostatic forces in view of the involvement of different interacting ions from the reagent used in surface treatment together with that of the polyelectrolyte dopant in PEDOT. The physics and chemistry of the interaction are thus very crucial to further understanding of this phenomenon as it would complement this initial result of the study.

4. Conclusions

From this preliminary investigation, it can be suggested that

1) Surface treatment, switching and surface roughness all appear to be key factors in increasing interface adhesion, hence device performance.

2) Switching an untreated PEDOT:PSS surface resulted in reduced interface adhesion.

3) Switching a surface-treated specimen significantly increased interface adhesion most probably because of reduced surface roughness.

These results require more detailed investigations, especially in relation to the actual device performance.

5. Acknowledgements

The authors wish to acknowledge the invaluable assistance offered by staff of the nano-fabrication laboratory, Department of Electrical Engineering, Princeton University, New Jersey, USA. BB acknowledges the authorities of Obafemi Awolowo University for kindly granting leave of absence for the Fellowship at Princeton University where the experimental work was carried out.

REFERENCES

[1] M. Grobosch and M. Knupfer, "Electronic Properties of

Organic Semiconductor/Electrode Interfaces: The Influence of Contact Contaminations on the Interface Energetic," *The Open Applied Physics Journal*, Vol. 4, No. 1, 2011, pp. 8-18.

[2] W. Brütting, "Physics of Organic Semiconductors," WILEY-VCH Verlag, Weinheim, 2005.

[3] H. Klauk, "Organic Electronics," WILEY-VCH Verlag, Weinheim, 2007.

[4] K, Walzer, B. Maennig, M. Pfeiffer and K. Leo, "Highly Efficient Organic Devices Based on Electrically Doped Transport Layers," *Chemical Reviews*, Vol. 107, No. 4, 2007, pp. 1233-1271.

[5] N. Koch, "Organic Electronic Devices and Their Functional Interfaces," *ChemPhysChem*, Vol. 8, No. 10, 2007, pp. 1438-1455.

[6] H. Ishii, K. Sugiyama, E. Ito and K. Seki, "Energy Level Alignment and Interfacial Electronic Structure at Organic/Metal and Organic/Organic Interfaces," *Advanced Materials*, Vol. 11, No. 8, 1999, pp. 605-608.

[7] A. Kahn, N. Koch and W. Gao, "Electronic Structure and Electrical Properties of Interfaces between Metals and π-Conjugated Molecular Films," *Journal of Polymer Science Part B*, Vol. 41, No. 21, 2003, pp. 2529-2548.

[8] J. C. Scott, "Metal-Organic Interface and Charge Injection in Organic Electronic Devices," *Journal of Vacuum Science & Technology A*, Vol. 21, No. 3, 2003, pp. 521-533.

[9] M. Knupfer and H. Peisert, "Electronic Properties of Interfaces between Model Organic Semiconductors and Metals," *Physica Status Solidi A*, Vol. 201, No. 6, 2004, pp. 1055-1074.

[10] M. Knupfer and G. Paasch, "Origin of the Interface Dipole at Interfaces between Undoped Organic Semiconductors and Metals," *Journal of Vacuum Science & Technology A*, Vol. 23, No. 4, 2005, pp. 1072-1077.

[11] N. Koch, "Energy Levels at Interfaces between Metals and Conjugated Organic Molecules," *Journal of Physics: Condensed Matter*, Vol. 20, No. 18, 2008, Article ID: 184008.

[12] S. Braun, W. R. Salaneck and M. Fahlman, "Energy-Level Alignment at Organic/Metal and Organic/Organic Interfaces," *Advanced Materials*, Vol. 21, No. 14-15, 2009, pp. 1450-1472.

[13] J. Hwang, A. Wan and A. Kahn, "Energetics of Metal-Organic Interfaces: New Experiments and Assessments of the Field," *Materials Science and Engineering R*, Vol. 64, No. 1-2, 2009, pp. 1-31.

[14] Y. Gao, "Surface Analytical Studies of Interfaces on Organic Semiconductor Devices," *Materials Science and Engineering R*, Vol. 68, No. 3, 2010, pp. 39-87.

[15] C. W. Tang and S. A. VanSlyke, "Organic Electroluminescent Diodes," *Applied Physics Letters*, Vol. 51, No. 12, 1987, pp. 913-915.

[16] J. Blochwitz, T. Fritz, M. Pfeiffer, *et al.*, "Interface Electronic Structure of Organic Semiconductors with Controlled Doping Levels," *Organic Electronics*, Vol. 2, 2001, pp. 97-104.

[17] K. Müllen and U. Scherf, "Organic Light Emitting Devices," WILEY-VCH Verlag, Weinheim, 2006.

[18] Y. Fukuda, W. Teruichi, W. Takeo, M. Satoshi and T. Masami, "An Organic LED Display Exhibiting Pure RGB Colors," *Synthetic Metals*, Vol. 111-112, 2000, pp. 1-6.

[19] A. Bandhopadhyay and A. J. Pal, "Large Conductance Switching and Memory Effects in Organic Molecules for Data—Storage Applications," *Applied Physics Letters*, Vol. 82, No. 8, 2003, pp. 1215-1217.

[20] A. Bandhopadhyay and A. J. Pal, "Large Conductance Switching and Binary Operation in Organic Devices: Role of Functional Groups," *Journal of Physical Chemistry B*, Vol. 107, No. 11, 2003, pp. 2531-2536.

[21] L. P. Ma, J. Liu and Y. Yang, "Organic Electrical Bistable Devices and Rewritable Memory Cells," *Applied Physics Letters*, Vol. 80, No. 16, 2002, pp. 2997-2999.

[22] G. Horowitz, "Organic Thin Film Transistors: From Theory to Real Devices," *Journal of Materials Research*, Vol. 19, No. 7, 2004, pp. 1946-1962.

[23] W. Clemens, I. Fix, J. Ficker, A. Knobloch and A. Ullmannn, "From Polymer Transistors toward Printed Electronics," *Journal of Materials Research*, Vol. 19, No. 7, 2004, pp. 1963-1971.

[24] S. Scheinert and G. Paasch, "Fabrication and Analysis of Polymer Field Effect Transistors," *Physica Status Solidi A*, Vol. 201, No. 6, 2004, pp. 1263-1301.

[25] F. Cicoira and C. Santato, "Organic Light Emitting Field Effect Transistors: Advances and Perspectives," *Advanced Functional Materials*, Vol. 17, No. 17, 2007, pp. 3421-3427.

[26] C. D. Dimtrakopoulos and D. J. Mascaro, "Organic Thin Film Transistors: A Review of Recent Advances," *IBM Journal of Research and Development*, Vol. 45, No. 1, 2001, pp. 11-27.

[27] X. Qui, Y. Hu, G. Dong, L. Wang, J. Xei and Y. Ma, "Water Effect on the Stability of Organic Thin Film Field Effect Transistors (OTF-FET)," *Applied Physics Letters*, Vol. 83, 2003, pp. 1644-1646.

[28] V. Dediu, M. Murgia, F. C. Matacotta, C. Taliani and S.

Barbanera, "Room Temperature Spin Polarized Injection in Organic Semiconductor," *Solid State Communications*, Vol. 122, No. 3, 2002, pp. 181-184.

[29] W. J. M. Naber, S. Faez and W. G. av der Wiel, "Organic Spintronics," *Journal of Physics D: Applied Physics*, Vol. 40, No. 12, 2007, pp. R205-R228.

[30] Z. H. Xiong, D. Wu, Z. V. Vardeny and J. Shi, "Giant Magneto-Resistance in Organic Spin-Valves," *Nature*, Vol. 427, 2004, pp. 821-824.

[31] C. Santato and F. Rosei, "Organic/Metal Interfaces Seeing Both Sides, News and Views," *Nature Chemistry*, Vol. 2, No. 5, 2010, p. 344.

[32] T. Tong, B. Babatope, S. Admassie, J. Meng, O. Akwogu, W. Akande and W. O. Soboyejo, "Adhesion in Organic Electronic Structures," *Journal of Applied Physics*, Vol. 106, No. 8, 2009, Article ID: 083708.

[33] J. R. de Lima, L. O. Péres, J. R. Garcia, J. Gruber and I. A. Hümmelgen, "Poly(acettoxy-p-phenylene vinylene) Based Diode with a Soft Breakdown," *Solid-State Electronics*, Vol. 44, No. 33, 2000, pp. 565-570.

[34] H. Carchano, R. Lacoste and Y. Segui, "Bistable Electrical Switches in Polymer Thin Films," *Applied Physics Letters*, Vol. 19, No. 19, 1971, pp. 414-415.

[35] B. Villeret and M. Nechtschein, "Memory Effects in Conducting Polymers," *Physical Review Letters*, Vol. 63, No. 12, 1989, pp. 1285-1287.

[36] T. C. Chung, J. H. Kaufman, A. J. Heeger and F. Wudl, "Charge Storage in Doped Poly(Thiophene): Optical and Electrochemical Studies," *Physical Review B*, Vol. 30, No. 2, 1984, pp. 702-710.

[37] V. Cimrová and D. Neher, "Anomalous Electrical Characteristics, Memory Phenomena and Microcavity Effects in Polymeric Light-Emitting Diodes," *Synthetic Metals*, Vol. 76, No. 1-3, 1996, pp. 125-128.

[38] D. Ma, M. Aguiar, J. A. Freire and I. A. Hümmelgen, "Organic Reversible Switching Devices for Memory Applications," *Advanced Materials*, Vol. 12, No. 14, 2000, pp. 1063-1066.

[39] D. M. Taylor and C. A. Mills, "Memory Effect of the Current-Voltage Characteristic of a Low-Band Gap Conjugated Polymer," *Journal of Applied Physics*, Vol. 90, No. 1, 2001, pp. 306-309.

[40] S. Moller, S. R. Forrest, C. Perlov, W. Jackson and C. Tausig, "Electrochromic Conductive Polymer Fuses for Hybrid Organic/Inorganic Semiconductor Memories," *Journal of Applied Physics*, Vol. 94, No. 12, 2003, pp. 7811-7819.

[41] H. L. Gomes, A. R. V. Benvenho, D. M. de Leeuw, M. Colle, P. Stallinga, F. Verbakel and D. M. Taylor, "Switching in Polymeric Resistance Random-Access Memories (RRAMS)," *Organic Electronics*, Vol. 9, No. 1, 2008, pp. 119-128.

[42] H. S. Majumdara, A. Bolognesi and A. J. Pal, "Switching and Memory Devices Based on a Polythiophene Derivative for Data-Storage Applications," *Synthetic Metals*, Vol. 140, No. 2-3, 2004, pp. 203-206.

[43] J. R. Koo, S. W. Pyo, J. H. Kim, S. Y. Jung, S. S. Yoon, T. W. Kim, Y. H. Choi and Y. K. Kim, "Current-Voltage (*I-V*) Characteristics of the Molecular Electronic Devices Using Various Organic Molecules," *Synthetic Metals*, Vol. 156, No. 2-4, 2006, pp. 298-301.

[44] J. S. Chen and D. G. Ma, "Single-Layer Organic Memory Devices Based on N,N′-Di(Naphthalene-l-yl)-N,N′-Diphenyl-benzidine," *Applied Physics Letters*, Vol. 87, No. 2, 2005, Article ID: 023505.

[45] T. Graves-Abe and J. C. Sturm, "Programmable Organic Thin-Film Devices with Extremely High Current Densities," *Applied Physics Letters*, Vol. 87, No. 13, 2005, pp. 133502-133503.

[46] Y. S. Lai, C.-H. Tu, D.-L. Kwong and J. S. Chen, "Bistable Resistance Switching of Poly(N-Vinylcarbazole) Films for Nonvolatile Memory Applications," *Applied Physics Letters*, Vol. 87, No. 12, 2005, p. 122101.

[47] T. Oyamada, H. Tanaka, K. Matsushige, H. Sasabe and C. Adachi, "Switching Effect in Cu:TCNQ Charge Transfer-Complex Thin Films by Vacuum Codeposition," *Applied Physics Letters*, Vol. 83, No. 6, 2003, pp. 1252-1255.

[48] Y. Yang, J. Ouyang, L. Ma, R. J.-H. Tseng and C.-W. Chu, "Electrical Switching and Bistability in Organic/Polymeric Thin Films and Memory Devices," *Advanced Functional Materials*, Vol. 16, No. 8, 2006, pp. 1001-1014.

[49] G. G. Wallace, P. C. Dastoor, D. L. Officer and C. O. Too, "Conjugated Polymers: New Materials for Photovoltaics," *Chemical Innovation-Enabling Science*, Vol. 30, No. 1, 2000, pp. 14-22.

[50] M. Fahlman and W. R. Salaneck, "Surfaces and Interfaces in Polymer-Based Electronics," *Surface Science*, Vol. 500, No. 1-3, 2002, pp. 904-922.

[51] F. Cacialli and M. Stoneham, "Polymer Electronics: The Skill Lies in the Blending," *Journal of Physics: Condensed Matter*, Vol. 14, No. 47, 2002, pp. 9-11.

[52] D. Kozanogiu, D. H. Apaydin, A. Cirpan and E. N. Esenturk, "Power Conversion Efficiency Enhancement of Organic Solar Cells by Addition of Gold Nanostars, Nanorods and Nanospheres," *Organic Electronics*, Vol. 14, No. 7, 2013, pp. 1720-1727.

[53] C. D. Muller, A. Falcou, N. Reckefuss, M. Rojahn, V. Wiederhirn, P. Rudati, H. Frohne, O. Nuyken, H. Becker

and K. Meerholz, "Multicolor Organic Light-Emitting Displays by Solution Processing," *Nature*, Vol. 421, 2003, pp. 829-833.

[54] S. Moller, C. Perlov, W. Jackson, C. Tausig and S. R. Forrest, "A Polymer/Semiconductor Write-Once-Read-Many Times Memory," *Nature*, Vol. 426, 2003, pp. 166-169.

[55] O. Akogwu, D. Kwabi, S. Midturi, M. Eleruja, B. Babatope and W. O. Soboyejo, "Large Strain Deformation and Cracking of Nano-Scale Gold Films on PDMS Substrate," *Materials Science and Engineering B*, Vol. 170, No. 1-3, 2010, pp. 32-40.

24

Ab Initio Study of the Electronic and Vibrational Properties of 1-nm-Diameter Single-Walled Nanotubes

Jesús Marquina[1], Chrystian Power[2], Jesús González[2,3], Jean-Marc Broto[4]
[1]Centro de Estudios Avanzado en Óptica, Facultad de Ciencias, Universidad de los Andes, Mérida, Venezuela
[2]Centro de Estudios de Semiconductores, Facultad de Ciencias, Universidad de los Andes, Mérida, Venezuela
[3]Malta Consolider Team, CITIMAC, Facultad de Ciencias, Universidad de Cantabria, Santander, España
[4]LNCMI, Paul Sabatier University, Toulouse, France

ABSTRACT

The electronic structure, band gap, density of states of the (8,8), (14,0) and (12,3) single-walled carbon nanotubes by the SIESTA (Spanish Initiative for Electronic Simulations with Thousands of Atoms) method in the framework density-functional theory (DFT) with the generalized gradients approximation (GGA) were studied. Also, we studied the vibrational properties of the (8,8) and (14,0) nanotubes. Only the calculated relaxed geometry for (12,3) nanotube show significant deviations from the ideal rolled graphene sheet configuration. The electronic transition energies of van Hove singularities were studied and compared with previous results. The calculated band structures, density of states and dispersion curves for all tubes were in good agreement with theoretical and experimental results.

Keywords: Carbon Nanotubes; Electronic Structure; Dispersion Curve

1. Introduction

Carbon nanotubes can be regarded as graphitic cylinders generated just by rolling one sheet of graphene with a finite width [1]. Also it is known that a nanotube can be metallic or semiconducting depending on their diameter and chirality [2-4]. In recent years, it has attracted great attention in many fields of science and technology because of its novel and unique structural and electronic properties [5]. Nanotubes are promising for applications, such as probes and sensors; tips for atomic force microscopy, gas sensors, biosensors and physical sensors [6].

There has been much theorical and experimental work on the vibrational and electronic properties of carbon nanotubes. The first theorical calculation of SWCNT's electronic structure was carried out using zone folding scheme [7,8]. Within this framework, all the armchair nanotubes (n,n) are expected to be metallic, while approximately one third of zigzag (n,0) and chiral (n,m) tubes should be metallic, with remainder being semiconducting [3,4]. The zone-folding scheme in general works reasonably well and has proven immense success in providing physically relevant information on the nature of electronic and vibrational interactions in carbon nanotubes [9,10]. However, previous studies show the effects of the σ-π hybridization in small diameter SWCNT's on their electronic properties [11,12]. An example of particular importance is the (5,0) nanotube which has been demonstrated by *ab initio* (using local density approximation in the framework of the density-functional theory) and symmetry-adapted non-orthogonal tight-binding calculations to be a metal [12-14] but not a semiconductor with a moderate gap as reported in the previous π-band tight-binding work [2,3].

The phonon dispersion curves over the full Brillouin zone are useful in such fields as the inelastic neutron scattering [15], Second-order Raman spectroscopy [16], double-resonance processes [17-20] and the electron-phonon interactions [21-24]. The calculations of the phonon dispersions have been made using both empirical methods and the accuracy *ab initio* calculations. Initial work on this problem using zone folding and force constant models [9,25-27] shows that the zone-folding method has two shortcomings: the structural relaxation effect has been completely ignored which makes the transferability of the phonon spectra between the graphene and the SWCNT's a serious question and that the vibrational eigenvectors are also different from the graphene to the SWCNT's. The *ab initio* phonon dispersion is advantageous because such calculations do not depend on any predefined parameter. There are different authors that have reported *ab initio* phonon dispersion of SWCNT's, all using the supercell approach [11,28,29].

One way to study the structural and mechanical properties of carbon nanotubes is by applying hydrostatic pressure. A recent study shows that in DWCNT's the outer tube acts as a protection shield for the inner tube (at least up 11 GPa) [30,31]. This means that the inner tube retains its properties as single-walled nanotube. In Figures 5 and 6 of Ref. [30], the Raman spectrums of the radial breathing modes (RBM) of the inner and the outer shells of pristine and filled DWCNT's have been observed, respectively. In both in pristine or Te-filled DWCNT's tubes with diameters around 1 nm for the inner tubes of DWCNT's are presented.

In this paper, we present first-principles calculations of the structural, electronic, and vibrational properties of SWCNT's about 1 nm in diameter to address the point mentioned above. We will discuss the relaxed geometry of the (8,8), (14,0) and (12,3) nanotubes comparing it to the ideal cylindrical configuration. We also examine the importance of the detailed structure in discussing electronic properties of nanotubes. Finally, we discuss about phonon dispersion curves for the (8,8) nanotube.

2. Computational Method

Our first-principles calculations are based on density functional theory (DFT) employ a numerical-atomic-orbital basis set. We used the package SIESTA *ab initio* simulation package [32]. We worked within the gradient generalized approximation (GGA) as parametrized by Perdew-Burke-Ernzerhof (PBE) [33]. The core electrons were replaced by pseudopotentials of the type Troullier-Martins [34], the valence electrons were described by localized pseudo-atomic orbitals [35]. The valence electrons were described by a double-ζ polarized basis set with a cutoff radius of 4.99 a.u. for the s and 6.25 for the p and the polarizing d orbitals, as is determined from an energy shift of 50 meV by localization. These values have been tested by M. Machón *et al.* [12]. They found that an increasing of the cutting of the orbitals only produces changes in the total energies less than 0.1 meV/atom and none significant effect neither on the structure nor on the energies of electronic states was observed. The dynamical matrix was found by a finite-difference approach, this is, calculating the Hellman-Feynmann forces for displaced atoms.

Calculations were performed for the following tubes: (8,8), (14,0) and (12,3). They were all considered as isolated. For that purpose, we used periodic-boundary conditions on supercell geometry with sufficient lateral separation among neighboring tubes. For the purpose of sampling the Brillouin zone in the direction of the tube axis, as well as for the computation of the force-constant matrix for phonon calculations, we used supercells consisted of five unit cells for the armchair (8,8) tube with 160 atoms and three unit cells for zigzag (14,0) tube with 162 atoms. The supercell length (about 1 nm) in the tube-axis direction is not too different from one another. We are aware that the supercell should be large enough so that the effect of the image displacement is negligible [11].

3. Results and Discussion

3.1. Dispersion Curve of Graphene

In **Figure 1** we show the phonon frequencies of graphene calculated from first principles. This will serve as a test of accuracy of the calculation method. In general, a good agreement between our calculations and experiment (see Ref. [36]) for the TO, LO and LA branches is found. This shows the accuracy of the basis set used in this work. We optimized the bond length of a graphene sheet in a supercell with the interlayer distance being 10 Å, which can be considered enough to neglect the interlayer interaction. The bond length of graphite obtained by this optimization is 1.437 Å, which is 1.5% higher than the experimental value (1.416 Å) [37]. The small splitting of the LO and LA phonons at the M point equal to 33 cm^{-1} coincides with the experimental value reported by inelastic x-ray scattering [36]. While, our calculated splitting between TO and LA modes at the K point 118 cm^{-1} is different from experimental value 71 cm^{-1}. Other theoretical values have been reported; Pavone *et al.* [38] obtained a splitting of 120 cm^{-1} and a value of 140 cm^{-1} by using LDA was also reported [28]. The zone center phonon value 1575 cm^{-1} is slightly below the experimental values for graphite and grapheme (1582 cm^{-1}) measured recently [39]. These differences can be explained by taking into account that other authors have estimated uncertainties of 10 cm^{-1} for the frequencies higher than 1300 cm^{-1} [11,40].

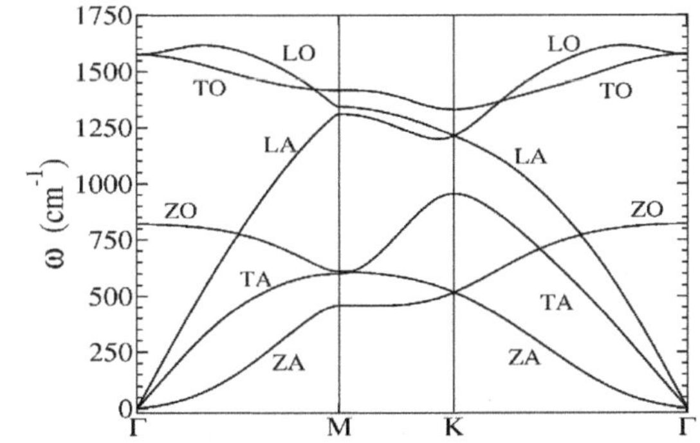

Figure 1. ***Ab initio*** **calculations of the phonon dispersion of graphene. We used a supercell with 98 atoms. The mesh cutoff was 350 Ry, with 3321** ***k*** **points and a 50 meV PAO energy shift.**

3.2. Geometries and Electronic Band Structure of Nanotubes

The closed distances between atoms on different tubes being as long as 10 Å; to suppress sufficiently the inter-tube interaction. We completely optimized the atomic positions of atoms by using the conjugate-gradient (CG) method. We also optimized the length of the unit cell along the axes of the tubes by changing the lattice constant (c) along the tube axis. The atomic positions were relaxed until the forces on the atoms were less than 0.04 eV/Å.

For the relaxed structure we calculated the theoretical lattice constant. In **Table 1** are shown the results of geometry parameters (**Figure 2**) relaxed together with the parameters corresponding to the ideal cylindrical geometry obtained by rolling up the graphene plane. To determine the magnitude of the translation vector (T) we used the values 1.44 Å and 1.437 Å for ideal cylindrical geometry and relaxed geometry, respectively. Hence, we find that the lattice constant of the nanotubes in the direction of the tube axis (z) changed by less than 0.3% during the relaxation, in agreement with the trend indicated by Machón *et al.* [12]. Also, the bonds and angles in the direction of the tube axis and in the direction of the circumference for the (8,8) and (14,0) nanotubes changed by less than 0.3% with respect to the ideal structure. Nevertheless, the bonds c and a for the (12,3) nanotube experience changes by up to 10%. In the (12,3) nanotube the bonds and angles tend to form a perfect hexagon for the relaxed geometry and the sum of their three bond angles is always smaller than 360° for all tubes.

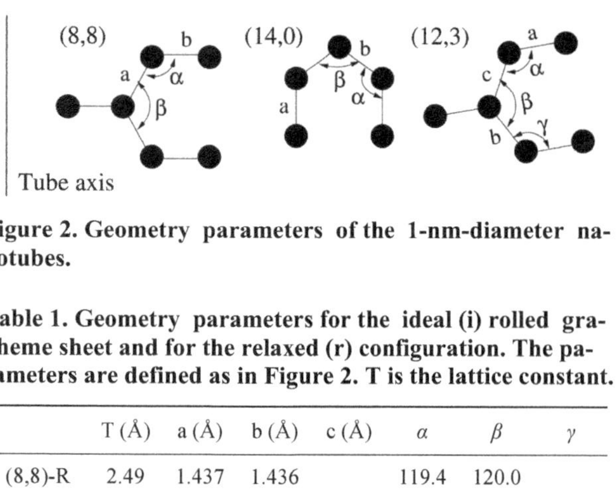

Figure 2. Geometry parameters of the 1-nm-diameter nanotubes.

Table 1. Geometry parameters for the ideal (i) rolled grapheme sheet and for the relaxed (r) configuration. The parameters are defined as in Figure 2. T is the lattice constant.

	T (Å)	a (Å)	b (Å)	c (Å)	α	β	γ
(8,8)-R	2.49	1.437	1.436		119.4	120.0	
(8,8)-I	2.49	1.437	1.433		119.3	120.0	
(14,0)-R	4.32	1.436	1.438		120.2	118.5	
(14,0)-I	4.31	1.437	1.435		120.1	118.7	
(12,3)-R	6.59	1.438	1.439	1.436	120.0	120.0	118.7
(12,3)-I	6.59	1.389	1.335	1.597	112.5	115.9	130.0

In **Table 1**, we observe that the values of the relaxed bond lengths are approximately the same for three nanotubes. Imtani & Jindal found three equal bond lengths and three unequal bond angles in the structure of different chiral tubes [41]. It is in agreement with the relaxed bond length for the chiral tube (12,3). However, is in disagreement with calculations obtained by Jiang *et al.* [42], they found three unequal bond lengths in the structure of chiral (4,2) and (9,3) tubes. According to [41], these differences were attributed to the unsatisfactory reproduction of the graphite sheet bond length (1.42 Å) from the beginning. On the other hand, Jiang *et al.* have also found three unequal bond angles in the structure of chiral (4,2) and (9,3) tubes. They obtained a value of β larger than the ideal value and the other two bond angles are smaller. Imtani & Jindal found three unequal bond angles in the structure of different chiral tubes. Our results show bond angles very close to 120° for any nanotubes. We think that these differences could be related to the radius of the nanotubes, *i.e.* our nanotubes do not show effects of confinement.

The electronic band structures were calculated using the relaxed geometries obtained above. The band structures obtained are given in **Figure 3**. The density of states for the (8,8) and (12,3) tubes shown a metallic character. This metallic character of both nanotubes is in good agreement with the classification as metal 2p reported by Saito *et al.* [4]. Experimental results also confirm the theoretical predictions about the metallicity of the nanotube (8,8). Ouyang *et al.* [43] have reported the tunneling spectrum of (8,8) nanotubes in bundles and isolated. They have finding a gap-like feature at E_F that is not present in spectra of the isolated (8,8) armchair, in good agreement with our result. Also, an *ab initio* calculation recent on a armchair (7,7) nanotube (using full potential-linarized augmented plane wave (FP-LAPW) in the GGA approximation) shown that it is a metallic nanotube with zero band gap and with the valence and conduction bands cross each other at the Fermi level [44], just as is observed in **Figure 3** for (8,8) and (12,3) nanotubes.

The nanotube (14,0) is clearly a semiconductor material with a direct energy gap, as expected. This is consistent with previous results obtained from *ab initio* code VASP [45] and with the classification as semiconductor type I given by Saito *et al.* [4].

In the Tight-Binding/Zone-Folding method, the transition energies between van Hove singularities in the electronic density of states is related to the diameter of the tube by

$$E_{ii} = 2n\gamma_0 a_{cc} / d \quad (1)$$

where $\gamma_0 = 2.9$ eV is the overlapping integral and

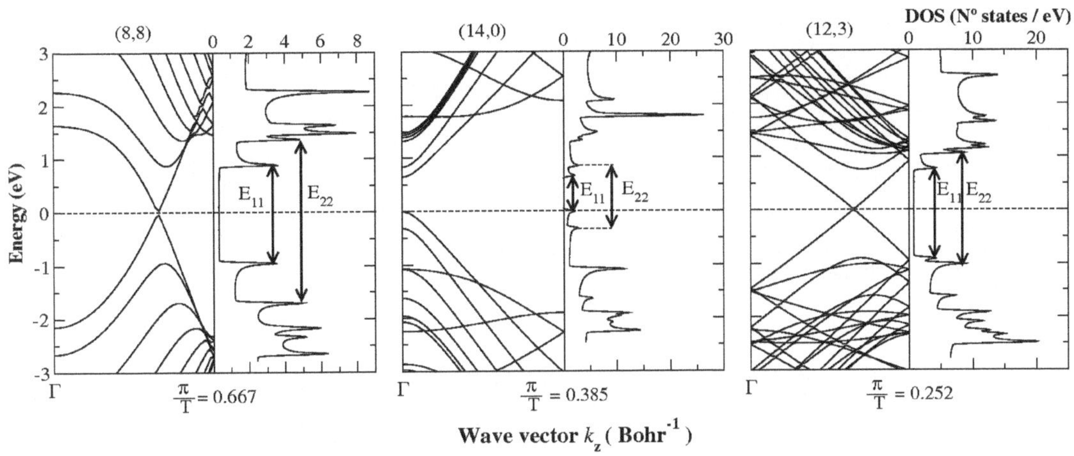

Figure 3. Calculated electronic band structures and DOS of the relaxed SWCNT's. The Fermi level is at 0 eV. One thousand *k* points for (8,8) nanotube and five thousand *k* points for (14,0) and (12,3) nanotubes were employed and the mesh cutoff was at 1600 Ry for (8,8) nanotube and at 1200 Ry for (14,0) and (12,3) nanotubes.

$a_{cc} = 1.437$ Å (**Table 1**) is the carbon-carbon distance mean of the nanotubes. The integer n has the values 1, 2, 4, 5 ... for the transitions E_{11}, E_{22}, E_{33}, E_{44}, of the semi-conducting tubes, respectively, and 3, 6, ... for the transitions E_{11}, E_{22}, of the metallic tubes, respectively [46].

In **Table 2** are shown electronic transition energies (E_{ii}). In this table is observed that the calculated values for the gap energy (E_g) are in good agreement with the previous report. On the contrary, the values for energies E_{11} and E_{22} differ from values reported usign different techniques. The energies E_{ii} obtained from Zone-Folding (Z-F) and Resonance Raman Spectroscopy (RRS) are based in Tight-Binding approximation [46,47]. Thus, the differences between our results and those obtained by Z-F or RSS are due to the type of approximation used for calculations of the energies E_{ii}, *i.e.* Tight-Binding or Density Functional Theory. On the other hand, they find that the experimental values 1.45 and 2.73 eV for the (8,8) nanotube shown in [43] are below our values. Although, even the approach used is better than those obtained by other methods, it does not reproduce well the valence bands in the electronic band structure. This is a problem inherent in all DFT-based methods.

The **Figure 4** shows phonon dispersion curves of (8,8) and (14,0) nanotubes. The phonon-dispersion curve for (10,10) nanotube reported by Ye *et al.* [29] is very similar to the nanotube (8,8). And also, the dispersion curve of (14,0) zigzag tube is in good qualitative agreement with a dispersion curve for the (11,0) nanotube reported in [29]. According to group theory there are eight modes Raman active for an armchair carbon nanotube, with symmetries and number of optically active phonon modes given by [48]:

$$\Gamma(R) = 2A_{1g} \otimes 2E_{1g} \otimes 4E_{2g}, \quad (2)$$

$$\Gamma(IR) = 3E_{1u}. \quad (3)$$

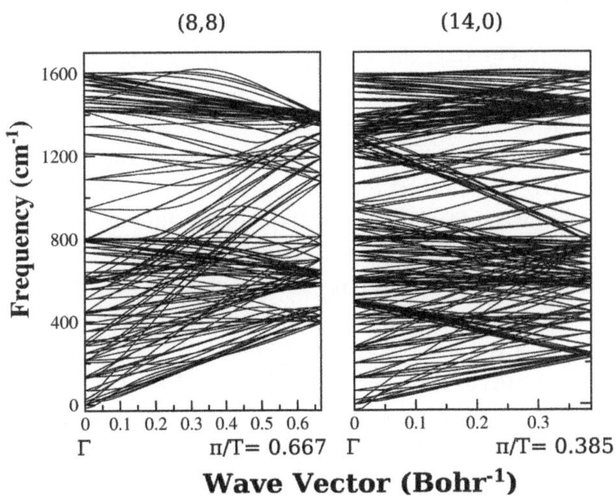

Figure 4. Phonon dispersion curves. We have used used a supercell with 160 atoms for the (8,8) nanotube and 162 atoms for the (14,0) nanotube. Seventy-five *k* points for (8,8) nanotube and fifty *k* points for (14,0) and (12,3) nanotubes were employed and the mesh cutoff was at 350 Ry for all nanotubes.

Table 2. Energy-gap values and electronic transitions energies values.

	E_g (eV)		E_{11} (eV)		E_{22} (eV)	
	Our	Rep.	Our	Rep.	Our	Rep.
(8,8)	0.00	0.00[c]	1.84	2.28[a] 2.22[b], 1.45[d]	3.08	4.56[a] 2.73[d]
(14,0)	0.67	0.62[e]	0.67	0.76[a]	1.16	1.51[a]
(12,3)	0.00	0.00[c]	1.69	2.31[a] 2.04[b]	2.08	4.62[a]

[a]Ref. [46]; [b]Ref [47]; [c]Ref. [4]; [d]Ref. [43]; [e]Ref. [45].

The A modes are non-degenerate and the E modes are doubly degenerate.

In **Table 3** we have assigned the symmetry of the different modes of vibration calculated to the (8,8) nanotube,

Table 3. Vibrational modes in zone center for (8,8) nanotubes. R and IR mean Raman-Active and infrared-active.

Symmetry	Our *Ab inicio* (GGA)	Ref. [29] *Ab initio* (LDA)	Ref. [54] Tight-Binding	Ref. [49] Exp.
E_{2g} (R)	1603	1593	1698	1609
A_{1g} (R)	1589	1580	1672	1593
E_{1u} (IR)	1583	1574	1683	-
E_{1g} (R)	1568	1561	1668	1567

following the assignment given in [29] taking into account the Equations (2) and (3). In a recent paper presented earlier by Rao *et al.* Raman bands were observed for an isolated nanotube with chirality that are consistent with an ensemble of (8,8), (9,9), (10,10), and (11,11) tubes [49]. They observed seven Raman peaks and failed to detect the lowest frequency mode in the experiment because of the strong Rayleigh scattering back- ground. These values are used for comparison with our calculations. From **Table 3** is clear that in the high frequency region the calculations made with the GGA approximation agree much better with experiment than those using LDA or Tight-binding. It is known that, the LDA and Tight-binding approximations tends to over estimate the high-energy phonons.

In the high-frequency region the phonon dipersions is well reproduced in our calculations. For example, the modes for A_{1g} symmetry are 1589 and 1593 cm^{-1} for (8,8) and (14,0) tubes, respectively. These values are just below the highest frequency mode. On the other hand, taking into account that the strongest G band (assigned to A_{1g} symmetry [50,51]) at about 1590 cm^{-1} does not depend on the tube diameter and that the G bands of the outer tubes (in pristine DWCNT's) fall in the same spectral range as the G^{+} of an isolated semiconducting SWCNT (1589 - 1593 cm^{-1}) [30,52,53]. We conclude that our results for the zone-center phonons in the high-frequency region are in very good agreement with the experimental values.

4. Conclusion

We have made a study of DFT as implemented in the code SIESTA, using localized atomic wave functions as a basis set with the GGA, and applied it to the study of SWCNT's about 1 nm in diameter. We found that the structure for (8,8) and (14,0) nanotubes does not show significant change compared with the rolled graphene sheet, while the (12,3) nanotube shows significant change. It has been observed a natural tendency to form regular hexagons that dominates in the later case. Our results predict that the electronic band structure for (8,8) and (12,3) nanotubes has metallic behavior, that is, behaves as metal at 0 K. Whereas the (14,0) nanotube is a semiconducting. Also, it is found that the accuracy of the method for predicting the transition energies of Van Hove singularities is insufficient. This is a problem inherent in DFT-based methods because it does not reproduce well the valence bands in the electronic band structure. Finally, we have shown that our results are comparable with the experimental values, this proves that the method used can be quite accurate for the high-energy vibrations because of their reliable description of the chemical bonds.

5. Acknowledgements

The authors are grateful to the PCP Carbon Nanotubes (France)-MCT (Venezuela) for financial support. We also would like to express our gratitude to Nicolas RENON and the CALMIP (Calcul en Midi-Midi-Pyrénées) team for giving us access to the supercomputer used for parallel computing. One of us (J. M.) would like to thank Dr. L. Rincón for his interest in this work and the useful suggestions.

REFERENCES

[1] S. Lijima, "Helical Microtubules of Graphitic Carbon," Nature, Vol. 354, 1991, pp. 56-58.

[2] N. Hamada, S.-I. Sawada and A. Oshiyama, "New One-Dimensional Conductors: Graphitic Microtubules," *Physical Review Letters*, Vol. 68, No. 10, 1992, pp. 1579-1581.

[3] R. Saito, M. Fujita, G. Dresselhaus and M. S. Dresselhaus, "Electronic Structure of Chiral Graphene Tubules," *Applied Physics Letters*, Vol. 60, No. 18, 1992, pp. 2204-2206.

[4] R. Saito, K. Sato, Y. Oyama, J. Jiang, G. G. Samsonidze, G. Dresselhaus and M. S. Dresselhaus, "Cutting Lines near the Fermi Energy of Single-Wall Carbon Nanotubes," *Physical Review B*, Vol. 72, No. 15, 2005, Article ID: 153413.

[5] M. S. Dresselhaus, G. Dresselhaus and P. C. Ecklund, "Science of Fullerene and Carbon Nanotubes," Academic Press, San Diego, 1996.

[6] P. J. F. Harris, "Carbon Nanotube Science," Cambridge University Press, London, 2009.

[7] R. Saito, M. Fujita, G. Dresselhaus and M. S. Dresselhaus, "Electronic Structure of Graphene Tubules Based on C60," *Physical Review B*, Vol. 46, No. 3, 1992, pp. 1804-1811.

[8] J. W. Mintmire and B. I. Dunlap, C. T. White. "Are Fullerene Tubules Metallic?" *Physical Review Letters*, Vol. 68, No. 5, 1992, pp. 631-634.

[9] R. Saito, T. Takeya, T. Kimura, G. Dresselhaus and M. S. Dresselhaus, "Raman Intensity of Single-Wall Carbon Nanotubes," *Physical Review B*, Vol. 57, No. 7, 1998, pp. 4145-4153.

[10] R. A. Jishi, L. Venkataraman, M. S. Dresselhaus and G. Dresselhaus, "Symmetry Properties of Chiral Carbon Nanotubes," *Physical Review B*, Vol. 51, No. 16, 1995, pp. 11176-11179.

[11] D. Sánchez-Portal, E. Artacho, J. M. Soler, A. Rubio and P. Ordejón, "*Ab Initio* Structural, Elastic, and Vibrational Properties of Carbon Nanotubes," *Physical Review B*, Vol. 59, No. 19, 1999, pp. 12678-12688.

[12] M. Machón, S. Reich, C. Thomsen, D. Sánchez-Portal and P. Ordejon, "*Ab Initio* Calculations of the Optical Properties of 4-Å-Diameter Single-Walled Nanotubes," *Physical Review B*, Vol. 66, No. 15, 2002, Article ID: 155410.

[13] K. Kanamitsu and S. Saito, "Geometries, Electronic Properties, and Energetics of Isolated Single Walled Carbon Nanotubes," *Journal of the Physical Society of Japan*, Vol. 71, 2002, pp. 483-486.

[14] V. N. Popov, "Curvature Effects on the Structural, Electronic and Optical Properties of Isolated Single-Walled Carbon Nanotubes within a Symmetry-Adapted Non-Orthogonal Tight-Binding Model," *New Journal of Physics*, Vol. 6, 2004, pp. 1-17.

[15] S. Rols, Z. Benes, E. Anglaret, J. L. Sauvajol, P. Papanek, J. E. Fischer, G. Coddens, H. Schober and A. J. Dianoux, "Phonon Density of States of Single-Wall Carbon Nanotubes," *Physical Review Letters*, Vol. 85, No. 24, 2000, pp. 5222-5225.

[16] C. Thomsen, "Second-Order Raman Spectra of Single and Multiwalled Carbon Nanotubes," *Physical Review B*, Vol. 61, No. 7, 2000, pp. 4542-4544.

[17] J. Maultzsch, S. Reich and C. Thomsen, "Chirality-Selective Raman Scattering of the D Mode in Carbon Nanotubes," *Physical Review B*, Vol. 64, No. 12, 2001, Article ID: 121407.

[18] J. Maultzsch, J. S. Reich and S. C. Thomsen, "Raman Scattering in Carbon Nanotubes Revisited," *Physical Review B*, Vol. 65, No. 23, 2002, Article ID: 233402.

[19] V. Zólyomi and J. Kiirti, "Calculating the Discrepancy between the Stokes and Anti-Stokes Raman D Band of Carbon Nanotubes Using Double Resonance Theory," *Physical Review B*, Vol. 66, No. 7, 2002, Article ID: 073418.

[20] G. G. Samsonidze, R. Saito, A. Jorio, A. G. S. Filho, A. Griineis, M. P. Pimenta, G. Dresselhaus and M. S. Dresselhaus, "Phonon Trigonal Warping Effect in Graphite and Carbon Nanotubes," *Physical Review Letters*, Vol. 90, No. 2, 2003, Article ID: 027403.

[21] G. L. Kane, E. J. Mele, R. S. Lee, J. E. Fischer, P. Petit, H. Dai, A. Thess, R. E. Smalley, A. R. M. Verschueren, S. J. Tans and G. Dekker, "Temperature-Dependent Resistivity of Single-Wall Carbon Nanotubes," *EPL* (*Europhysics Letters*), Vol. 41, No. 6, 1998, pp. 683-689.

[22] Z. Yao, G. L. Kane and G. Dekker, "High-Field Electrical Transport in Single-Wall Carbon Nanotubes," *Physical Review Letters*, Vol. 84, No. 13, 2000, pp. 2941-2944.

[23] H. Suzuura and T. Ando, "Phonons and Electron-Phonon Scattering in Carbon Nanotubes," *Physical Review B*, Vol. 65, No. 23, 2002, Article ID: 235412.

[24] A. Sédéki, L. G. Caron and G. Bourbonnais, "Electron-Phonon Coupling and Peierls Transition in Metallic Carbon Nanotubes," *Physical Review B*, Vol. 62, No. 11, 2000, pp. 6975-6978.

[25] Z. J. Q. Ding, Q. Jiang and D. Tian, "Vibrational Spectrum of Carbon Nanotubes," *Fullerene Science and Technology*, Vol. 4, No. 1, 1996, pp. 31-37.

[26] A. Charlier, E. McRae, M.-F. Charlier, A. Spire and S. Forster, "Lattice Dynamics Study of Zigzag and Armchair Carbon Nanotubes," *Physical Review B*, Vol. 57, No. 11, 1998, pp. 6689-6696.

[27] R. A. Jishi, L. Venkataraman, M. S. Dresselhaus and G. Dresselhaus, "Phonon Modes in Carbon Nanotubules," *Chemical Physics Letters*, Vol. 209, No. 1-2, 1993, pp. 77-82.

[28] O. Dubay and G. Kresse, "Accurate Density Functional Calculations for the Phonon Dispersion Relations of Graphite Layer and Carbon Nanotubes," *Physical Review B*, Vol. 67, No. 3, 2003, Article ID: 035401.

[29] L. H. Ye, B. G. Liu, D. S. Wang and R. Han, "*Ab Initio* Phonon Dispersions of Single-Wall Carbon Nanotubes," *Physical Review B*, Vol. 69, No. 23, 2004, Article ID: 235409.

[30] E. Belandria, M. Millot, J. M. Broto, E. Flahaut, F. Rodriguez, R. Valiente and J. González, "Pressure Dependence of Raman Modes in Double Wall Carbon Nanotubes Filled with 1D Tellurium," *Carbon*, Vol. 48, No. 9, 2010, pp. 2566-2572.

[31] J. Marquina, E. Flahaut and J. González, "High-Pressure Optical Absorption Studies of Double-Walled Carbon Nanotubes," *Physica Status Solidi* (*RRL*)—*Rapid Research Letters*, Vol. 6, No. 9-10, 2012, pp. 382-384.

[32] D. Sánchez-Portal, P. Ordejón, E. Artachoa and J. M. Soler, "Density-Functional Method for Very Large Systems with LCAO Basis Sets," *International Journal of Quantum Chemistry*, Vol. 65, No. 5, 1997, pp. 453-461.

[33] J. P. Perdew, K. Burke and M. Ernzerhof, "Generalized Gradient Approximation Made Simple," *Physical Review Letters*, Vol. 77, No. 18, 1996, pp. 3865-3868.

[34] N. Troullier and J. L. Martins, "Efficient Pseudopotentials for Plane-Wave Calculations," *Physical Review B*, Vol. 43, No. 3, 1991, pp. 1993-2006.

[35] J. Junquera, O. Paz, D. Sánchez-Portal and E. Artacho, "Numerical Atomic Orbitals for Linear-Scaling Calcula-

tions," *Physical Review B*, Vol. 64, No. 23, 2001, Article ID: 235111.

[36] J. Maultzsch, S. Reich, C. Thomsen, H. Requardt and P. Ordejón, "Phonon Dispersion in Graphite," *Physical Review Letters*, Vol. 92, No. 7, 2004, Article ID: 075501.

[37] M. Cardona, "In Electrons and Photons in Solids," Scuola Normale, Pisa, 2001.

[38] P. Pavone, R. Bauer, K. Karch, O. Schtt, S. Vent, W. Windl, D. Strauch, S. Baroni and S. de Gironcoli, "*Ab Initio* Phonon Calculations in Solids," *Physica B: Condensed Matter*, Vol. 219-220, 1996, pp. 439-441.

[39] J. Marquina, Ch. Power and J. González, "Espectroscopia Raman del Grafeno Monocapa y el Grafito: Acoplamiento Electrón Fonón y Efectos no Adiabáticos," *Revista Tumbaga*, Vol. 1, No. 5, 2010, pp. 183-194.

[40] Y. Miyamoto, M. L. Cohen and S. G. Louie, "*Ab Initio* Calculation of Phonon Spectra for Graphite, BN, and BC_2N Sheets," *Physical Review B*, Vol. 52, No. 20, 1995, pp. 14971-14975.

[41] A. N. Imtani and V. Jindal, "Structure of Chiral Single-Walled Carbon Nanotubes under Hydrostatic Pressure," *Computational Materials Science*, Vol. 46, No. 2, 2009, pp. 297-302.

[42] H. Jiang, P. Zhang, B. Liu, Y. Huang, P. H. Geubelle, H. Gao and K. C. Hwang, "The Effect of Nanotube Radius on the Constitutive Model for Carbon Nanotubes," *Computational Materials Science*, Vol. 28, No. 3-4, 2003, pp. 429-442.

[43] M. Ouyang, J. L. Huang, C. L. Cheung and C. M. Lieber, "Energy Gaps in 'Metallic' Single-Walled Carbon Nanotubes," *Science*, Vol. 292, No. 5517, 2001, pp. 702-705.

[44] H. S. Khadigeh, "*Ab Initio* Study of Electronic Properties of a Armchair (7,7) Carbon Nanotube," *Advances in Materials Physics and Chemistry*, Vol. 2, 2012, pp. 159-162.

[45] T. Ito, K. Nishidate, M. Baba and M. Hasegawa, "First Principles Calculations for Electronic Band Structure of Single-Walled Carbon Nanotube under Uniaxial Strain," *Surface Science*, Vol. 514, No. 1-3, 2002, pp. 222-226.

[46] M. Hulman, R. Pfeiffer and H. Kuzmany, "Raman Spectroscopy of Small-Diameter Nanotubes," *New Journal of Physics*, Vol. 6, 2004, pp. 1-17.

[47] C. Fantini, A. Jorio, M. Souza, M. S. Strano, M. S. Dresselhaus and M. A. Pimenta, "Optical Transition Energies for Carbon Nanotubes from Resonant Raman Spectroscopy: Environment and Temperature Effects," *Physical. Review Letters*, Vol. 93, No. 14, 2004, Article ID: 147406.

[48] O. E. Alon, "Number of Raman- and Infrared-Active Vibrations in Single-Walled Carbon Nanotubes," *Physical Review B*, Vol. 63, No. 20, 2001, Article ID: 201403.

[49] A. M. Rao, E. Richter, S. Bandow, B. Chase, P. C. Eklund, K. A. Williams, S. Fang, K. R. Subbaswamy, M. Menon, A. Thess, R. E. Smalley, G. Dresselhaus and M. S. Dresselhaus, "Diameter-Selective Raman Scattering from Vibrational Modes in Carbon Nanotubes," *Science*, Vol. 275, No. 5297, 1997, pp. 187-191.

[50] H. D. Sun, Z. K. Tang, J. Chen and G. Li, "Polarized Raman Spectra of Single-Wall Carbon Nanotubes Mono-Dispersed in Channels of $ALPO_4$-5 Single Crystals," *Solid State Communications*, Vol. 109, No. 6, 1999, pp. 365-369.

[51] C. Thomsen, "Second-Order Raman Spectra of Single and Multiwalled Carbon Nanotubes," *Physical Review B*, Vol. 61, No. 7, 2000, pp. 4542-4544.

[52] A. Jorio, A. G. Souza Filho, G. Dresselhaus, M. S. Dresselhaus, A. K. Swan, M. S. Ünlü, B. B. Goldberg, M. A. Pimenta, J. H. Hafner, C. M. Lieber and R. Saito, "G-Band Resonant Raman Study of 62 Isolated Single-Wall Carbon Nanotubes," *Physical Review B*, Vol. 65, No. 15, 2002, Article ID: 155412.

[53] M. Dresselhaus, G. Dresselhaus, R. Saito and A. Jorio, "Raman Spectroscopy of Carbon Nanotubes," *Physics Reports*, Vol. 409, No. 2, 2005, pp. 47-99.

[54] D. Kahn and J. P. Lu, "Vibrational Modes of Carbon Nanotubes and Nanoropes," *Physical Review B*, Vol. 60, No. 9, 1999, pp. 6535-6540.

25

Recalculate Structural, Elastic, Electronic, and Thermal Properties in $LaAlO_3$ Rhombohedral Perovskite

Abdelkader Boudali, Fatiha Saadaoui, Mostefa Zemouli, Mohamed Driss Khodja, Kadda Amara
Laboratory of Physico-chemical Studies, University of Saida, Saida, Algeria

ABSTRACT

We study the structural, elastic and electronic properties of perovskite insulator $LaAlO_3$ using two different methods: the full-potential linearized augmented plane wave method and the pseudo-potential plane wave scheme in the frame of generalized gradient approximation and local density approximation GGA + mBJ. We have evaluated the ground state quantities. Also, we have presented the results of the band structure and densities of states. These results are in favourable agreement with previous theoretical works and the existing experimental data. To complete the fundamental characteristics of this compound, we have analyzed the thermodynamic properties.

Keywords: Ab-Initio Calculations; Elastic Properties; Electronic Structure; Thermodynamic

1. Introduction

Lanthanum aluminate ($LaAlO_3$) is a promising alternative gate dielectric for the replacement of SiO_2 in silicon MOSFETs due to its high-*k* value (~25) [1], wide energy band gap (~5 - 6 eV) [2,3], large optical band gap 6.2 eV [2-4] and good thermal stability in contact with Si [5,6]. Being a perovskite (ABO_3), $LaAlO_3$ is also structurally compatible with many functional compounds, like manganites, superconductors and ferroelectrics. There have been many studies of the electronic and structural properties of $LaAlO_3$, both of experimental [7-16] and theoretical aspects of bulk [17-22] as well as surface [23-26]. The structure of $LaAlO_3$ is rhombohedral at room temperature and usually defined as hexagonal with space group R-3c (No.167). The lattice parameters are a = 0.536 nm and c = 1.311 nm [27] **Figure 1**.

To the best of our knowledge, there are no theoretical reports on the thermal behaviour of $LaAlO_3$ in the literature. Consequently, the primary purpose of this work is to provide some additional information to the existing data on the physical properties of $LaAlO_3$ with state-of the-art first-principles calculations. After introducing the problem, the remaining paper is organized as follows: in Section 2, we describe the calculation procedure. Results are presented in Section 3. Section 4 contains the conclusion.

[*]Corresponding author.

2. Theoretical Method

The zero-temperature energy calculations are performed using both the all-electron full-potential linear augmented-plane-wave method (FP-LAPW) [28] and the plane-wave pseudopotential (PPs-PW) [29] method. In both cases exchange-correlation effects were described using the Perdew *et al.* (GGA96) form [30] and local density approximation Tr-Blaha LDA + mBJ [31]. The FP-LAPW calculations have been carried out using the WIEN2k code [28]. The sphere radii used in the calculations are 2.32, 1.74 and 1.60 *a.u.* for La, Al and O, respectively. Within these spheres, the charge density and potential are expanded in terms of crystal harmonics up to angular momenta $L = 10$, and a plane wave expansion has been used in the interstitial region. The Brillouin zone integrations for the total energy have been carried out using 35 special k-points in the irreducible Brillouin zone. Well converged solutions were found for $R_{cut}K_{max} = 8$, where K_{max} is the plane wave cut-off and R_{cut} is the atomic sphere radii. Both the muffin-tin radius and the number of *k*-points are varied to ensure convergence. Core states are treated fully relativistic but for valence states relativistic effects have been included in a scalar relativistic treatment.

The PP-PW calculations were performed using the CASTEP computer code [29,32] and the thermodynamic properties of $LaAlO_3$, we apply the quasi-harmonic Debye model [33-41], in which the non-equilibrium Gibbs

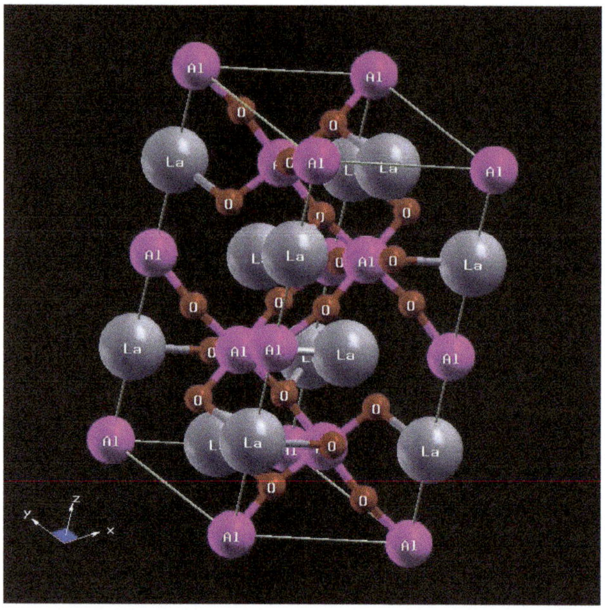

(a)

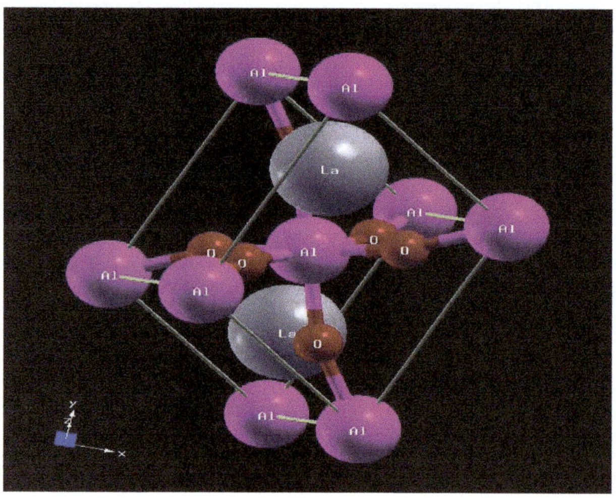

(b)

Figure 1. Structure of rhombohedral hexagonal (a cell conventional and *b* cell primitive).

function $G^*(V; P, T)$ details can be found in my article Cubic $LaAlO_3$ [33].

3. Results and Discussions

3.1. Structure and Elastic Constants

We firstly calculate the structural and elastic properties using FP-LAPW and PPs-PW methods. **Figure 2** shows the variation of the total energy as a function of volume and fitted to a Murnaghan equation of state [42] to obtain the equilibrium lattice constant, bulk modulus, and its pressure derivative. The results are summarized in **Table 1**, together with some theoretical results and the available experimental data. Obviously, the obtained results by FP-LAPW method are in better agreement with the

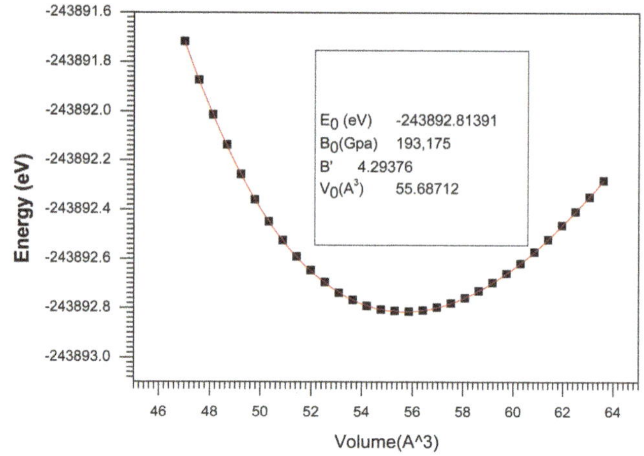

(a)

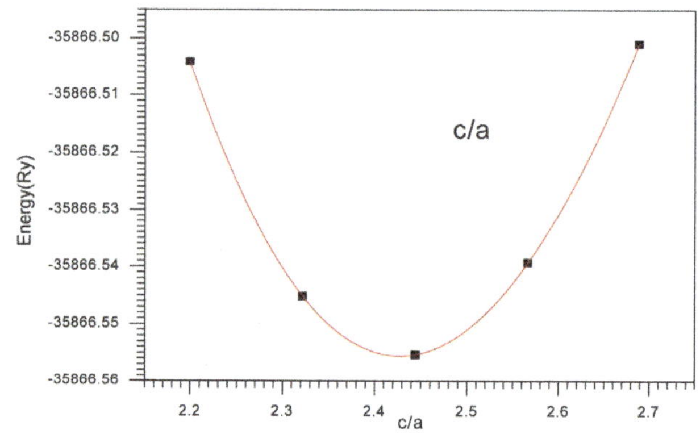

(b)

Figure 2. Variations of the energy with the unit cell volume and c/a.

Table 1. Lattice constant *a*, bulk modulus *B*, pressure derivative of bulk modulus *B'* and elastic constants parameters.

Present work		Experiment	Other theoretical works	
(FP-LAPW)		(PW-PP)	[50]	[51]
a (Å)	5.46	5.43	5.37	5.37
C (Å)	13.13	13.30	13.110	13.14
B (GPa)	189.30	184.26	195.76	185.63
B'	4.06			
C11 (GPa)		389.56	337	381.56
C33 (GPa)		410.89	411	411.17
C44 (GPa)		118.14	121	115.33
C66		129	93	140.63
C12		135.45	151	100.30
C13		92.20	93	72.54
C14		43.12	46	29.66

experiment compared to those calculated by PP-PW. The elastic constants are summarized in **Table 1**, which are in good agreement with the theoretical results and the available experimental data. It is clearly seen that the results we have calculated by FP-LAPW are in better agreement with experiment compared to those calculated by PP-PW. Our data will be beneficial to future investigation. We have performed elastic constants, calculations with the PW-PP method using the CASTEP code. The obtained values are presented in **Table 1**. Unfortunately, to our knowledge, no other results are available in literature for comparison. In **Table 2**, we report CASTEP calculations of the atomic positions after relaxation.

3.2. Electronic Properties

Density of states and electronic band structure often provide sufficient information for a thorough characterization of the electronic properties of a material. The energy band structure, total and partial density of states of $LaAlO_3$ were calculated by FP-LAPW method scheme in the frame of local density approximation Tr-Blaha + mBJ. From **Figures 3** and **4**, it appears that the main effect of TB-mBJ potential, with respect to GGA [43], is to shift up the unoccupied La *d* states. This is due to the kinetic-energy density term which increases the energy of unoccupied states and changes the corresponding band structure and density of states [44]. In PBE calculation [43], the states at the bottom of the conduction band derive from La *d* states hybridized with the states of the La *f* band which occur just above the conduction band edge. In our TB-mBJ calculations, the La *d* unoccupied states are shifted to higher energies. This give rise to pure and narrower band of La *f* localized states at the bottom of the conduction band. Similar observations had been made by D. J. Singh for La_2O_3 [45]. Our calculations show that the valence band maximum is located at Γ and the conduction band is minimum at Γ point.

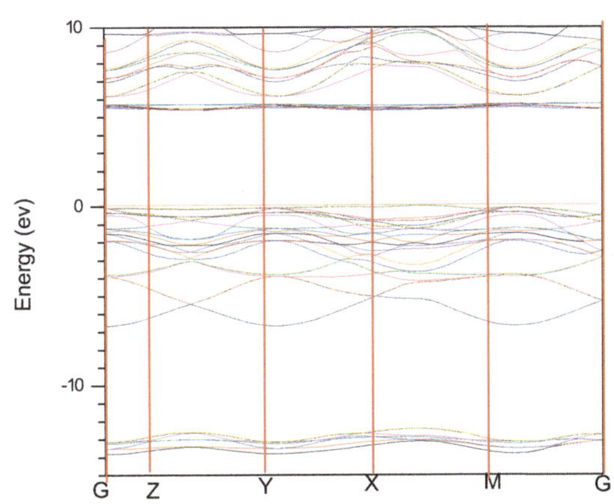

Figure 3. B and structure for high-symmetry directions in the Brillouin zone. The zero of energy corresponds to the fermi level.

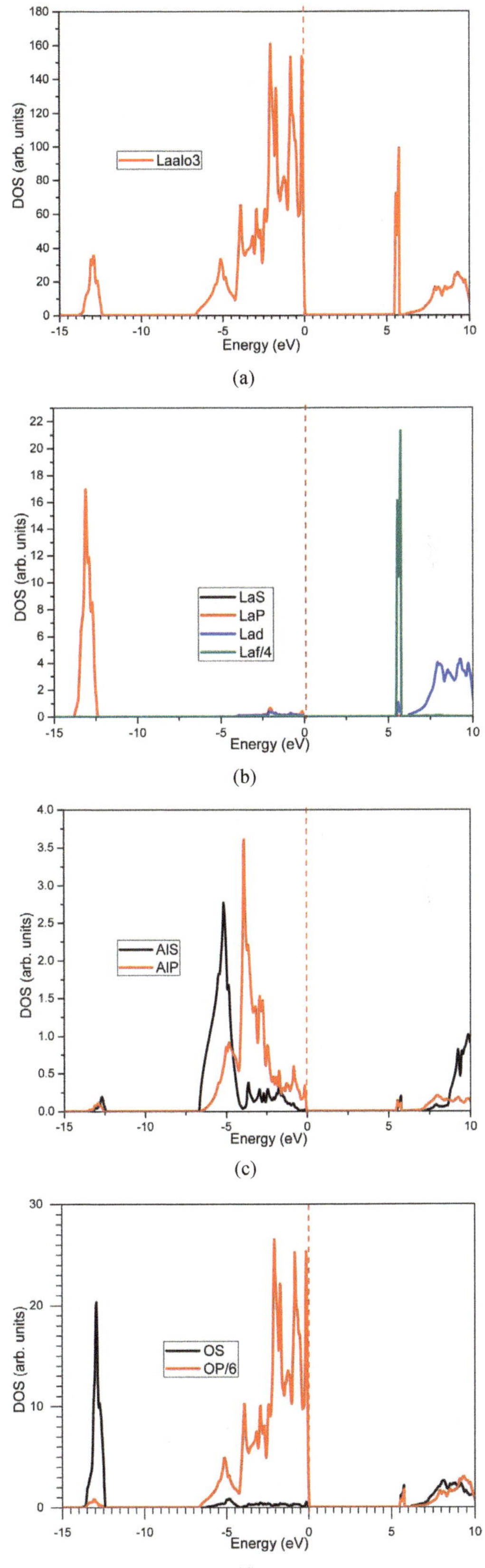

Figure 4. Calculated partial density of states and total density of states of $LaAlO_3$.

Table 2. Positions of the atoms obtained by relaxation.

		Fractional coordinates		
Atom	Number	x	y	z
O	1	0.224639	0.275361	0.750000
O	2	0.750000	0.224639	0.275361
O	3	0.275361	0.750000	0.224639
O	4	−0.224639	−0.275361	−0.75000
O	5	−0.750000	−0.224639	−0.275361
O	6	−0.275361	−0.750000	−0.224639
Al	1	0.000000	0.000000	0.000000
Al	1	0.500000	0.500000	0.500000
La	1	0.250000	0.250000	0.250000
La	1	−0.250000	−0.250000	−0.250000

From **Figure 4**, we can find that the DOS can be mainly divided into four parts. The first part extending from −13.89 to −12.34 eV is of the combination of Al *s*, Al *p*, O *s* and La *p* states, and it is mainly due to the contribution of O *s* states; the second part from −6.73 to - 0 eV is mainly made from La *p* states, with slight contributions of O *s*; the third part from −7.3 to 0 eV is mainly the contribution of O 2*p* states; the forth part extending from 5.40 to 10 eV is mainly composed of La *d* and Al *s*, *p* states.

Our calculation show that the valence band maximum is located at Γ and the conduction band minimum is at Γ resulting in an direct gap of 5.4 eV in FP-LAPW, and 3.4 eV in PPs-PW, compared to 5.6 eV experimentally [2,3]. We have to remember that GGA is known to underestimate band gaps in semiconductors. Robertson *et al.* [22]. Who used a screen exchange (SX) GGA [23] found that $LaAlO_3$ was a semiconductor with an indirect gap of 4.4 eV. Our calculation with LDA + mBJ shows that $LaAlO_3$ is an insulator. This modification of the electronic properties implies automatically changes in the optical properties.

3.3. Thermodynamic Properties

The thermal properties are determined in the temperature range from 0 to 600 K, where the quasi-harmonic model remains fully valid. The pressure effect is studied in the 0 - 30 GPa range. The temperature effects on the lattice parameters are shown in **Figure 5**. The Volume parameter increases with increasing temperature but the rate of increase is very moderate. On the other hand, it is noted from **Figure 6** that the relationships between bulk modulus and pressure are all nearly linear at various temperatures of 0, 100, 200, 300, 400, 500 and 600 K, respectively. The bulk modulus increases with pressure at a given temperature and decreases with temperature at given pressure. These results are due to the fact that the effect of increasing pressure on the material is the same as decreasing temperature of the material.

The investigation on the heat capacity of crystals is an old topic of condensed matter physics with which illustrious names are associated [46-48]. Knowledge of the heat capacity of a substance not only provides essential insight into its vibrational properties but is also mandatory for many applications. Two famous limiting cases are correctly predicted by the standard elastic continuum theory [48]. At high temperatures, the constant volume heat capacity Cv tends to the Petit and Dulong limit [49]. At sufficiently low temperatures, Cv is proportional to T^3 [48]. At intermediate temperatures, however, the temperature dependence of Cv is governed by the details of vibrations of the atoms and for a long time could only be determined from experiments. **Figures 7** and **8** represent the variation of the heat capacity, Cv (T), and volume expansion coefficient, α(T) as function of the temperature, respectively. These two quantities indicate a sharp increase up to ~400 K which is due to the anharmonic approximation of the Debye model used here. However, at higher temperature, the anharmonic effect on Cv is suppressed, and Cv is very close to the Dulong-Pettit

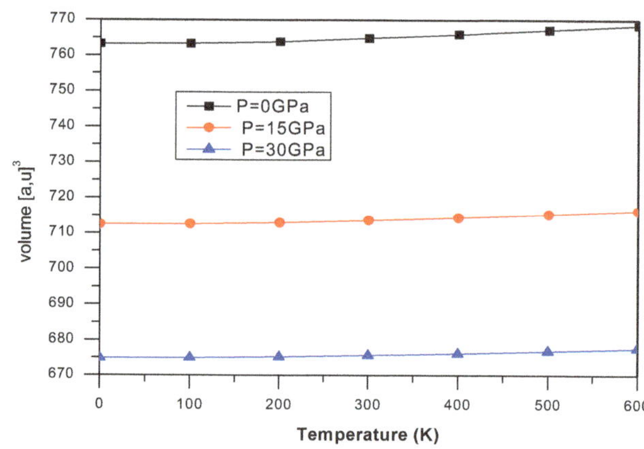

Figure 5. The relationships between lattice volumes and temperature at pressures of 0, 15 and 30 GPa, respectively.

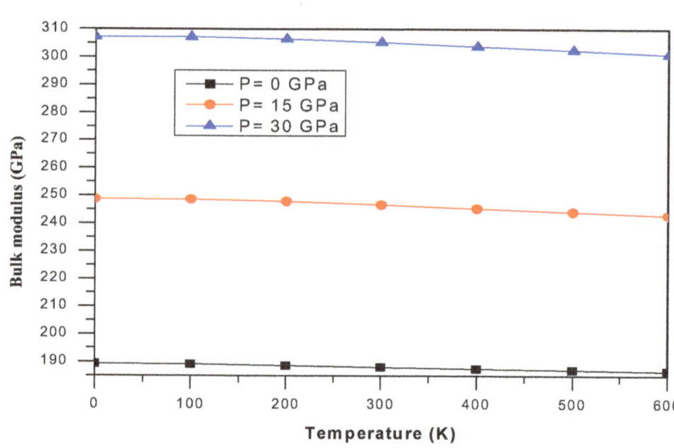

Figure 6. The relationships between bulk modulus and temperature at pressures of 0, 15 and 30 GPa, respectively.

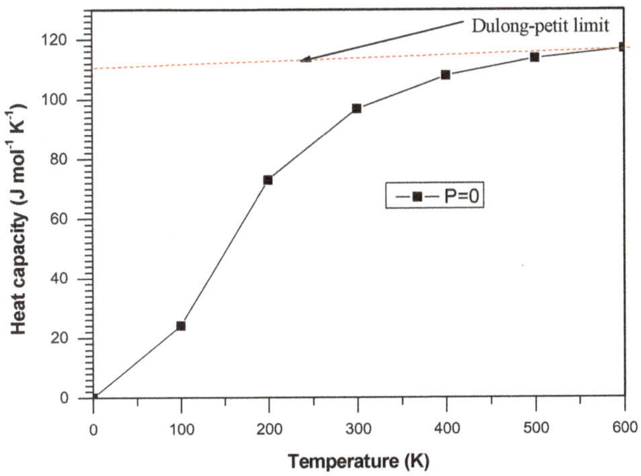

Figure 7. The heat capacity versus temperature at pressures of 0 GPa, respectively.

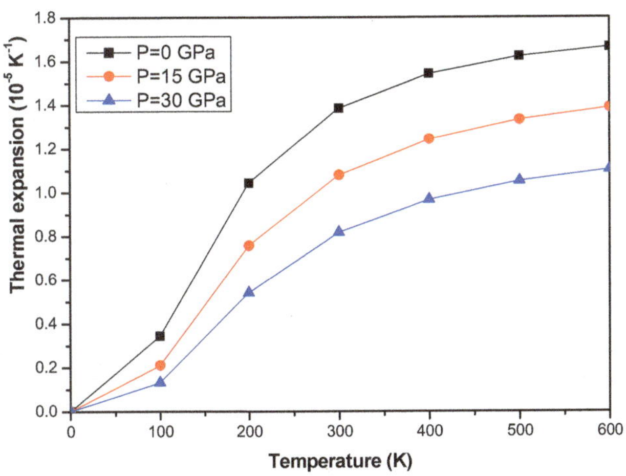

Figure 8. The thermal expansion versus temperature at pressures of 0, 15 and 30 GPa, respectively.

limit [49] ($Cv(T)$-3R for mono-atomic solids), which is common to all solids at high temperatures.

Finally, in **Figure 8**, we have plotted our results for the thermal expansion coefficient α of $LaAlO_3$. It is shown that, for a given pressure, α increases with temperature at low temperatures especially at zero pressure and gradually tends to a linear increase at higher temperatures. As pressure increases, the increase of α with temperature becomes smaller. While for a given temperature, α decreases strongly with increasing pressure, and it is very small at higher temperatures and higher pressures.

4. Conclusion

In summary, the structural, elastic and electronic properties of perovskite isulator $LaAlO_3$ at temperature $T = 0$ K were extensively studied using full-potential linearized augmented plane wave (FP-LAPW) as well as pseudo-potential plane-wave (PPs-PW) method. The exchange-correlation potential was calculated with the frame of generalized gradient approximation (GGA) and local density approximation Tr-Blaha + mBJ. The ground-state parameters, such as lattice parameter, bulk modulus and its pressure derivative and the elastic constants are in good agreement with some theoretical results and the available experimental data. The partial contribution of each atom to the total density of states was calculated. From the band structure, we find that $LaAlO_3$ is a direct $\Gamma \rightarrow \Gamma$ gap close about 98% to the experimental value indicating that the material is an insulator. With this change of electronic properties, the optical properties must be recalculated. To complete the basic characteristics of this compound, we analyzed their thermodynamic properties using the quasiharmonic model of Debye. We see that the predicted heat capacity Cv is close to the Dulong-Petit limit, which is common to all solids at high temperatures.

REFERENCES

[1] B.-E. Park and H. Ishiwara, "Formation of $LaAlO_3$ Films on Si (100) Substrates Using Molecular Beam Deposition," *Applied Physics Letters*, Vol. 82, No. 8, 2003 pp. 1197-1199.

[2] S. G. Lim, S. Kriventsov, T. N. Jackson, J. H. Haeni, D. G. Schlom, A. M. Balbashov, R. Uecker, P. Reiche, J. L. Freeouf and G. Lucovsky, "Dielectric Functions and Optical Bandgaps of High-K Dielectrics for Metal-Oxide-Semiconductor Field-Effect Transistors by far Ultraviolet Spectroscopic Ellipsometry," *Journal of Applied Physics*, Vol. 91, No. 7, 2002, pp. 4500-4505.

[3] X. Lu, Z. G. Liu, Y. P. Wang, Y. Yang, X. P. Wang, H. W. Zhou and B. Y. Nguyen, "Lattice Symmetry Applied in Transfer-Matrix Methods for Photonic Crystals," *Journal of Applied Physics*, Vol. 94, No. 2, 2003, pp. 811-821.

[4] P. Delugas, V. Fiorentini and A. Filippetti, "Dielectric Properties and Long-Wavelength Optical Modes of the High-κ Oxide $LaAlO_3$," *Physical Review B*, Vol. 71, No. 13, 2005, pp. 134302-134307.

[5] L. F. Edge, D. G. S4chlom, S. A. Chambers, E. Cicerrella, J. L. Freeouf, B. Holländer and J. Schubert, "Measurement of the Band Offsets between Amorphous $LaAlO_3$ and Silicon," *Applied Physics Letters*, Vol. 84, No. 2, 2004, pp. 726-728 .

[6] P. Sivasubramani, M. J. Kim, B. E. Gnade, R. M. Wallace, L. F. Edge, D. G. Schlom, H. S. Craft and J.-P. Maria, "Outdiffusion of La and Al from Amorphous $LaAlO_3$ in Direct Contact with Si (001)," *Applied Physics Letters*, Vol. 86, No. 20, 2005, Article ID: 201901.

[7] X. H. Zeng, L. H. Zhang, G. J. Zhao, J. Xu, Y. Hang, H. Y. Pang, M. Y. Jie, C. F. Yan and X. M. He, "Crystal Growth and Optical properties of $LaAlO_3$ and Ce-Doped $LaAlO_3$ Single Crystals," *Journal of Crystal Growth*, Vol.

271, No. 1-2, 2004, pp. 319-324.

[8] M. Losurdo, M. M. Giangregorio, M. Luchena, P. Capezzuto, G. Bruno, R. G. Toro, G. Malandrino, I. L. Fragalà and R. Lo Nigro, "Structural-Optical Study of High-Dielectric-Constant Oxide Films," *Applied Surface Science*, Vol. 253, No. 1, 2006, pp. 322-327.

[9] A. A. Demkov, R. Lui, X. Zhang and H. Loechelt, "Study of Aniline on a Si(111)7×7 Surface by Scanning Tunneling Microscopy," *Journal of Vacuum Science & Technology B*, Vol. 18, No. 5, 2000, pp. 2335-2338.

[10] H. Wadati, Y. Hotta, A. Fujimori, T. Susaki, H. Y. Hwang, Y. Takata, K. Horiba, M. Matsunami, S. Shin, M. Yabashi, K. Tamasaku, Y. Nishino and T. Ishikawa, "Hard X-Ray Photoemission Study of $LaAlO_3/LaVO_3$ Multilayers," *Physical Review B*, Vol. 77, No. 4, 2008, Article ID: 045122.

[11] R. A. B. Devine, "Infrared and Electrical Properties of Amorphous Sputtered $(La_xAl_{1-x})_2O_3$ Films," *Journal of Applied Physics*, Vol. 93, No. 12, 2003, pp. 9938-9942.

[12] V. V. Afanas'ev, A. Stesmans, C. Zhao, M. Caymax, T. Heeg and J. Schubert, Y. Jia, D. G. Schlom and G. Lucovsky, "Band Alignment between (100) Si and Complex Rare Earth/Transition Metal Oxides," *Applied Physics Letters*, Vol. 85, No. 24, 2004, pp. 5917-5919.

[13] B.-E. Park and H. Ishiwara, "Electrical Properties of $LaAlO_3$/Si and $Sr_{0.8}Bi_{2.2}Ta_2O_9/LaAlO_3$/Si Structures," *Applied Physics Letters*, Vol. 79, No. 6, 2001, pp. 806-808.

[14] Y. Y. Mi, Z. Yu, S. J. Wang, P. C. Lim, Y. L. Foo, A. C. H. Huan and C. K. Ong, "Epitaxial $LaAlO_3$ Thin Film on Silicon: Structure and Electronic Properties," *Applied Physics Letters*, Vol. 90, No. 18, 2007, Article ID: 181925.

[15] V. Garcia, M. Bibes, J.-L. Maurice, E. Jacquet, K. Bouzehouane, J.-P. Contour and A. Barthélémy, "Spin-Dependent Tunneling through High-k $LaAlO_3$," *Applied Physics Letters*, Vol. 87, No. 21, 2005, Article ID: 212501.

[16] E. Cicerrella, J. L. Freeouf, L. F. Edge, D. G. Schlom, T. Heeg, J. Schubert and S. A. Chambers, "Optical Properties of La-Based High-K Dielectric Films," *Journal of Vacuum Science & Technology A*, Vol. 23, No. 6, 2005, pp. 1676-1680.

[17] A. A. Knizhnik, I. M. Iskandarova, A. A. Bagatur'yants, B. V. Potapkin, L. R. C. Fonseca and A. Korkin, "First-Principles Calculations of the Electrical Properties of $LaAlO_3$ and Its Interface with Si," *Physical Review B*, Vol. 72, No. 23, 2005, Article ID: 235329.

[18] P. Delugas, V. Fiorentini and A. Filippetti, "Dielectric Properties and Long-Wavelength Optical Modes of the High-κ Oxide $LaAlO_3$," *Physical Review B*, Vol. 71, No. 13, 2005, Article ID: 134302.

[19] P. W. Peacock and J. Robertson, "Band Offsets and Schottky Barrier Heights of High Dielectric Constant Oxides," *Journal of Applied Physics*, Vol. 92, No. 8, 2002, pp. 4712-4721.

[20] K. Xiong, J. Robertson and S. J. Clark, "Defect States in the High-Dielectric-Constant Gate Oxide $LaAlO_3$," *Applied Physics Letters*, Vol. 89, No. 2, 2006, Article ID: 022907.

[21] K. Xiong, J. Robertson and S. J. Clark, "Electronic Defects in $LaAlO_3$," *Microelectronic Engineering*, Vol. 85, No. 1, 2008, pp. 65-69.

[22] C. J. Först, K. Schwarz and P. E. Blöchl, "Structural and Electronic Properties of the Interface between the High-*k* Oxide LaAlO3 and Si (001)," *Physical Review Letters*, Vol. 95, No. 13, 2005, Article ID: 137602.

[23] J. L.Tang, J. Zhu, W. F. Qin, J. Xiong, Y. Zhang and Y. R. Li, "Atomic Relaxation and Electronic Redistribution of $LaAlO_3$ (001) Surfaces," *Physics Letters A*, Vol. 365, No. 1-2, 2007, pp. 149-155.

[24] R. Pentcheva and W. E. Pickett, "Charge Localization or Itineracy at $LaAlO_3/SrTiO_3$ Interfaces: Hole Polarons, Oxygen Vacancies, and Mobile Electrons," *Physical Review B*, Vol. 74, No. 3, 2006, Article ID: 035112.

[25] C. H. Lanier, J. M. Rondinelli, B. Deng, R. Kilaas, K. R. Poeppelmeier and L. D. Marks, "Surface Reconstruction with a Fractional Hole: $(\sqrt{5}\times\sqrt{5})R26.6^{\circ}$ $LaAlO_3$ (001)," *Physical Review Letters*, Vol. 98, No. 8, 2007, Article ID: 086102.

[26] S. Gemming and G. Seifert, "$SrTiO_3$ (001) $LaAlO_3$ (001) Multilayers: A Density-Functional Investigation," *Acta Materialia*, Vol. 54, No. 16, 2006, pp. 4299-4306

[27] C. J. Howard, B. J. Kennedy and B. C. Chakoumakos, "Neutron Powder Diffraction Study of Rhombohedral Rare-Earth Aluminates and the Rhombohedral to Cubic Phase Transition," *Journal of Physics: Condensed Matter*, Vol. 12, No. 4, 2000, pp. 349-365.

[28] P. Blaha, K. Schwarz and P. Sorantin "Full-Potential, Linearized Augmented Plane Wave Programs for Crystalline Systems," *Computer Physics Communications*, Vol. 59, No. 2, 1990, pp 399-415.

[29] M. D. Segall, P. J. D. Lindan, M. J. Probert, C. J. Pickard, P. J. Hasnip, S. J. Clark and M. C. Payne, "First-Principles Simulation: Ideas, Illustrations and the CASTEP Code," *Journal of Physics: Condensed Matter*, Vol. 14, No. 11, 2002. pp. 2717-2744.

[30] J. P. Perdew, S. Burke and M. Ernzerhof, "Generalized Gradient Approximation Made Simple," *Physical Review Letters*, Vol. 77, No. 18, 1996, pp. 3865-3868.

[31] F. Tran and P. Blaha, "Accurate Band Gaps of Semiconductors and Insulators with a Semilocal Exchange-Correlation Potential," *Physical Review Letters*, Vol. 102, No.

22, 2009, Article ID: 226401.

[32] L. Kleinman and D. M. Bylander, "Efficacious Form for Model Pseudopotentials," *Physical Review Letters*, Vol. 48, No. 20, 1982, pp. 1425-1428.

[33] A. Boudali, B. Amrani , M. Driss khodja, A. Abada and K. Amara, "First-Principles Study of Structural, Elastic, Electronic, and Thermal Properties of $LaAlO_3$ Perovskite," *Computational Materials Science*, Vol. 45, No. 4, 2009, pp. 1068-1072.

[34] M. A. Blanco, E. Francisco and V. Luaña, "GIBBS: Isothermal-Isobaric Thermodynamics of Solids from Energy Curves Using a Quasi-Harmonic Debye Model," *Computer Physics Communications*, Vol. 158, No. 1, 2004, pp. 57-72.

[35] M. A. Blanco, A. Martín Pendás, E. Francisco, J. M. Recio and R. Franco, "Thermodynamical Properties of Solids from Microscopic Theory: Applications to MgF_2 and Al_2O_3," *Journal of Molecular Structure*: *THEOCHEM*, Vol. 368, 1996, pp. 245-255.

[36] M. Flórez, J. M. Recio, E. Francisco, M. A. Blanco and A. M. Pendás, "First-Principles Study of the Rocksalt-Cesium Chloride Relative Phase Stability in Alkali Halides," *Physical Review*, Vol. 66, No. 14, 2002, pp. 144112-144719.

[37] E. Francisco, J. M. Recio, M. A. Blanco and A. M. Pendás, "Quantum-Mechanical Study of Thermodynamic and Bonding Properties of MgF_2," *The Journal of Physical Chemistry*, Vol. 102, No. 9, 1998, pp. 1595-1601.

[38] E. Francisco, M. A. Blanco and G. Sanjurjo, "Atomistic simulation of SrF_2 polymorphs," *Physical Review*, Vol. 63, No. 9, 2001, Article ID: 094107.

[39] J. P. Poirier, "Introduction to the Physics of the Earth's Interior," Cambridge University Press, Oxford, 2000.

[40] R. Hill, "The Elastic Behaviour of a Crystalline Aggregate," *Proceedings of the Physical Society*, Vol. 65, No. 5, 1952, pp. 349-354.

[41] N. Iles, A. Kellou, K. D. Khodja, B. Amrani, F. Lemoigno, D. Bourbie and H. Aourag, "Atomistic Study of Structural, Elastic, Electronic and Thermal Properties of Perovskites Ba(Ti,Zr,Nb)O3," *Computational Materials Science*, Vol. 39, No. 4, 2007, pp. 896-902.

[42] V. G. Tyuterev and N. Vast, "Murnaghan's Equation of State for the Electronic Ground State Energy," *Computational Materials Science*, Vol. 38, No. 2, 2006, pp. 350-353.

[43] X. Luo and B. Wang, "First-Principles Study of the Electronic and Optical Properties in Rhombohedral $LaAlO_3$," *Journal of Applied Physics*, Vol. 104, No. 5, 2008, Article ID: 053503.

[44] D. Koller, F. Tran and P. Blaha, "Merits and Limits of the Modified Becke-Johnson Exchange Potential," *Physical Review*, Vol. 83, No. 19, 2011, pp. 195134-195143.

[45] D. J. Singh, "Electronic Structure Calculations with the Tran-Blaha Modified Becke-Johnson Density Functional," *Physical Review*, Vol. 82, No. 20, 2010, pp. 205102-205119.

[46] A. Einstein, "Die Plackschen Theorie der Strahlung und die Theorie der spezifischen Warme," *Annalen der Physik*, Vol. 22, No. 2, 1907, pp. 180-190.

[47] W. Nernst, A. F. Lindemann and Z. Elektrochem, "Angew," *The Journal of Physical Chemistry*, Vol. 17, No. 817, 1977.

[48] P. Debye, "The Theory of Specific Warmth," *Annalen Der Physik*, Vol. 39, No. 14, 1912, pp. 789-839.

[49] A. T. Petit and P. L. Dulong, "Recherches sur Quelques Points Importants de la Théorie de la Chaleur," *Annales de Chimie et de Physique*, Vol. 10, 1819, pp. 395-413.

[50] M. A. Carpenter, T. W. Darling, J. D. Bass, D. L. Lakshtanov, S. V. Sinogeikin and S. D. Jacobsen, "American Geophysical Union Fall Meeting," 2006 (unpublished).

[51] X. Luo and B. Wang, "Structural and Elastic Properties of $LaAlO_3$ from First-Principles Calculations," *Journal of Applied Physics*, Vol. 104, No. 7, 2008, Article ID: 073518.

Permissions

The contributors of this book come from diverse backgrounds, making this book a truly international effort. This book will bring forth new frontiers with its revolutionizing research information and detailed analysis of the nascent developments around the world.

We would like to thank all the contributing authors for lending their expertise to make the book truly unique. They have played a crucial role in the development of this book. Without their invaluable contributions this book wouldn't have been possible. They have made vital efforts to compile up to date information on the varied aspects of this subject to make this book a valuable addition to the collection of many professionals and students.

This book was conceptualized with the vision of imparting up-to-date information and advanced data in this field. To ensure the same, a matchless editorial board was set up. Every individual on the board went through rigorous rounds of assessment to prove their worth. After which they invested a large part of their time researching and compiling the most relevant data for our readers. Conferences and sessions were held from time to time between the editorial board and the contributing authors to present the data in the most comprehensible form. The editorial team has worked tirelessly to provide valuable and valid information to help people across the globe.

Every chapter published in this book has been scrutinized by our experts. Their significance has been extensively debated. The topics covered herein carry significant findings which will fuel the growth of the discipline. They may even be implemented as practical applications or may be referred to as a beginning point for another development. Chapters in this book were first published by Scientific Research Publishing Inc.; hereby published with permission under the Creative Commons Attribution License or equivalent.

The editorial board has been involved in producing this book since its inception. They have spent rigorous hours researching and exploring the diverse topics which have resulted in the successful publishing of this book. They have passed on their knowledge of decades through this book. To expedite this challenging task, the publisher supported the team at every step. A small team of assistant editors was also appointed to further simplify the editing procedure and attain best results for the readers.

Our editorial team has been hand-picked from every corner of the world. Their multi-ethnicity adds dynamic inputs to the discussions which result in innovative outcomes. These outcomes are then further discussed with the researchers and contributors who give their valuable feedback and opinion regarding the same. The feedback is then collaborated with the researches and they are edited in a comprehensive manner to aid the understanding of the subject.

Apart from the editorial board, the designing team has also invested a significant amount of their time in understanding the subject and creating the most relevant covers. They scrutinized every image to scout for the most suitable representation of the subject and create an appropriate cover for the book.

The publishing team has been involved in this book since its early stages. They were actively engaged in every process, be it collecting the data, connecting with the contributors or procuring relevant information. The team has been an ardent support to the editorial, designing and production team. Their endless efforts to recruit the best for this project, has resulted in the accomplishment of this book. They are a veteran in the field of academics and their pool of knowledge is as vast as their experience in printing. Their expertise and guidance has proved useful at every step. Their uncompromising quality standards have made this book an exceptional effort. Their encouragement from time to time has been an inspiration for everyone.

The publisher and the editorial board hope that this book will prove to be a valuable piece of knowledge for researchers, students, practitioners and scholars across the globe.

List of Contributors

Ambran Hartono and Edi Sanjaya
Department of Physics, Faculty of Science and Technology, Islamic State University Syarif Hidayatullah, Jakarta, Indonesia

Suparno Satira, Mitra Djamal and Herman Bahar
Theoretical High Energy Physics and Instrumentation Research Group, Faculty of Mathematics and Natural Sciences, Institut Teknologi Bandung, Bandung, Indonesia

Ramli Ramli
Department of Physics, Faculty of Mathematics and Natural Sciences, Universitas Negeri Padang, Padang, Indonesia

Avijit Nandi, Chandana Sen, Randhir K. Sinha and Chittaranjan Sinha
Department of Chemistry, Inorganic Chemistry Section, Jadavpur University, Kolkata, India

Debashis Mallick
Department of Chemistry, Mrinalini Datta Mahavidyapith, Birati, India

Noureddine Selmi and Ali Sari
Nuclear Research Center of Birine, Birine, Algeria

Prasanna Chandrasekhar, Petar Pirgov, Brian J. Zay and David Lawrence
Ashwin-Ushas Corporation, Marlboro, NJ, USA

Sean Morefield
US Army Engineer Research and Development Center, Construction Engineering Research Laboratory (ERDC-CERL), Champaign, IL, USA

Tilghman L. Rittenhouse and Salvatore G. Clementi
Chemical and Biological Technologies Directorate, US Defense Threat Reduction Agency (DTRA), Fort Belvoir, VA, USA

Quoc Truong
Warrior Science, Technology and Applied Research Directorate, US Army Natick Soldier Research, Development and Engineering (RD&E) Center, Natick, MA, USA

Russell R. Greene
Hazardous Materials Research Center (HMRC), Battelle Columbus Laboratories, Battelle Memorial Institute, Columbus, OH, USA

Muna Moussa Abbas
Department of Physics, College of Science, Baghdad University, Baghdad, Iraq

Zhixiang Han
School of the Environment and Safety Engineering, Jiangsu University, Zhenjiang, China
State Key Laboratory of Chemo/Biosensing and Chemometrics, College of Chemistry & Chemical Engineering, Hunan University, Changsha, China

Lihui Liang
Hunan Provincial People's Hospital, Changsha, China

Xiaobing Zhang
State Key Laboratory of Chemo/Biosensing and Chemometrics, College of Chemistry & Chemical Engineering, Hunan University, Changsha, China

Qingqing Yang
School of the Environment and Safety Engineering, Jiangsu University, Zhenjiang, China

Qun Ma, Yinchang Li, Tingting Jiang, Junxia Wang, Fei Li, Hongyun Jin and Yongqian Wang
Engineering Research Center of Nano-Geo Materials of Education China University of Geosciences, Wuhan, China

Lili Ji
Jiangcheng College of China University of Geosciences, Wuhan, China

Claudia A. Cortés-Escobedo and Elizabeth P. Juárez-Camacho
CIITEC, Instituto Politécnico Nacional, Mexico City, Mexico

Ana M. Bolarín-Miró, Félix Sánchez-De Jesús and Iván L. Samperio-Gómez
Área Académica de Ciencias de la Tierra y Materiales, UAEH, Hidalgo, Mexico

Raúl Valenzuela
Departamento de Materiales Metálicos y Cerámicos, IIM, UNAM, Mexico City, Mexico

Souad Ammar
ITODYS, UMR, CNRS 7086, Université de Paris-Diderot, Paris, France

Nadia Febiana Djaja, Dionisius Agung Montja and Rosari Saleh
Departemen Fisika, FMIPA-Universitas Indonesia, Depok, Indonesia

Hiroaki Onoda and Taisuke Yamaguchi
Department of Informatics and Environmental Sciences, Kyoto Prefectural University, Kyoto, Japan

Osamu Takakuwa and Hitoshi Soyama
Department of Nanomechanics, Graduate School of Engineering, Tohoku University, Sendai, Japan

Udumula Subbarao and Sebastian C. Peter
New Chemistry Unit, Jawaharlal Nehru Centre for Advanced Scientific Research, Bangalore, India

M. Raghasudha
Department of Chemistry, Jayaprakash Narayan College of Engineering, Mahabubnagar, India

D. Ravinder
Department of Physics, Nizam College, Basheerbagh Osmania University, Hyderabad, India

P. Veerasomaiah
Department of Chemistry, Osmania University, Hyderabad, India

Katsuhiko Inaba and Shintaro Kobayashi
X-Ray Research Laboratory, Rigaku Corporation, Tokyo, Japan

Akira Okada and Tamio Endo
Graduate School of Engineering, Mie University, Mie, Japan

Kenichi Uehara
Graduate School of Engineering, Mie University, Mie, Japan
Adwel R&D Co. Ltd., Shiga, Japan

Sanapa Lakshmi Reddy
Department of Physics, Sri Venkateswara Degree College, Kadapa, India

Ghazala Y. Hermiz, Bushra A. Aljurani and Hassan A. Thabit
Physics Department, College of Science, University of Baghdad, Baghdad, Iraq

Longlong Chen and Jian Li
School of Physical Science & Technology, Southwest University, Chongqing, China

Pengfei Hu, Yanyan Lou, Bo Lu, Min Shao and Jiansen Ni
Laboratory for Microstructures, Shanghai University, Shanghai, China

Yali Cao
Institute of Applied Chemistry, Xinjiang University, Urumqi, China

Meng Cao
School of Materials Science and Engineering, Shanghai University, Shanghai, China

Mohammed T. Hussein, Eman M. Nasir and Addnan H. Al-Aarajiy
Department of Physics, University of Baghdad, Baghdad, Iraq

Witsarut Sriratana and Riichi Murayama
Department of Intelligent Mechanical Engineering, Faculty of Engineering, Fukuoka Institute of Technology, Fukuoka, Japan

Lerdlekha Tanachaikhan
Faculty of Engineering, Ramkhamhaeng University, Bangkok, Thailand

Kasimayan Uma
Centre for Social Contribution and Collaboration, Nagoya Institute of Technology, Nagoya, Japan

Solaiappan Ananthakumar
Materials and Minerals Division, National Institute for Interdisciplinary Science and Technology (CSIR), Kerala, India

Ramalinga Viswanathan Mangalaraja
Department of Materials Engineering, University of Concepcion, Concepcion, Chile

Takashi Jimbo
Department of Frontier Materials, Nagoya Institute of Technology, Nagoya, Japan

Tetsuo Soga
Centre for Social Contribution and Collaboration, Nagoya Institute of Technology, Nagoya, Japan
Department of Frontier Materials, Nagoya Institute of Technology, Nagoya, Japan

Wan Azelee Wan Abu Bakar, Susilawati Toemen, Rusmidah Ali and Hazwan Faiz Abd Rahim
Department of Chemistry, Faculty of Science, Universiti Teknologi Malaysia, Johor Bahru, Malaysia

Fawaz Nasro Saiof
Higher Institute for Laser Research, Damascus University, Damascus, Syrian

Mohammed Alwan Hamza
College of Science, Baghdad University, Baghdad, Iraq

Babaniyi Babatope
Department of Physics, Obafemi Awolowo University, Ile-Ife, Nigeria

Akogwu Onobu
Department of Mechanical and Aerospace Engineering, Princeton University, Princeton, USA

Winston O. Soboyejo
Department of Mechanical and Aerospace Engineering, Princeton University, Princeton, USA
Materials Science and Engineering Department, African University of Science and Technology, Abuja, Nigeria

Olusegun O. Adewoye
Materials Science and Engineering Department, African University of Science and Technology, Abuja, Nigeria

Jesús Marquina
Centro de Estudios Avanzado en Óptica, Facultad de Ciencias, Universidad de los Andes, Mérida, Venezuela

Chrystian Power
Centro de Estudios de Semiconductores, Facultad de Ciencias, Universidad de los Andes, Mérida, Venezuela

Jean-Marc Broto
LNCMI, Paul Sabatier University, Toulouse, France

Jesús González
Centro de Estudios de Semiconductores, Facultad de Ciencias, Universidad de los Andes, Mérida, Venezuela
Malta Consolider Team, CITIMAC, Facultad de Ciencias, Universidad de Cantabria, Santander, España

Abdelkader Boudali, Fatiha Saadaoui, Mostefa Zemouli, Mohamed Driss Khodja and Kadda Amara
Laboratory of Physico-chemical Studies, University of Saida, Saida, Algeria